Developments in Environmental Modelling, 9

# Fundamentals of Ecological Modelling

Developments in Environmental Modelling

Series Editor: S.E. Jørgensen
Langkaer Vaenge 9,
3500 Vaerløse,
Copenhagen,
Denmark

1. ENERGY AND ECOLOGICAL MODELLING
   edited by W.J. Mitsch, R.W. Bossermann and J.M. Klopatek

2. WATER MANAGEMENT MODELS IN PRACTICE:
   A CASE STUDY OF THE ASWAN HIGH DAM
   by D. Whittington and G. Guariso
   1983 xxii + 246 pp.

3. NUMERICAL ECOLOGY
   by L. Legendre and P. Legendre
   1983 xvi + 419 pp.

4A. APPLICATION OF ECOLOGICAL MODELLING IN ENVIRONMENTAL
    MANAGEMENT, PART A
    edited by S.E. Jørgensen
    1983 viii + 735 pp.

4B. APPLICATION OF ECOLOGICAL MODELLING IN ENVIRONMENTAL
    MANAGEMENT, PART B
    edited by S.E. Jørgensen and W.J. Mitsch
    1983 viii + 438 pp.

5. ANALYSIS OF ECOLOGICAL SYSTEMS: STATE-OF-THE-ART IN
   ECOLOGICAL MODELLING
   edited by W.K. Lauenroth, G.V. Skogerboe and M. Flug
   1983 992 pp.

6. MODELLING THE FATE AND EFFECT OF TOXIC SUBSTANCES IN THE
   ENVIRONMENT
   edited by S.E. Jørgensen
   1984 viii + 342 pp.

7. MATHEMATICAL MODELS IN BIOLOGICAL WASTE WATER TREATMENT
   edited by S.E. Jørgensen and M.J. Gromiec
   In preparation

8. FRESHWATER ECOSYSTEMS: MODELLING AND SIMULATION
   by M. Straškraba and A.H. Gnauck
   1985 309 pp.

Developments in Environmental Modelling, 9

# Fundamentals of Ecological Modelling

By

**Professor S.E. JØRGENSEN**
*Langkaer Vaenge 9, 3500 Vaerløse, Copenhagen, Denmark*

**ELSEVIER**
Amsterdam — Oxford — New York — Tokyo   1986

*Published jointly by*
Professor S.E. JØRGENSEN
Langkaer Vaenge 9, 3500 Vaerløse, Copenhagen, Denmark

*and*

ELSEVIER SCIENCE PUBLISHERS B.V.
Sara Burgerhartstraat 25
P.O. Box 211, 1000 AE Amsterdam, The Netherlands

*Distributors for the United States and Canada:*

ELSEVIER SCIENCE PUBLISHING COMPANY INC.
52 Vanderbilt Avenue
New York, NY 10017 U.S.A.

Library of Congress Cataloging-in-Publication Data
Jørgensen, Sven Erik, 1934–
   Fundamentals of ecological modelling.
   (Developments in environmental modelling; 9)
   1. Ecology-Simulation methods. I. Title.
II. Series.
QH541.15.S5J67  1986    574.5'0724    85-20582
ISBN 0-444-99535-8

ISBN 0-444-99535-8 (Vol. 9)
ISBN 0-444-41948-9 (Series)

Printed in Denmark by Fair-Print AS, Roskilde

# Contents

PREFACE ................................................................................................ 6

**1. INTRODUCTION** ............................................................................ 9
    1.1 Physical and Mathematical Models ................................................ 9
    1.2 Models as a Management Tool ...................................................... 9
    1.3 Models as a Scientific Tool .......................................................... 10

**2. CONCEPTS OF MODELLING** ...................................................... 12
    2.1 Elements of Modelling ................................................................ 12
    2.2 Modelling Procedure .................................................................. 15
    2.3 Classes of Ecological Models ...................................................... 19
    2.4 Selection of Models Complexity and Structure ............................ 24
    2.5 Verification ................................................................................ 39
    2.6 Sensitivity Analysis .................................................................... 48
    2.7 Parameter Estimation ................................................................ 52
    2.8 Validation .................................................................................. 74
    2.9 Constraints on Models ................................................................ 75
    2.10 Computers and Ecological Modelling ........................................ 83

**3. ECOLOGICAL MODELLING** ........................................................ 90
    3.1 Application of Unit Processes in Ecological Modelling ................ 90
    3.2 Physical Processes ...................................................................... 93
        3.2.1 Transport Processes ............................................................ 93
    3.3 Sorption .................................................................................... 98
        3.3.1 Temperature dependence .................................................... 100
        3.3.2 Evaporation ........................................................................ 101
    3.4 Chemical Processes .................................................................... 101
        3.4.1 Chemical oxidation ............................................................ 102
        3.4.2 Photolysis .......................................................................... 102
        3.4.3 Hydrolysis .......................................................................... 103
        3.4.4 Ionization, Complexation and Precipitation ........................ 103
    3.5 Photosynthesis .......................................................................... 104
        3.5.1 Production at Secondary and Higher Trophic Levels ............ 112
        3.5.2 Energy Flow in Secondary Production ................................ 113
        3.5.3 Decomposition .................................................................. 117
        3.5.4 Adaption ............................................................................ 120

**4. CONCEPTUAL MODELS** .............................................................. 123
    4.1 Application of Conceptual Models .............................................. 123
    4.2 Types of Conceptual Diagrams .................................................. 125
    4.3 The Conceptual Diagram as Modelling Tool .............................. 132

**5. STATIC MODELS** .......................................................................... 143
    5.1 Application of Static Models ...................................................... 143
    5.2 Input/output Environ Analysis .................................................... 143
    5.3 Response Models ........................................................................ 151

**6. MODELLING POPULATION DYNAMICS** .................................... 156
    6.1 Basic Concepts .......................................................................... 156
    6.2 Growth Models .......................................................................... 157
    6.3 Interactions between Population .................................................. 162
    6.4 Matrix Models .......................................................................... 170
    6.5 Harvest Models .......................................................................... 183

**7. DYNAMIC BIOGEOCHEMICAL MODELS** ................................................................ 193

7.1 Application of Dynamic Models ........................................................................ 193
7.2 BOD/DO Models ............................................................................................. 196
   7.2.1 Simple BOD/DO Models ........................................................................ 196
   7.2.2 Complex BOD/DO Models ..................................................................... 201
7.3 Application of Hydrodynamics in Biogeochemical Models ................................. 204
   7.3.1 Introduction .......................................................................................... 204
   7.3.2 The Hydrodynamic of complete mixed Systems ...................................... 205
   7.3.3 Hydrodynamic Models of Streams and Rivers ........................................ 208
   7.3.4 Estuarine Models .................................................................................. 211
   7.3.5 Multidimensional Models and numerical Models ..................................... 213
   7.3.6 Modelling a Stratified Lake ................................................................... 216
7.4 Eutrophication Models .................................................................................... 220
   7.4.1 Eutrophication ...................................................................................... 220
   7.4.2 Eutrophication Models an Overview ....................................................... 224
   7.4.3 Some relative simple Eutrophication Models ........................................... 225
   7.4.4 Complex Eutrophication Model .............................................................. 233
7.5 Wetland Models .............................................................................................. 246
   7.5.1 Introduction .......................................................................................... 246
   7.5.2 Cypress Dome Simulation Model ........................................................... 247
7.6 Models in Ecotoxicology ................................................................................. 252
   7.6.1 Introduction .......................................................................................... 252
   7.6.2 Principles of Modelling the Distribution and Effects of Toxic Substances .............. 253
   7.6.3 Simplifications in Ecotoxicological Models ............................................. 257
7.7 Models in Toxicology ...................................................................................... 265
   7.7.1 Introduction .......................................................................................... 265
   7.7.2 Accumulation ....................................................................................... 266
   7.7.3 Multiple Compartment Models ............................................................... 268
7.8 Distribution of Air Pollutants .......................................................................... 270
   7.8.1 General about Air Pollution ................................................................... 270
   7.8.2 Basic Equations for Long-Range Transport Models of Air Pollution .................. 272
   7.8.3 Models of the Distribution and Effect of Acidic Rain .............................. 273
   7.8.4 Inclusion of Vertical Transport in Air Pollution Models .......................... 277
   7.8.5 Models of Air Chemistry ....................................................................... 281
   7.8.6 Plume Dispersion .................................................................................. 282
7.9 Models of Soil Processes, Plant Growth and Crop Production ........................... 290
   7.9.1 Introduction .......................................................................................... 290
   7.9.2 Mass and Heat Transfer in Soil ............................................................. 291
   7.9.3 The Inflix of Plants on the Water Balance .............................................. 299
   7.9.4 How to consider the Climatic Influence on Plant Growth and Water Balance ........ 302
   7.9.5 Models of Plant Growth and Crop Production ......................................... 308
   7.9.6 Models of the Nitrogen Processes in Soil ............................................... 312

**8. APPLICATION OF ECOLOGICAL MODELS IN ENVIRONMENTAL MANAGEMENT ....** 321

8.1 Environmental Management Models .................................................................. 321
8.2 Environmental Problems and Models ................................................................ 327
8.3 Management Examples ..................................................................................... 329
   8.3.1 Validation of a Prognosis, based upon a Management Model ..................... 329
   8.3.2 Model for evaluating Human Carrying Capacity of a Recreational Area .............. 336
   8.3.3 Simulation of Management Alternatives in Wetland Forests ...................... 344

**9. ECOSYSTEMS CHARACTERISTICS AND MODELS** .................................................. 349

9.1 Characteristic Features of Ecosystems .............................................................. 349
9.2 Ecosystem Dynamics ...................................................................................... 351
9.3 Ecological Models with Goal Functions ........................................................... 353
9.4 Application of Catastrophy Theory to Ecological Modelling .............................. 356

**10. REFERENCES** ................................................................ 361

**10. INDEX** ......................................................................... 388

# PREFACE

It is the intention that this book be suitable for a variety of engineers and ecologists, who may wish to gain an introduction to the rapidly growing field of ecological and environmental modelling. An understanding of the fundamentals of environmental problems and ecology, as is presented in the textbook "Principles of Enviromental Science and Technology" is assumed. Furthermore it is assumed that the reader has a fundamental knowledge of differential equations, and matrix calculations.

Only very few books, giving an introduction into ecological modelling, have been published. There seems therefore to be a need for a book to be applicable to courses in this subject. Although many books have been published on the topic they require in most cases that the reader already has an understanding of the field or at least has had some experience in development of ecological models. It is the aim of this book to bridge this gap.

It has been the scope of the author to give on the one hand an overview of the field and on the other to enable the reader to develop his own models. An attempt has been made to meet these objectives by covering the following points:

1) Discussion of the modelling procedure in detail and presentation of the development of models step by step. Advantages and shortcomings of each step are discussed and simple examples are used to illustrate all the steps.

2) Presentation of most model types by use of theory, overview tables on applications, complexity, examples and illustrations.

3) Presentation of both simple and complex models. As a presentation of a complex model is rather space consuming, only the characteristic features and problems of the most complex models are given. The alternative would have been to present fewer models, but this would not have given the reader an idea of the wide spectrum of models available to day.

Emphasis has been laid on understanding the nature of models. Models are very useful tools in ecology and environmental management, but if they are developed and used carelessly, they might do more harm than good. Modelling is not just a mathematical exercise, but requires a profound knowledge to the modelled system. This is illustrated several times throughout the book.

After the introduction in chapter one, chapter two deals with the modelling procedure in all phases. The author attempts to give the reader a complete answer to the question: How to model a biological system? Chapter three gives an overview of applicable submodels or unit-processes i.e. elements in models. The same process can be modelled in several ways depending on the application of the submodel. Which one to select is discussed in chapter two and in relation to the presented models in chapter four to nine. Chapter four reviews different method of model conceptualization. As different modellers prefer different methods, the author has found it of importance to present all available methods.

The ambitious modeller would go for a dynamic model, but often the problem, the system and/or the data might require that a simpler static model is applied. In many context, a static model is completely satisfactory. Chapter five presents therefore the development and the use of static models including management case studies. In principles there is no difference between population models and other models, but the former have had a different history and are used to solve different problems. Chapter six gives an overview of population models, but a more comprehensive treatment of this subject must be found in books focusing entirely on this type of models. Ecological models in their broadest sense comprise also population dynamic models and ecological applications of such models are therefore included in this chapter.

Chapter seven covers dynamic biogeochemical models. They are illustrated by the presentation of BOD/DO models, hydrodynamic models, eutrophication models, models of toxic substances in the environment and in the organism, air pollution models and soil pollution models. This type of models have found a very wide use and it was therefore considered of importance to give a rather comprehensive treatment of the development and application of biogeochemical models.

Chapter eight illustrates the application of models in environmental management. Problems in this context are discussed and three examples are presented in detail to illustrate the application in practice.

Chapter nine, the final chapter, presents a recent development in ecological modelling: how to give the models the property of softness and flexible structure, which we know that ecosystems have? Different approaches to this question are presented and discussed in this chapter.

The individual chapters can to a certain extent be read independently, which renders it feasible to tailor the book to any course in ecological modelling. Chapter one and two are compulsory, as they cover the general theory behind ecological modelling. Chapter three can be considered as a "handbook" in submodels. Chapters four to nine can be combined to cover the subjects of almost any course in modelling biological systems.

We have planned to use the following parts of the book in a course of modelling for biologists:chapters 1 and 2 will be read entirely,while chapter 3 only will be used as a handbook.Chapters 4,5,6,8 and 9 will be used almost entirely and sections 7.1, 7.2, 7.4, 7.5 and 7.6 will be read entirely. In another course aimed for agriculture engineers it is the intention to replace 7.4 and 7.5 by 7.9,but also to use less time on chapters 4,5,6 and 9.

The book contains illustrations,which are models presented in detail, and examples, where a problem is formulated and the solution to the problem presented. It is hoped that this will give the reader the knowledge that is required to construct models. Overviews of available models applicable to solving a certain problem complex are presented in several chapters. At the beginning of every chapter a summary of the content is given to facilitate 1) a selection of the most relevant parts of the chapter and 2) an understanding of the sequence.

# ACKNOWLEDGEMENT

The author would like to express his appreciation to Leif Albert Jørgensen, Henning F. Mejer and Billie Vestergaard for their constructive advices and encouragement during the preparation of this book. Furthermore, I am very grateful that it has been possible in preparation of the book to reproduce models, figures and tables from Dansk Hydraulisk Institut's report on a model for transport of material in the unsaturated zone (mainly carried out by Karsten Hoegh Jensen and Hans Chr. Ammentorp), from Hydrotekninsk

Institut's report on the model Nitcros and from contributions to the book "Application of Ecological Modelling in Environmental Management" (part A and B), written by Ole Stig Jacobsen, William J. Mitsch, Martha W. Gilliland and R.L.Phipps and L.H.Applegate.

Copenhagen, July 1985

Sven Erik Jørgensen

## SYMBOLS AND MATHEMATICAL NOTATION

A wide spectrum of symbols and mathematical notations is used in ecological modelling. A variable f.inst. dissolved oxygen might have different symbols in different context. If it is used in a mathematical equation D is often used, while it in a computer equation might be called OXYG or DISS, because here symbols of 2-4 letters are often used in a computer program.

Various mathematical notations are applied in ecological modelling literature; f.inst. Bold-Faced types or a letter surmounted by a line or an arrow for a matrix.

This book has not attemped to homogenize the use of symbols and mathematical notations in ecological modelling, but the symbols and notations used in the original references have been applied. The disadvantage is that a general list of symbols and notations cannot be given, but on the other side the reader will be acquainted with the wide spectrum of symbols and notations found in the literature of ecological modelling. Hopefully, this will not make it too difficult for the readers to follow the meaning of the equations, as explanations to symbols and notations are given together with the equations.

# 1. INTRODUCTION

## 1.1. PHYSICAL AND MATHEMATICAL MODELS

Mankind has always used models, it means a simplified picture of reality, as a tool to solve problems. The model will of course never contain *all* the features of the real system, because, then it would be the real system itself. But it is of importance that the model contains the *characteristic* features, that are essential in the context of the problem to be solved or described.

The philosophy behind the use of model might be best illustrated by use of an example. We have for many years used physical models of ships to determine the profile, that gives a ship the smallest resistance in water. Such a model will have the shape and the relative main dimensions of the real ship, but will not contain all the details such as e.g. the instrumentation, the lay-out of the cabins etc. Such details are of course irrelevant to the objectives of that model. Other models of the ship serve other aims: blue prints of the electrical wiring, lay-out of the various cabins, drawings of pipes etc.

Correspondingly, an ecological model must contain the features, that are of interest for the management or scientific problem,that we wish to solve by use of the model. An ecosystem is a much more complex system than a ship, this implies that it is a far more complicated matter to capture the main features of importance for an ecological problem. However, intense research during the last decades, has made it possible to-day to set up workable ecological models.

The model might be physical,such as the ship model used for the resistance measurements,which may be called microcosmos or it might be a mathematical model, which describes the main characteristics of the ecosystem and the related problems in mathematical terms.

Physical models will only be touched on very briefly in this book,which will focus entirely on the construction of mathematical models.

The field of ecological modelling has developed rapidly during the last decade due essentially to two factors:

1) the development of computer technology, which has enabled us to handle very complex mathematical systems.

2) a general understanding of pollution problems, including that a complete elimination of pollution is not feasible ("zero discharge"), but that a proper pollution control with the limited economical resources available requires serious considerations of the influence of pollution impacts on ecosystems.

## 1.2. MODELS AS A MANAGEMENT TOOL

The idea behind the use of ecological management models is demonstrated in fig. 1.1. Urbanization and technological development has had an increasing impact on the

9

environment. Energy and pollutants are released into ecosystems, where they may cause more rapid growth of algae or bacteria, damage species, or alter the entire ecological structure. Now, an ecosystem is extremely complex, and so it is an overwhelming task to predict the environmental effects that such an emission will have. It is here that the model comes into the picture. With sound ecological knowledge it is possible to extract the features of the ecosystem that are involved in the pollution problem under consideration, to form the basis of the ecological model (see also the discussion in chapter 2). As indicated in fig. 1.1, the model resulting can be used to select the environmental technology best suited for the solution of specific environmental problems, or legislation reducing or eliminating the emission set up.

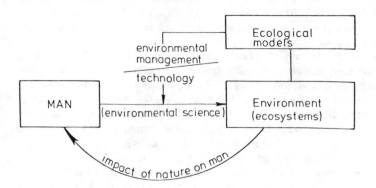

Fig. 1.1: Relations between environmental science, ecology, ecological modelling and environmental management and technology.

### 1.3. MODELS AS A SCIENTIFIC TOOL

Models are a widely used instrument in science. The scientist often uses physical models to carry out experiments in situ or in the laboratory to eliminate disturbance from processes irrelevant to his investigation. Chemostats are used e.g. to measure algal growth as function of nutrient concentrations. Sediment cores are examined in the laboratory to investigate sediment-water interactions without disturbance from other ecosystems components. Reaction chambers are used to find reaction rates for chemical processes etc.

But mathematical models are also widely applied in science. Newton's laws are relatively simple mathematical models of the influence of gravity on bodies, but they do not account for frictional forces, influence of wind etc. Ecological models do not differ essentially from other scientific models not even by their complexity, as many models used in nuclear physics during the last decades might be even more complex than ecological models.

The application of models in ecology is almost cumpulsory, if we want to understand the function of such a complex system as an ecosystem. It is simply not possible to survey the many components and their reactions in an ecosystem without the use of a model as synthesis tool. The reactions of the system might not necessarily be the sum of all the individual reactions; this implies that the properties of the ecosystem as a system will not be revealed without the use of a model of the entire system.

It is therefore not surprising that ecological models have been used increasingly in ecology as an instrument to understand the properties of the ecosystem. This application has clearly revealed the advantages of models as a useful tool in ecology; this might be summarized in the following points:

1) Models are useful instruments in **survey** of complex systems.

2) Models can be used to reveal **system properties**.

3) Models reveal **the weakness in our knowledge** and can therefore be used to set up research priorities.

4) Models are useful in tests of **scientific hypotheses,** as the model can simulate ecosystem reactions, which can be compared with observations.

# 2. CONCEPTS OF MODELLING

Chapter two covers the topic of modelling theory and its application in practice. After the definitions of model components and modelling steps have been presented, a tentative modelling procedure is given. The steps in the presented modelling procedure are discussed in detail. The chapter focuses furthermore on model selection - meaning the selection of model components, processes and in particular model complexity. Various methods to select "closed to the right" complexity of the model are presented. The conceptual diagram or model is the first presentation of the model, but due to the great number of possibilities, this step is only mentioned briefly in this chapter. The topic is covered in detail in chapter 4. The following steps, however, are discussed in detail in this chapter: verification, parameter estimation and validation. Illustrations are included to show the reader how these steps are carried out in practical model building.

Several model formulations are always at hand and to choose among these will require that sound scientific constraints are imposed on the model. Possible constraints are introduced and discussed. A mathematical model will almost always require the use of a computer and thereby a computer language. The chapter is terminated by a discussion of this problem.

## 2.1. ELEMENTS OF MODELLING

An ecological model consists, in its mathematical formulation, of five components:

(1) **Forcing functions or external variables,** which are functions or variables of an external nature that influence the state of the ecosystem. In a management context the problem to solve can often be reformulated as follows: if certain forcing functions are varied what will be the influence on the state of the ecosystem? The model is, in other words, used to predict what will change in the ecosystem, when forcing functions are varied with the time. Examples of forcing functions are the input of pollutants to the ecosystem, the consumption of fossil fuel, or a fishery policy, but temperature, solar radiation, and precipitation are also forcing functions (which, however, we cannot at present manipulate). Forcing functions, which are controllable by man, are often named **control functions.**

(2) **State variables** describe, as the name indicates, the state of the ecosystem. The selection of the variables is crucial for the model structure, but in most cases the choice is obvious. If, for instance, we want to model the eutrophication of a lake it is natural to include the phytoplankton concentration and the concentrations of nutrients. When the model is used in a management context the values of the state variables predicted by changing the forcing functions can be considered as the result of the model, because the model will contain relations between the forcing functions and the state variables. Most models will contain more state variables than are **directly required** for purpose of management, because the relations are so complex that they require the introduction of additional state variables. For instance, it would be sufficient in many eutrophication models to relate the input of one nutrient with the phytoplankton concentration, but as

this variable is influenced by more than one nutrient (it is influenced by other nutrient concentrations, temperature, hydrology of the water body, zooplankton concentration, solar radiation, transparency of the water etc.) a eutrophication model will most often contain a number of state variables.

(3) The biological, chemical and physical processes in the ecosystem are represented in the model by means of **mathematical equations.** They are the relations between forcing functions and state variables. The same type of processes can be found in many ecosystems, which implies that the same equations can be used in different models. Chapter 3 has therefore been devoted to summarizing the mathematical representations of these unit processes. It is, however, not possible to-day to have one equation that represents a given process in all ecological contexts. Most of the processes have several mathematical representations, which are *equally* valid either because the process is too complex to be understood in sufficient detail at present, or because some specified circumstances allow us to use simplifications.

(4) The mathematical representation of processes in the ecosystem contains coefficients or **parameters.** They can be considered constant for a specific ecosystem or part of ecosystem (see however the discussion of distributed models and lumped models in 2.9). In the casual models the parameters will have a scientific definition, e.g. the maximum growth rate of phytoplankton. Many parameter values are known within limits. In Jørgensen et al. (1979) can be found a comprehensive collection of ecological parameters. However, only a few parameters are known exactly and so it is necessary to calibrate the others.

By the **calibration** it is attempted to find the best accordance between computed and observed state variables by variation of a number of parameters. The calibration might be carried out by trial and error procedure or by use of software developed to find the parameters that give the best fit.

In many static models, where process rates are given as average values in a given time interval and in many simple models, which contain only a few and well defined or directly measured parameters, calibration is not required. In models aimed at simulating the dynamics of the ecological processes, the calibration is crucial for the quality of the model due to the reasons summarized below:

a)  As indicated above the parameters are in most cases known only within limits.

b)  Different species of animals and plants have different parameters, which can be found in the literature (see Jørgensen et al. 1979). However, most ecological models do not distinguish between different species of phytoplankton, but consider them as one state variable. In this case it is possible to find limits for the phytoplankton parameters, but as the composition of the phytoplankton varies throughout the year an exact average value cannot be found.

c)  The influence of the ecological processes which are of minor importance to the state variables under consideration, and therefore not included in the model, can to a certain extent be considered by the calibration, where the results of the model are compared with the observation, from the ecosystem. This might also explain why the parameters have different values in the same model when used for different ecosystems. The calibration can, in other words, take the site differences and the ecological processes of minor importance into account, but obviously it is essential to reduce the use of calibration for this purpose. The calibration must never be used to force the model to fit observations if this implies, that unrealistic parameters are obtained. If a reasonable fit cannot be achieved with realistic parameters the entire model should be questioned. It is therefore extremely important to have realistic ranges for all parameters or at least for the most sensitive. This implies that a sensitive analysis must be carried out, whereby the influence of changes in submodels, parameters or forcing function on the most crucial state variables is found. In section 2.5 will be shown how it is possible to make a sensitivity analysis and for what purpose it can be used.

(5) Most models will also contain universal constants such as the gas constant, molecular weights etc. Such constants are of course not subject to a calibration.

Models can be defined as formal expressions of the essential elements of a problem in either physical or mathematical terms. The first recognition of the problem is often, and most likely, to be, expressed *verbally*. This can be recognized as an essential preliminary step in the modelling procedure, but the term formal expressions implies that a translation into physical or mathematical terms must take place before we have a model in the sense we are applying throughout this book.

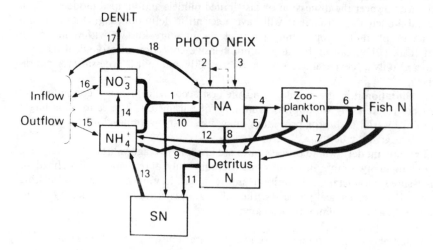

Fig. 2.1: The nitrogen cycle in an aquatic ecosystem. The processes are:
(1) Uptake of $NO_3^-$ and $NH_4^+$ by algae, (2) Photosynthesis, (3) Nitrogen fixation, (4) Grazing with loss of undigested matter, (5), (6) and (7) are predation and loss of undigested matter by predation, (8) Mortality, (9) Mineralization, (10) Settling, (11) Settling of ditritus, (12) Settling, (13) Release from sediment, (14) Nitrification, (15), (16) and (18) Input/output, (17) Denitrification.

The verbal model is difficult to visualize and it is therefore conveniently translated into a **conceptual diagram,** which contains the state variables, the forcing functions and how these components are interrelated by processes. The conceptual diagram can be considered as a model, and is named a conceptual model. A number of models in the ecological literature stop at this stage due to lack of knowledge about the mathematical formulation of processes. They can, however, be used to illustrate the relationships qualitatively. Chapter 4 deals with conceptual models and the various methods applied to set up this type of models.

Fig. 2.1 illustrates a conceptual model of the nitrogen cycle in a lake. The state variables are nitrate, ammonium, nitrogen in phytoplankton, nitrogen in zooplankton, nitrogen in fish, nitrogen in sediment and nitrogen in detritus. In the diagram are shown the following forcing functions: inflow, outflow and the concentrations of nitrate and ammonium in the in- and outflow. Other forcing functions not shown are: solar radiation and temperature. The arrows in the diagram illustrate the processes numbered 1 to 18. If we want to proceed to a quantitative model, it is necessary to formulate these processes by use of mathematical expressions (equations).

It is of great importance to verify and validate models. **Verification** is a test of the *internal logic* of the model. Typical questions in the verification phase are: Does the model react as expected? Will f.inst. increased discharge of organic matter give lower concentration of oxygen in a river model concerned with the oxygen balance of the system? Is the model long-term stable? Does the model follow the law of mass conservation? etc.

Verification is therefore largely a subjective assessment of the behaviour of the model. To a large extent verification will inevitably go on during the use of the model before the calibration phase, which has been mentioned above.

**Validation** must be distinguished from verification, but previous use of the words has not been consistent. Validation consists of an objective test on how well the model outputs fit the data. The selection of possible objective tests will be discussed in section 2.9.


## 2.2. MODELLING PROCEDURE

The primary focus of all research at all times is to define the problem. Only in that way can it be ensured that limited research resources can be correctly allocated and not dispersed into irrelevant activities.

The definition of the actual problem will need to be bound by the constituents of space, time and subsystems. The bounding of the problem in space and time is usually easier, and consequently more explicit, than the identification of the ecological subsystems to be incorporated in the models.

Some of the projects of the International Biological Programme (IBP) assumed that it was necessary to model the whole ecosystem and that it was unnecessary to define subsystems of that ecosystem. When the final synthesis was attempted, major gaps were found in many of the projects, which could not be filled by any of the experimental or survey results, and these gaps were frequently emphasized by the absence of any preliminary synthesis (Jeffers, 1978).

The experience of IBP has led many ecologists to question the need for studies of whole ecosystems and to focus their attention on carefully designed sets of subsystems. In the synthesis of the eutrophication of lakes for example, most attention must be concentrated to algal growth and nutrient cycles as a basis for the prediction of the effect of the nutrients on the eutrophication process.

The use of models in the ecological context is relatively new and few guides are therefore available for the construction of ecological management models. A tentative guideline is presented in fig. 2.2.

In addition to defining the problem and its parameters in space and time, it is important to emphasize that this procedure is unlikely to be correct at the first attempt, and so there is no need to aim at perfection in one step. The main requirement is to get started (Jeffers, 1978). All ecosystems have a distinctive character and a comprehensive knowledge of the system that is going to be modelled is often needed to get a good start.

It is difficult to determine the optimum number of subsystems to be included in the model for an acceptable level of accuracy, and often it is necessary to accept a lower level than intended at the start due to a lack of data.

It has been argued that a more complex model should be able to account more accurately for the complexity of the real system but this is not true. Some additional factors have to be induced in this consideration. As increasing number of parameters are added to the model there will be an increase in uncertainty. The parameters have to be estimated either by observations in the field, by laboratory experiments or by calibrations, which again are based on field measurements. Parameter estimations are therefore never error free. As these errors are carried through into the model they will contribute to the uncertainty of the prediction derived from the model, and there seems therefore to be some great advantage in reducing the complexity of models.

Some ecologists argue that ignorance of species diversity increases the risk of neglecting important elements of their dynamics. However, comparison of models with different complexities (Jørgensen et al., 1978 and 1981) demonstrates that the deviations of simpler models from alternative models, which take biological diversity into accound, might be negligible for the purpose of the model. This trade off between complexity and simplicity in the choice of model is one of the most difficult modelling problems. Some attempts have been made to provide some general rules. The method published by Jørgensen et al. (1977) measures the *response* of the model to more state variables and concludes that only the major influences of importance for the problem in focus should be included in the model. The method might also be interpreted as a sensitivity test on the addition of state variables. The selection of the model complexity will be discussed further in section 2.6.

Once the model complexity, at least at the first attempt, has been selected, it is possible to conceptualize the model (e.g. in the form of a diagram such as those shown for the nitrogen cycles in fig. 2.1). This will give information on, which state variables and processes are required in the model. For most processes a mathematical description is available, and most of the parameters have, at least within limits, known values from the literature. Tables of parameters used in ecological models can be found in Jørgensen et al. (1979).

It is possible at this stage to set up alternative equations for the same process and apply the model to test the equations against each other. However, the many ecological processes not included in the model, have some influence on the processes in the model. Furthermore, the parameter values used from the literature are often not fixed numbers, but are rather indicated as intervals. Biological parameters can most often not be determined with the same accuracy as chemical or physical parameters due to changing and uncontrolled experimental conditions. Consequently, calibration by the application of a set of measured data (see also the discussion in 2.1) is almost always required. However, the calibration of several parameters is not realistic. Mathematical calibration procedures for ten or more parameters are not available for most problems. Therefore, it is recommended that sound values from the literature are used for all parameters, and that a sensitivity analysis of the parameters (see fig. 2.2) is made before the calibration. The most sensitive parameters should be selected, as an acceptable calibration of 4-8 parameters is possible with the present techniques.

If it is necessary to calibrate 10 parameters or more it is advantageous to use two different series of measurements for the calibration of five parameters each, preferably by selecting measuring periods where the state variables are most sensitive to the parameters calibrated (Mejer et al., 1980, Jørgensen et al., 1981). It is of great importance to make the calibration on the basis of reliable data; unfortunately, many ecological models have been calibrated against inaccurate information.

It is characteristic of most ecological models that analysis and calibration of submodels are required. If ecological models are build without the knowledge of the ecosystem and its subsystems, they are often not realistic. These considerations are included in the procedure indicated above for the calibration and also in the modelling procedure presented in fig. 2.2.

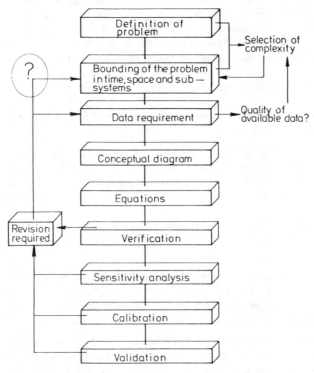

Fig. 2.2: A tentative modelling procedure.

After the calibration it is important to validate the model, preferably against a series of measurements from a period with changed conditions, e.g. with changed external loading or climatic conditions.

The right complexity cannot, as already discussed, be selected generally. It seems that there has been a tendency rather to choose too complex rather than too simple a model, probably because it is too easy to add to the complexity: It is far more troublesome to obtain the data that are necessary to calibrate and validate a more complex model. As we have repeated here several times, it is necessary to select complexity on the basis of **the problem, the system and the data available.**

The tentative modelling procedure presented in fig. 2.2 is only one among many workable procedures. However, the components of other possible procedures are approximately the same. The goals and the objectives of the model determine its nature . The steps: setting up a conceptual diagram, verification and validation are repeated in all procedures. As mentioned in 2.1 calibration and thereby sensitivity analysis might sometimes be redundant, when the parameters already are known with sufficient accuracy.

Modelling should, however, be considered an iterative process. When the model in the first instance has been verified, calibrated or validated, new ideas will emerge on how to improve the model. The modeller will again and again wish to build new data, knowledge and experience either from his own experiments or from the scientific literature into the model; this implies that he at least to a certain extent must go through the entire procedure again to come up with a better model. The modeller knows that he can always build a better model, which has higher accuracy, is a better prognosis tool or contains more relevant details than the previous model. He will approach the ideal model asymtotically,but will never reach it. However, *limited resources* will sooner or later stop the iteration and the modeller will declare his model to be good enough within the *given limitations*.

An example might be used to describe a model development:

1) A conceptual model was set up based upon a comprehensive study of Lake Glumsø in 1973 (Jørgensen et al., 1973).

2) This first model could not be used to set up a prognosis, as it was only a conceptuali-zation of the ecological knowledge on the lake as an ecosystem. The objectives for the next step was to establish a relationship between eutrophication (measured by the concentrations of phytoplankton and by the transparency) and the nutrient input to the lake.
   Consequently a **mathematical model,** which focused on eutrophication, was made on basis of additional data. The model was calibrated and validated satisfactorily (Jørgensen, 1976).

3) During the following two years several **possible improvements** of the model were tested, which resulted in a slightly changed model and improved calibration and validation. That model was used **to set up prognosis** for various management strate-gies, which could be compared more objectively (Jørgensen et al., 1978).

4) The experience with model calibration and validation made it clear, that the available data did not reflect adequately the dynamics of the most important state variable in the model: phytoplankton. The sampling frequency during the period 1973-76 was monthly, and it was therefore attempted to increase the sampling frequency to 3-4 times a week during the spring- and summer bloom period to obtain better data for **a more accurate calibration** (Jørgensen et al., 1981).

5) A diversion of the waste water took place in 1981, which made it possible to validate the prognosis set up in 1978. Previously the model was validated under unchanged nutrient loadings, but it was an open question, whether a significant reduction in the nutrient loadings would cause such changes in the ecosystem, that the previously developed model would be invalid. Simultaneously with the prognosis validation, **further possible improvements** were tested and a few changes adopted. The validation of the prognosis gave under all circumstances the clear result, that the model was a

good prognosis tool, but further improvements could be induced by use of a current change of the most crucial parameters **to account for the changes in species composition** observed by the nutrient reduction (Jørgensen et al., 1985 and Jørgensen, 1985).

6) Further improvements seem therefore to take the direction of incorporation of **dynamic structure** into the model to improve the model ability to make predictions **under radically changed circumstances.**

## 2.3. CLASSES OF ECOLOGICAL MODELS

It is useful to distinguish beteen various types of models and discuss the selection of model types briefly. In fact in the introduction was made a division of models into two groups:

research or scientific models and management models.

In Table 2.1a. other pairs of model types are shown. A stochastic model contains *stochastic input disturbances and random measurement errors,* see fig. 2.3. If they are both assumed to be zero the **stochastic model** would reduce to a **deterministic model,** provided that the parameters are known exactly and not estimated in terms of statistical distributions. It is worth underlining, that a deterministic model is tantamount to the assumption, that one has perfect knowledge of the behaviour of the system. It implies that the future response of the system is completely determined by a knowledge of the present state and future measured inputs. Stochastic models are only touched on in this book in a few cases.

The application of the expressions **compartment and matrix models** is not consistent, but some modellers distinguish between these two classes of models entirely by the mathematical formulation as indicated in Table 2.1a. Both mathematical formulations will be used in this book, but the classification will not be widely applied.

The classification **reductionistic and holistic models** is based upon a difference in the scientific ideas behind the model. The reductionistic modeller will attempt to incorporate as many details of the system as possible to be able to capture its behaviour. He believes that the properties of the system are the sum of all the details. The holistic modeller on the other side attempts to include in the model properties of the ecosystem working as a system by use of general system principles. In this case are the properties of the system, not the sum of all the details   considered, but the system possesses some additional properties because the subsystems are working as a unit.

**Dynamic systems** might have four classes of states. **The initial state** changes through **transient states** to a state, where the system **oscillates round a steady state,** as shown in fig. 2.4. The transient phase can only be described by use of a dynamic model, which uses differential or difference equations to describe the system response to external factors. Differential equations are used to represent continuous changes of state with time, while difference equations use discrete time steps. The steady state corresponds to all derivatives are equal to zero. The oscillations round the steady state are described by use of a dynamic model, while **steady state** itself can be described by use of a **static model.** As all derivatives are equal to zero in steady states *the static model is reduced to algebraic equations.* Some dynamic systems have no **steady state** f.inst. systems which show **limit cycles.** This forth state possibility obviously requires a dynamic model to describe the system behaviour. In this case the system is always non-linear, although there are non-linear systems, which have steady states.

19

A **static model** assumes consequently that all variables and parameters are independent of time. The advantage of the static model is its potential for simplifying subsequent computational effort through the elimination of one of the independent variables in the model relationships. A typical example is a model which computes an average of spatial variations of quality in a river system for an average time-in-variant set of waste water discharge, temperature and stream flow rate conditions. The model can be used as management tool by comparison of various steady state situations, but the model cannot be used to predict when these situations will occur. If forecasting systems must be applied it is necessary to use dynamic models, which are characterized by time-variant state variables. The eutrophication model presented in section 7.4 is a typical example of dynamic models. The model can be applied to predict the response of a lake ecosystem after initiation of nutrient removal treatment. In this case a prognosis will be able to describe the successive improvement of the water quality as function of time.

A **distributed model** accounts for variations of variables in time and space. A typical example would be an advection-diffusion model for the transport of dissolved substance along a stream. It might include variations in the three orthogonal directions. The analyst might, however, decide on basis of prior observations, that gradients of dissolved material along one or two directions are not sufficiently large to merit inclusion in the model. He therefore, a the water quality to be uniform and independent on position within a defined volume. The distributed model would then be reduced by that assumption to **a lumped parameter model.** A typical example of this kind of model is the continuously stirred tank reactor idealization of lake water quality dynamics. Whereas the lumped model is frequently based upon ordinary differential equations, the *distributed parameter model is usually defined by partial differential equations.*

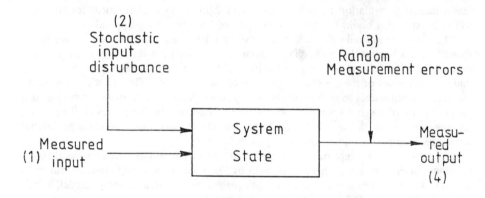

Fig. 2.3: A stochastic model considers (1) (2) and (3) while a deterministic model assumes that (2) and (3) are zero.

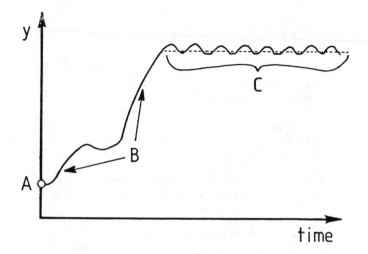

Fig. 2.4: Y is a state variable expressed as a function of time. A is the initial state, B transient states and C oscillation round steady state. The dotted line corresponds the steady state, which can be described by use of a static model.

Most distributed and lumped models are nonlinear and a special case of this general class of nonlinear models is the linear model. The great advantage of the linear model is that it obeys the principle of superposition. If the input forcing function IF gives the output response OR, and likewise the forcing function IFF is related to the output ORR, then the combinations of inputs (a IF + b IFF) wil produce the model response (a OR + b ORR), where a and b are constants.

**The casual or internally descriptive model** characterizes how the inputs are connected to the states and how the states are connected to each other and to the outputs of the system, wheras the **black box model** reflects only what changes the input will effect in the output responses. In other words the causal model provides a description of the internal mechanisms of process behaviour. The black box model deals only with what is measurable: the input and the output.

## TABLE 2.1a:

## Classification of Models (pairs of model types).

| Type of models | Characterization |
|---|---|
| Research models | used as a research tool |
| Management models | used as a management tool |
| Deterministic models | the predicted values are computed exactly |
| Stochastic models | the predicted values depend on probability distribution |
| Compartment models | the variables defining the system are quantified by means of time-dependent differential equations |
| Matrix models | use matrices in the mathematical formulation |
| Reductionistic models | include as many relevant details as possible |
| Holistic models | use general principles |
| Static models | the variables defining the system are not dependent on time |
| Dynamic models | the variables defining the system are a function of time (or perhaps of space) |
| Distributed models | the parameters are considered functions of time and space |
| Lumped models | the parameters are within certain prescribed spatial locations and time, considered as constants |
| Linear models | first degree equations are used consecutively |
| Nonlinear models | one or more of the equations are not first degree |
| Causal models | the inputs, the states and the outputs are interrelated by use of causal relations |
| Black box models | the input disturbances affect only the output responses. No causality is required |
| Autonomous models | the derivatives are not explicitly dependent on the independent variable (time) |
| Nonautonomous models | the derivatives are explicitly dependent on the independent variable (time) |

**TABLE 2.1b:**

**Identification of Models.**

| Type of models | Organization | Pattern | Measurements | Examples |
|---|---|---|---|---|
| Biodemographic | Conservation of species or genetic information | Life cycles | Number of individuals or species | Chapter 6 |
| Bioenergetic | Conservation energy | Energy flow | Energy | Chapter 4 |
| Biogeochemical | Conservation of mass | Element cycles | Mass or concentra- tions | Chapt. 7 |

A model, which relates the input of nutrient with the phytoplankton concentration in a reservoir directly, is an example of a black box model. The relationship might be found on basis of a statistical analysis of the forcing function (nutrient input) and the phytoplankton concentration measured in the reservoir water. If on the other hand the processes are described in the model by use of equations, which give the relationship, the model will be causal.

The modeller might prefer to the use of black box descriptions in the cases, where his knowledge about the processes is rather limited. The disadvantage by the the black box model is however, that it is limited in application to the considered ecosystem or at least to a similar ecosystem. If a general applicability is needed it is necessary to set up a causal model. This latter type is much more widely used in ecology than the black box model, mainly due to the understanding that the causal model gives the model user of the function of the ecosystem.

Autonomous models are not explicitly dependent on time (the independent variable):

$$dy/dt = ay^b + ct^d + e \tag{2.1}$$

Nonautonomous models contain terms, g(t), that make the derivatives dependent on time, f.inst.:

$$dy/dt = ay^b + cy^d + e + g(t) \tag{2.2}$$

The expressions homogeneous and nonhomogeneous models are often used to cover autonomous respectively nonautonomous models, when the derivatives are linear functions.

In Table 2.1b. another classification of models is shown. The differences between the three types of models are the choice of components used as state variables. If the model aims for a description of a number of individuals, species or classes of species the model will be called **biodemographic**. A model, which describes the energy flows is named **bioenergetic** and the state variables will typically be expressed in kW or kW per unit of volume or area. **The biogeochemical models** consider the flow of material and the state

23

variables are indicated as kg, kg/m³ or kg/m². Often this type of models will include one or more element cycles.

Biodemographic models are mainly treated in chapter 6 on Modelling the Population Dynamics, while the chapter on "Bio-geochemical Dynamic Models" focus on biogeo-chemical models. Energy can to a certain extent replace organic matter as 1 kg of biological material can be given a content of energy. It is therefore often quite simple to transfer a biogeochemical model to a bioenergetic model, that describes the energy cyclus. The difference between these two model types is therefore minor and is often related to the conceptual phase, see chapter 4.

## 2.4. SELECTION OF MODEL COMPLEXITY AND STRUCTURE

When the modeller has clarified the scope of the model, the basic properties of the ecosystem, that he is going to model and the data availability, the next step will be to set up a conceptual diagram for the model. As modelling is an iterative process he might be forced to go back and *redefine the problem or expand the data requirements*, perhaps already after the conceptualization phase. But sometimes it may be impossible to expand the amount of available data and the modeller is then forced to simplify the model already at this stage. The model is determined by the problem, the ecosystem and the data. And even for the most enthusiastic modeller the resources are limited.

A mathematical model will therefore always be a result of *several simplifications and assumptions* and it is a difficult task to make the right ones. A ecosystem can be model-led in several ways according to the purpose of the project . The choice of subsystems or model compartments is arbitrary. Thus, several alternative models can be derived for the same environment and usually no objective method is used to select one particular model in stead of another, given the modelling goals. The choice of the compartments involves a conceptualization of the system under study so that the right information can be obtained from the model. The process of conceptualization is the  most fundamental, because once decisions are made at this level, all results and conclusions will be depen-dent on this choice.

Various methods of constructing a conceptual model will be discussed in chapter 4, but here the considerations involved, when the modeller **selects the complexity and the structure** of the model, will be touched on. Only a few theoretical approaches are available to solve this crucial problem, but anyhow they will be able to give some guidelines on model selection.

A model which will be capable of accounting for the complete input-output beha-viour of the real ecosystem and be valid in all experimental frames can never be fully known (Zeigler 1976). This model is called the base model by Zeigler, and it would be very complex and require such a great number of computational resource, that it would be almost impossible to simulate. The base model of an ecosystem will never be fully known, because of the complexity of the system and the impossibility to observe all states. However, given an experimental frame of current interest, a modeller is likely to find it possible to construct a relatively simple model, that is workable in that frame. This is a lumped model and it is the modellers image of the ecosystem with the compo-nents lumped together and with simplified interactions (Zeigler 1976).

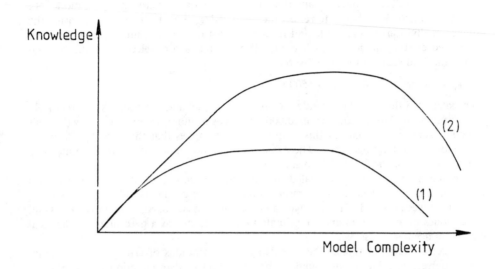

Fig. 2.5: Knowledge plotted versus model complexity measured f.inst. by the number of state variables. The knowledge increases up to a certain level. Increased complexity beyond this level will not add to ones knowledge about the modelled system. At a certain level ones knowledge might be decreased. (2) corresponds to an available data set, which is more comprehensive or has a better quality than (1).

It is a general assumption that a model may be made more realistic by adding more and more connections, up to a point. Addition of new parameters after that point does not contribute further to improved simulations, but on the contrary more parameters imply more uncertainty, because of the possible lack of information about the flows, which the parameters quantify, Given a certain amount of data, the addition of new state variables or parameters beyond a certain model complexity does not add to our ability to model the ecosystem, but only adds unaccounted uncertainty. These ideas are visualized in fig.2.5. The relationship between knowledge gained through a model and its complexity (f.inst. measured as the number of state variables or the number of connectivities) is shown for two levels of data quality and quantity.The question under discussion can be formulated with relation to this figure∗: how can we select the complexity and the structure of the model to assure that we are on the optimum for "knowledge gained", or the best answer to the question posed to the model. Costanza and Sklar (1985) have examined articulation, accuracy and effectivenes of 87 different mathematical models of Wetlands. They use the following equation for the articulation index:

$$A_i = \frac{N_i - 1}{k_i + (N_i - 1)} \times 100 \tag{2.3}$$

Where $A_i$ is the articulation index for mode i, $N_i$ is the number of divisions in mode i, and $k_1$ is a scaling factor for mode i. The number of divisions in each mode are: the number of components or state variables for the component mode, $N_c$, the number of time steps for the time mode, $N_t$, and the number of spatial units for the space mode, $N_s$. The scaling factor was chosen to reflect the relative degree of difficulty of increasing the number of divisions in the mode and to give an idea of the maximum size of the most articulated existing models in each mode. The scaling factors selected for the components, time and space respectively were:

$$k_c = 50, \ k_t = 1000 \text{ and } k_s = 5000.$$

They calculated the articulation index for both the model and the data, as it is relatively easy to run a simulation model with 10.000 time steps or more, whereas it is very difficult to collect supportive data at this frequency. This implies that the articulation of the data is often the limiting factor. The average articulation index of the three modes was found for each of the 87 models axamined.

An index of accuracy was calculated as the percentage of the total (historical) variation that was explained by the model, averaged over all three modes and stated as a fraction between 0 and 1. The average value was used to standardize the index across all three modes of articulation and to estimate model accuracy as a percentage of the total maximum accuracy possible.

Costanza and Sklar ranked the models by use of an index of effectiveness or explanatory power. This index was found as the coefficient of determination for each mode multiplied by the minimum of the data or model articulation index for that mode, and then averaged over the three modes. The most effective model under this scheme is one that balances the costs of added articulation against the benefits of increased accuracy to do the best job of explaining all the models of the system.

The results of this review are summarized in Table 2.2 and figs. 2.6 and 2.7. In the table only the 26 models appears, that were documented sufficiently to allow to make all the calculations. Fig.2.7 indicates an interesting result as it supports the more philosophical statement shown in fig.2.5. Note that the effectiveness index used in fig.2.7 is different from that used in Table 2.2. It is roughly twice as high. Of course all indices must be considered relatively and the authors state that the results should be considered as a hypothesis due to the small amount of supporting data.

The results from this model review have some interesting scientific perspectives. In the past, scientists have tended to narrow their questions in order to achieve higher accuracy. This leads to models with low complexity but high descriptive accuracy. The results say much about little.

**Nature is however complex and it becomes impossible to describe reactions of all species to the combinations of all possible impacts** (forcing functions) **by use of accurate answers to narrow questions.** In physics it is impossible to know simultaneously the accurate location and velocity of a particle. This is in accordance with the uncertainty relations by Heisenberg, and can be explained by Bohr's complementarity principle: the observer effects the object. In ecology we have a similar uncertainty relation: all components and processes cannot be described accurately at the same time. **The product of the number of elements in the model and the descriptive accuracy of the model has an upper limit and the trade off for the modeller is between knowing much about little and little about much.** The complexity of nature can only be described by statistical high number of elements and therefore scientists have recently been looking at nature more comprehensively from a viewpoint of models and systems. This has enabled us to know a little more about much. In accordance with Costanza and Sklar, we therefore have **to construct models with high effectiveness, this is a function of both how much it attempts**

to explain (articulation) and how well it explains that, which was attempted (descriptive accuracy).

The same problem might also be approached in a more pragmatic way. If we estimate that the maximum amount of resources, that can be devoted to one project corresponds to $10^8$ measurements or determinations, the two extremes would be to apply these measurements to get one piece of information, but with very high accuracy or to attempt to squeeze as much information out of the data as possible for a very complex system. If we estimate the accuracy obtainable for one measurement to be 0.1 relatively - it is 10% standard deviation - we can in the first case obtain an accuracy of $0.1/\sqrt{10^8} = 10^{-5}$. In the latter case the question is how many state variables (components) can be considered in our model and we have still a fairly good picture of the system problem under consideration. If we have two dependent state variables and want to get a picture of their relation we need at least 3 measurements. With two measurements, we cannot decide whether the function is linear or nonlinear. Correspondingly, if we have three dependent variables and want to get a picture of their variations and interactions we need to describe the shape of a plane. Consequently we need at least $3^2 = 9$ measurements. Finally, if we consider n state variables and cannot exclude any interrelationships, it would be necessary to have $3^{n-1}$ measurements. From $3^{n-1} < 10^8$, we can see that n < 18.

In accordance with these considerations we might formulate a first approach to an approximate ecological (biological) uncertainty relation:

$$\frac{10^5 \cdot \Delta x}{3^{n-1}} = 1 \tag{2.4}$$

where $\Delta x$ is the relative accuracy (standard deviation) of one "situation" and n is the number of components in the model, time and space coordinates. Note that we have presumed that we consider dependent variables, for which a relationship is valid. Doubtless a model might often attempt relationships which can be omitted for the considered problem.

It is obvious that an increase in the number of dependent state variables, will very rapidly require that many measurements, that it becomes impossible to validate such a model due to shortage in human resources.

**TABLE 2.2:**

**Summary of Model Characteristics**

| Model reference Author | Date | PCNL | Articulation | Descriptive accurcy | Effectivenes |
|---|---|---|---|---|---|
| Walters et al. | 1980.2 | 50.0 | 20.5837 | 0.253333 | 7.82181 |
| Gardner et al. | 1980.0 | 50.0 | 7.1701 | 0.593333 | 6.38183 |
| Jørgensen | 1982.0 | 35.8 | 9.1769 | 0.460000 | 6.33204 |
| Huff and Young | 1980.0 | 0.0 | 6.4299 | 0.286667 | 5.52973 |
| Mitsch | 1976.1 | 47.8 | 7.5038 | 0.466667 | 5.24333 |
| Miller et al. | 1976.0 | 50.0 | 4.8895 | 0.560000 | 4.10720 |
| Brown | 1978.1 | 41.6 | 3.8755 | 0.626667 | 3.64298 |
| Ondok and Pokorny | 1982.0 | 14.3 | 3.7226 | 0.586667 | 3.27587 |
| Burns and Taylor | 1979.1 | 8.0 | 3.5714 | 0.293333 | 3.14286 |
| Wheeler et al. | 1978.0 | 0.0 | 3.0303 | 0.273333 | 2.48485 |
| Wiegert | 1971.3 | 50.0 | 3.0436 | 0.260000 | 2.36364 |
| Wiegert | 1971.2 | 50.0 | 3.0436 | 0.226667 | 2.06061 |
| Walters et al. | 1980.1 | 50.02 | 0.5837 | 0.066667 | 2.05837 |
| Botkin et al. | 1972.0 | 20.03 | 8.1976 | 0.100000 | 1.93548 |
| Richey | 1977.0 | 33.3 | 3.7706 | 0.333333 | 1.79904 |
| Sklar | 1983.1 | 45.2 | 3.7797 | 0.266667 | 1.51189 |
| Halfon | 1979.0 | 0.0 | 1.4983 | 0.633333 | 1.42338 |
| Hopkinson and Day | 1980.0 | 0.0 | 3.3801 | 0.453333 | 0.93508 |
| Paschal et al. | 1979.0 | 50.0 | 6.3024 | 0.106667 | 0.90484 |
| Stone and McHugh | 1979.0 | 0.0 | 4.8535 | 0.230000 | 0.88462 |
| Nyholm | 1978.1 | 50.0 | 3.8360 | 0.153333 | 0.88228 |
| Huff et al. | 1973.0 | 50.0 | 4.8430 | 0.246667 | 0.55458 |
| White et al. | 1978.0 | 0.0 | 0.9096 | 0.386667 | 0.52759 |
| Wiegert | 1971.5 | 50.0 | 3.0436 | 0.230000 | 0.45098 |
| Sklar | 1983.0 | 45.2 | 0.7694 | 0.333333 | 0.37471 |
| Wiegert | 1971.4 | 50.0 | 3.0436 | 0.063333 | 0.35849 |

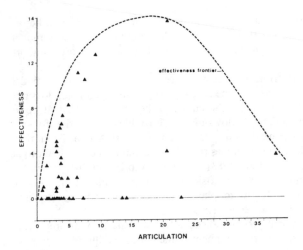

Fig.2.6: Plot of articulation index vs. descriptive accuracy index for the models reviewed in this study, showing the current accuracy frontier.

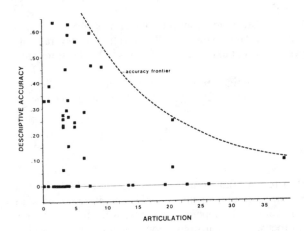

Fig.2.7: Plot of articulation index vs. effectiveness index showing the current effectiveness frontier.

There are only two possibilities to get around this dilemma: either to limit the number of state variables in the model to a number close to 15-20, or to describe the system by use of holistic methods and models, preferably by use of higher level scientific laws. See also the discussion about holistic and reductionistic approaches in 2.3.

Through a good knowledge to the system, it is possible to set up a mass- or energy-flow diagram. It might be considered a cenceptual model of its own but in this context the idea is to use it to recognize the most important flows for the model in question. Let us use energy flow diagram for Silver Springs, see fig. 2.8 as an example. If the goal of the model is to make predictions as to the net primary production for various conditions of temperature and input of fertilizers, it seems of importance to include plants, herbi-vores, carnivores and decomposers (as they mineralize the organic matter). A model consisting of these 4 state variables might be sufficient and the top carnivores, import and export can be deleted.

As energy flows are different from ecosystem to ecosystem, the selected model should also be different. A general model for one type of ecosystem, e.g. a lake, does not exist, on the contrary: it is necessary to adopt the model to the characteristic feature of the ecosystem. Figs.2.9 and 2.10 show the P-flows of two eutrophication models for two different lakes: a shallow lake in Denmark and Lake Victoria in East Africa. From time to time the latter has a thermocline, which implies that the lake should be divided into at least two horizontal layers, (Jørgensen et al., 1983). The food web is also different in the two lakes in that in Lake Victoria herbivorous fish, graze on phytoplankton, while in the danish lake the grazing is entirely by zooplankton. These differences were also reflected in the models set up for the two ecosystems.

In many shallow lakes the physical processes caused by the wind play an important role. In Lake Balaton the wind stirs up the sediment, which consists almost entirely of calcium compounds, having a high adsorption capacity for phosphorous compounds. Consequently studies on Lake Balaton have shown that the mass flows of phosphorous compounds from the water column to the sediment due to this effect is significant. Therefore an adequate description of the stirring up of the sediment, the adsorption of phosporous compounds on the suspended matter and sedimentation must be included in an eutrophication model for this lake.

Halfon et al (1978 and 1979) use another approach. They examine models of dif-ferent complexity and order them in **accordance with validation results**. As several state variables are compared, it is necessary to compare vectors or to weight the state vari-ables.

Jørgensen and Mejer (1979) use an examination of the inverse sensitivity called the ecological buffer capacity to select the number of state variables. The concept ecological buffer capacity is defined as:

$$\beta = dF/d(St) \tag{2.5}$$

where St is a state variable and F a forcing function. It is of course possible to define many different buffer capacities corresponding to all possible combinations of state variables and forcing functions. However, the scope of the model will often point out, which buffer capacity should be in focus. For a eutrophication model f.inst. it would be the change in input of phosphorus (or nitrogen) to the concentration of phytoplankton. Now the modeller examines the relationship between the buffer capacity in focus and the number of state variables. As long as the buffer capacity is changed significantly by addition of an extra state variable, the model complexity should be increased. But if additional state variables only change the buffer capacity insignificantly an increased model complexity will only augment the number of parameters and thereby add to the uncertainty.

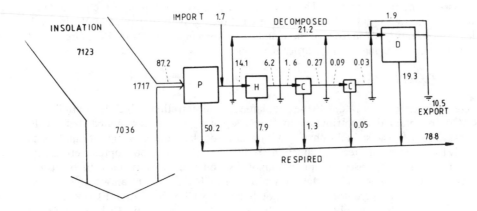

Fig.2.8: Energy flow diagram for Silver Springs, Florida.
Figures in kcal/m²/year. (Adapted from Odum, 1957).

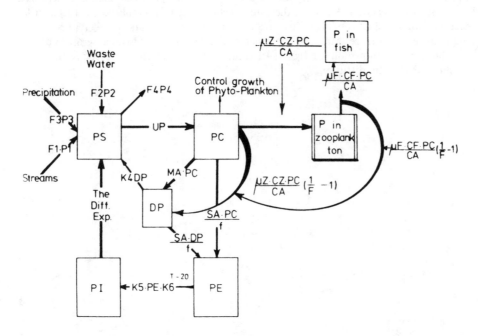

Fig.2.9: The P-cycle. The equations for the processes are included in the diagram.

P-inputs to the lake are streams F1 × P1, waste water F2 × P2 and precipitation F3 × P3. F represents flows and P phosphorus concentrations in the flows. F4 × P4 is output (flow times P-conc.). P4 is total phosphorus PS is soluble phosphorus, which is taken up by

31

phytoplankton by a rate of UP. PC represents the concentration of P in phytoplankton expressed as mg per liter water. The model uses a 2-steps description of the phytoplankton dynamics: first uptake of nutrient (here UP), then growth controlled by the intracellular concentration of nutrients, here indicated by an arrow from PC. Growth of zooplankton by grazing is a first order reaction:

$$\mu Z \times CZ \times PC/CA,$$

where $\mu Z$ is the growth rate, CZ the zooplankton concentration, which is multiplied by PC/CA to obtain the contribution to P in zooplankton as it is assumed that the P: biomass ration is the same in phytoplankton and zooplankton. Grazing rate multiplied by ((1/F)-1) represents fecal production by grazers. F is the food uptake efficiency. Similar equations are used for predation of fish on zooplankton. Growth of fish is $\mu F \times CF \times PC/CA$ where $\mu F$ is the growth of fish, CF the concentrations of fish. The fecal production by predators is correspondingly $\mu F \times CF \times PC/CA((1/F)-1) \times$ PC/CA is as for zooplankton to obtain P in fish, assuming the same ratio P: biomass as for phytoplankton. SA $\times$ PC is settling of phytoplankton. f represents the ratio of PC which goes to PE, exchangeable P in sediment. SA $\times$ PC/(1-f) goes to non-exchangeable P, which will never released to the water again. MA $\times$ PC represents a first order decay of P in phytoplankton. MA is mortality rate. DP is detritus phosphorus, which is mineralized by a first order reaction, K4 being the rate constant. PE is mineralized by a first order reaction with a rate $K5 \times PE.K6^{T-20}$, where K5 is a rate constant and $K6^{T-20}$ represents the temperature dependence. Many of the other rate constants are also temperature dependent, but it is not shown on the figure. PI is interstitial P, which is released to the water as soluble P (PS) by diffusion.

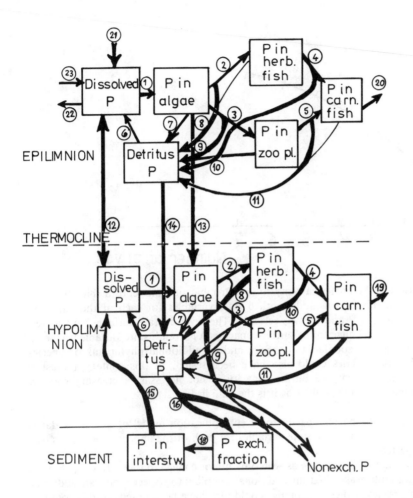

Fig.2.10: Eutrophication model illustrated by use of P-cycling. Arrows indicate processes. A thermocline is considered. Explanation of numbers are as follows: 1) uptake of phosphorus by algae 2) grazing by herbivorous fish 3) grazing by zooplankton 4), 5) predation on fish and zooplankton respectively by carnivorous fish. 6) mineralization 7) mortality of algae 8), 9), 10), 11) grazing and predation loss 12) exchange of P between epilimnion and hypolimnion 13) settling of algae (epilimnion - hypolimnion) 14) settling of detritus (epilimnion - hypolimnion) 15) diffusion of P from interstitial - to lake water 16) settling of detritus (hypolimnion - sediment) (a part goes to the non-exchangeable fraction) 17) settling of algae (hypolimnion - sediment) (a part goes to the non-exchangeable fraction) 18) mineralization of P in exchangeable fraction 19), 20) fishery 21) precipitation 22) outflows 23) inflows (tributaries).

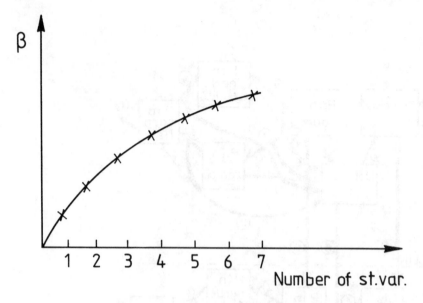

β

Number of st.var.

Fig.2.11: Illustrates the buffer capacity for a eutrophication model of a shallow danish lake. In this case a model with 6 state variables for each of the important nutrients (C, P and N) was selected. The seventh state variable gave as seen only minor change to the buffer capacity. As seventh state variable was tested an additional zooplankton species and an additional phytoplankton species. Other possibilities could also have been tested. In this context it must be pointed out that the buffer capacity is not necessary increasing with the number of state variables as it is the case in fig. 2.11.

Fig. 2.11 illustrates the buffercapacity for a eutrophication model of a shallow danish lake. In this case a model with 6 state variables for each of the important nutrients (C,P and N) was selected.

The seventh state variables gave as seen only minor change to the buffer capacity. As seventh state variable was tested an additional zooplankton species and an additional phytoplankton species. Other possibilities could also have been tested. In this context it must be pointed out that the buffercapacity is not necessary increasing with the number of state variables as it is the case in fig. 2.11.

Halfon (1983) uses Bossermann's measure of recycling (see Bossermann 1980 and 1982) as an index of connectivity as criteria for the selection of model structure. Ecosystems have a certain amount of recycling and an ecological model must mimic this recycling. If the model structure is too loose and not much recycling can be simulated structural uncertainty is introduced into the model. Adding links or state variables improves the model connectivity and thus recycling. At a certain point additions of new links will, however, not improve the model behavior much and therefore these additional links are useless from a model performance point of view. An example should be quoted to illustrate this method of selection model structure.

The pattern of interconnections among state variables can be described with an adjacency matrix A. An adjacency matric element $A_{ij} = 1$ if a direct link i-j exists and $= 0$ if no direct link exists. The direct connectivity of a model is the number of ones in the adjacency matric divided by $n^2$, where n is the number of rows or columns. Multi-lenghts links of order k can be studied by looking at the elements of the matrix $A^k$. For example the matrix $A^2$ shows the position and numbers of all 2-lenghts paths. The recycling measure, c, introduced by Bossermann is the number of ones in the first n matrices of the power series divided by $n^3$, which is equal to the number of total possible ones. c will vares between 0 and 1, when there are no paths respectively when all paths are realized.

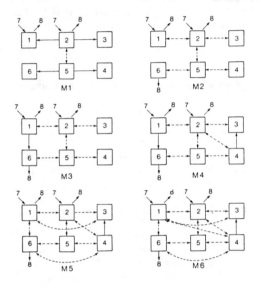

Fig.2.12: Model structures for first set of models with six state variables. Suspended sediments (1), water (2), fish (3) benthos (4), pore water (5), bottom sediments (6), inputs (7), outputs to the environment (8).

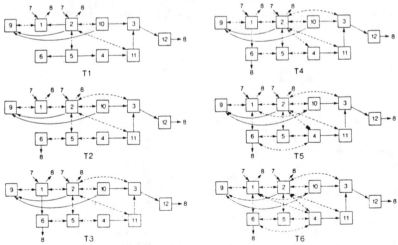

Fig.2.13: Model structures for second set of models with ten state variables. Suspended sediments (1), water (2), fish (3), benthos (4), pore water (5), bottom sediments (6), inputs (7), outputs t the environment (8), detritus (9), plankton (10), benthic fish (11), sea gulls (12).

Halfon (1983) used two sets of models, one with six (M-models) and one with ten state variables (T-models). Each set has six model configurations of increasing complexity (connectivity). The state variables are: suspended matter (1), water (2), fish (3), benthos (4), pore water (5) and bottom sediment (6). Fig. 2.10 shows the M-models and fig. 2.13 illustrates the T-models. The latter serie has the same state variables as the M-model with addition of detritus (9) phytoplankton (10), benthic fish (11) and sea gulls (12). The numbers 7 and 8 represent inputs and outputs respectively.

**TABLE 2.3**

**Adjacency matrix of model M2. Element $a_{ij}, j = 1,6$ may be zero (no internal recycling) or one (internal recycling)**

|   |   |            | TO |   |   |   |   |   |   |   |
|---|---|------------|----|---|---|---|---|---|---|---|
|   |   |            | 1  | 2 | 3 | 4 | 5 | 6 | 7 | 8 |
|   | 1 | Susp. sed   | 0 | 1 | 0 | 0 | 0 | 0 | 0 | 1 |
|   | 2 | Water       | 1 | 0 | 1 | 0 | 1 | 0 | 0 | 1 |
| F | 3 | Fish        | 0 | 1 | 0 | 0 | 0 | 0 | 0 | 0 |
| R | 4 | Benthos     | 0 | 0 | 0 | 0 | 1 | 0 | 0 | 0 |
| O | 5 | Pore water  | 0 | 1 | 0 | 1 | 0 | 1 | 0 | 0 |
| M | 6 | Bottom sed. | 0 | 0 | 0 | 0 | 1 | 0 | 0 | 1 |
|   | 7 | Inputs      | 1 | 1 | 0 | 0 | 0 | 0 | 0 | 0 |
|   | 8 | Outputs     | 0 | 0 | 0 | 0 | 0 | 0 | 0 | 0 |

Direct connectivity = 16/64 = 0.156.

**TABLE 2.4**

**Adjacency matrix of model T2. Element $a_{ij}, j = 1,12$, $j = 7$, $j = 8$ may be equal to zero (no internal recycling) on one (internal recycling)**

|   |    |             | TO |   |   |   |   |   |   |   |   |    |    |    |
|---|----|-------------|----|---|---|---|---|---|---|---|---|----|----|----|
|   |    |             | 1  | 2 | 3 | 4 | 5 | 6 | 7 | 8 | 9 | 10 | 11 | 12 |
|   | 1  | Susp. sed    | 0 | 1 | 0 | 0 | 0 | 0 | 0 | 1 | 1 | 0 | 0 | 0 |
|   | 2  | Water        | 1 | 0 | 1 | 0 | 1 | 0 | 0 | 1 | 1 | 1 | 1 | 0 |
|   | 3  | Fish         | 0 | 1 | 0 | 0 | 0 | 0 | 0 | 0 | 0 | 0 | 0 | 1 |
|   | 4  | Benthos      | 0 | 0 | 0 | 0 | 1 | 0 | 0 | 0 | 0 | 0 | 1 | 0 |
| F | 5  | Pore water   | 0 | 1 | 0 | 1 | 0 | 1 | 0 | 0 | 0 | 0 | 0 | 0 |
| R | 6  | Bottom sed.  | 0 | 0 | 0 | 0 | 1 | 0 | 0 | 1 | 0 | 0 | 0 | 0 |
| O | 7  | Inputs       | 1 | 1 | 0 | 0 | 0 | 0 | 0 | 0 | 0 | 0 | 0 | 0 |
| M | 8  | Outputs      | 0 | 0 | 0 | 0 | 0 | 0 | 0 | 0 | 0 | 0 | 0 | 0 |
|   | 9  | Detritus     | 1 | 0 | 0 | 0 | 0 | 0 | 0 | 0 | 0 | 0 | 0 | 0 |
|   | 10 | Plankton     | 0 | 1 | 1 | 0 | 0 | 0 | 0 | 0 | 1 | 0 | 0 | 0 |
|   | 11 | Benthic fish | 0 | 1 | 1 | 0 | 0 | 0 | 0 | 0 | 0 | 0 | 0 | 0 |
|   | 12 | Sea gulls    | 0 | 0 | 0 | 0 | 0 | 0 | 0 | 1 | 0 | 0 | 0 | 0 |

Direct connectivity = 29/144 = 0.194.

Table 2.3 shows the adjacency matrix of M2 and table 2.4 of T2. For each set of models two analysis were done: no considerable recycling within each state variables, i.e. $a_{jj} = 0$

or some recycling $a_{jj} = 1$.

Table 2.5 shows the complete calculation for the index c of model M4. c is found as $19 + 39 + 46 + 49 + 4 \times 49/8^3 = 0.682$.

## TABLE 2.5

**Boolean powers of the M4 model adjacency matrix and their first four sums. Calculation of $\bar{c}$.**

```
A¹                              A¹
0  1  0  0  0  1  0  1          0  1  0  0  0  1  0  1
1  0  1  1  1  0  0  1          1  0  1  1  1  0  0  1
0  1  0  0  0  0  0  0          0  1  0  0  0  0  0  0
0  1  1  0  1  0  0  0          0  1  1  0  1  0  0  0
0  1  0  1  0  1  0  0          0  1  0  1  0  1  0  0
0  0  0  0  1  0  0  1          0  0  0  0  1  0  0  1
1  1  0  0  0  0  0  0          1  1  0  0  0  0  0  0
0  0  0  0  0  0  0  0          0  0  0  0  0  0  0  0

A²                              A¹ + A²
1  0  1  1  1  0  0  1          1  1  1  1  1  1  0  1
0  1  1  1  1  1  0  1          1  1  1  1  1  1  0  1
1  0  1  1  1  0  0  1          1  1  1  1  1  0  0  1
1  1  1  1  1  1  0  1          1  1  1  1  1  1  0  1
1  1  1  1  1  0  0  1          1  1  1  1  1  1  0  1
0  1  0  1  0  1  0  0          0  1  0  1  1  1  0  1
1  1  1  1  1  1  0  1          1  1  1  1  1  1  0  1
0  0  0  0  0  0  0  0          0  0  0  0  0  0  0  0

A³                              A¹ + A² + A³ *
0  1  1  1  1  1  0  1          1  1  1  1  1  1  0  1
1  1  1  1  1  1  0  1          1  1  1  1  1  1  0  1
0  1  1  1  1  1  0  1          1  1  1  1  1  1  0  1
1  1  1  1  1  1  0  1          1  1  1  1  1  1  0  1
1  1  1  1  1  1  0  1          1  1  1  1  1  1  0  1
1  1  1  1  1  0  0  1          1  1  1  1  1  1  0  1
1  1  1  1  1  1  0  1          1  1  1  1  1  1  0  1
0  0  0  0  0  0  0  0          0  0  0  0  0  0  0  0

A⁴                              A¹ + A² + A³ + A⁴
1  1  1  1  1  1  0  1          1  1  1  1  1  1  0  1
1  1  1  1  1  1  0  1          1  1  1  1  1  1  0  1
1  1  1  1  1  1  0  1          1  1  1  1  1  1  0  1
1  1  1  1  1  1  0  1          1  1  1  1  1  1  0  1
1  1  1  1  1  1  0  1          1  1  1  1  1  1  0  1
1  1  1  1  1  1  0  1          1  1  1  1  1  1  0  1
1  1  1  1  1  1  0  1          1  1  1  1  1  1  0  1
0  0  0  0  0  0  0  0          0  0  0  0  0  0  0  0
```

$A^5$ through $A^8$ are the same as $A^4$. All further sums are the same. $\bar{c} = $ sum of number of ones in the first eight matrices of Boolean series/$n^3 = 0.682$.

\* This matrix is the reachability matrix.

In table 2.6 and 2.7 are summarized the results of the computations for the six M-models and six T-models both with and without internal recycling.

By looking over the results from the M-models in table 2.6 we see a marked change between models M3 and M4, as c increases from 0.449 to 0.682. Furthermore it has been attempted to add and delete paths to the six M-models and it was found that M4 was much less sensitive to addition and delection of any paths than model M3. Model M5 is still less sensitive to individual structural pertubations. It means that an inappropriate parameterization may have less crucial effect on the model behavior for model M4 (or M5 and M6) than for M3. The improved structural properties of M5 and M6 are not so much beter to overcome the fact that they have more parameters and therefore more uncertain flow rates than M4. Among the M-series M4 should be preferred.

The same formal reasoning is valid for the T-series and it is concluded that T2 or T3 should be used as structural models, depending on the information one has from the system of interest.

Such a structural analysis of a model cannot be dome completely in vacuum, but must be related to the system, when an application is sought.

The analysis can however diminish the number of arbitrary choices, as they are usually done. The method should also be used parallel to other possible approaches and can then be considered a very useful tool.

**TABLE 2.6:**

**Direct and indirect connectivity of the adjacency matrices for the first set of models with six state variables.**

| Model | Direct connectivity | $\bar{c}$ |
|---|---|---|
| Without internal recycling ($a_{jj} = 0$) | | |
| M1 | 0.15625 | 0.18359 |
| M2 | 0.23438 | 0.44531 |
| M3 | 0.25000 | 0.44922 |
| M4 | 0.29688 | 0.68164 |
| M5 | 0.37500 | 0.71289 |
| M6 | 0.40625 | 0.72070 |
| With internal recycling ($a_{jj} = 1$) | | |
| M1 | 0.25000 | 0.38281 |
| M2 | 0.32813 | 0.68945 |
| M3 | 0.34375 | 0.69531 |
| M4 | 0.39063 | 0.71289 |
| M5 | 0.46875 | 0.72852 |
| M6 | 0.50000 | 0.73243 |

**TABLE 2.7:**

**Direct and indirect connectivity of the adjacency matrices for the second set of models with ten state variables.**

| Model | Direct connectivity | $\bar{c}$ |
|-------|---------------------|-----------|
| Without internal recycling ($a_{jj} = 0$) | | |
| T1 | 0.15972 | 0.33391 |
| T2 | 0.19444 | 0.66898 |
| T3 | 0.20139 | 0.67419 |
| T4 | 0.21528 | 0.69734 |
| T5 | 0.25000 | 0.71065 |
| T6 | 0.26389 | 0.71412 |
| With internal recycling ($a_{jj} = 1$) | | |
| T1 | 0.22917 | 0.50637 |
| T2 | 0.26389 | 0.71470 |
| T3 | 0.27083 | 0.71759 |
| T4 | 0.28472 | 0.72454 |
| T5 | 0.31944 | 0.73264 |
| T6 | 0.33333 | 0.73438 |

The selection of complexity and structure of models is close to the **aggregation problem.** Aggregation is unification of system components that are homogeneous in some properties into blocks, each beeing a new component with properties defined by the aggregation laws. However, up to now, the theory of aggregation is still poorly developped. If the model is nonlinear the sole method to examine whether aggregation is possible or not is to compare the model outputs of two model versions.

It can be concluded from the various presented method that the model structure should not be selected randomly or arbitrary, but that the modeller should use the presented approaches to the problem to get a certain objectivity into this phase of modelling. As the entire model result is greatly dependent on the model structure and complexity as pointed out a few times above, it pays for the modeller to invest a little time in a proper and more objective selection of the model complexity and structure at this stage of the modelling procedure. Experience shows, that it will save some model corrections at a later stage, if the model has been calibrated and the validation phase indicates, that improvements might be needed. This does, however, not involve that corrections of the model structure at a later stage can be omitted. The methods presented for the selection of model structure are not so rigorous, that the very best model is always selected at the first instance. The method will assist the modeller to exclude some not-workable models, but not necessarily to choose the very best and only right model. But even that should be considered a great advantage.

### 2.5. VERIFICATION

The ecosystem and the problem are basis for the conceptual diagram, which already might be considered to be a model of its own. Therefore chapter 4 will be devoted to various forms of conceptual models. As it will be demonstrated in the chapter on conceptual models, it will be possible to use such models both as a management and scientific tool. In accordance with fig. 2.2 the conceptualization is followed by a mathematical formulation of the processes. Chapter 3 will give a survey of possible formulations of various ecological processes. Having made these two steps of the modelling procedure, the verification follows, see again fig.2.2.

Findeisen et al. (1979) gives the following definition of verification: **A model is said to be verified, if it behaves in the way the model builder wanted it to behave.** This definition implies that there is a model to be verified, which means that not only the model equations have been set up, but also that the parameters have been given reasonable realistic values. The sequence verification, sensitivity analysis and calibration must consequently not be considered a rigid step by step procedure, but rather as a iterative operation. The model is first given realistic parameters from the literature, then it is calibrated coarsely and finally the model can be verified, followed by a sensitivity analysis and a finer calibration. The model builder will have to go through this procedure several times, before the verification and the model output in the calibration phase will satisfy him.

Almost inevitably it will be necessary at some stage during this operation to make assumptions about the statistical properties of the noise sequences idealized in the model. To conform with the properties of white noise any error sequence should broadly satisfy the following constraints: that its mean value is zero, that it is not correlated with any other error sequence and that it is not correlated with the sequences of measured input forcing functions. Evaluation of the error sequences in this fashion can therefore essentially provide a check on whether the final model invalidates some of the assumptions inherent in the model. Should the error sequences not conform to their desired properties, this suggests that the model does not adequately characterize all of the more deterministic features of the observed dynamic behaviour. Consequently, the model structure should be modified to accomodate additional relationships.

To summarize this part of the verification:

1) the errors (comparison model output/observations) must have mean values of approximately zero.

2) the errors are not mutually cross related.

3) the errors are not correlated with the measured input forcing functions.

Results of this kind of analysis are given very illustratively in Beck (1978). Notice that this analysis requires good estimates of standard deviations in sampling and analysis (observations).

In addition and of equal importance to points 1-3 mentioned above  the verification requires a test of the internal logic of the model: does the model have the foreseen causality? And are the responses to pertubations as expected?

This part of the verification is to a certain extent based upon more subjective criteria. Typically the model builder formulates several questions about the reaction of the model. He provokes changes in forcing functions or initial conditions and simulates by use of the model responses to those changes. If the responses are not as expected, he will have to change the structure of the model or the equations, provided that the parameter space is approved. Examples of typical questions will illustrate this operation: Will increased BOD5-loading in a stream model imply decreased oxygen concentration?

Will increased temperature in the same model imply decreased oxygen concentration? Will the oxygen concentration be at a minimum at sun-rise when photosynthesis is included in the model? Will decreased predator concentration in a prey-predator model, in the first instance imply increased prey concentration? Will increased nutrient loadings in a eutrophication model give increased concentration of phytoplankton? etc. Numerous other examples can be given.

Finally, the long term stability of the model should be examined in the verification phase. The model is run for a long period using a certain pattern in the fluctuations of the forcing functions. It should then be expected that the state variables, too, show a certain pattern in their fluctuations. Fig. 2.14 illustrates an example. It is a ten year simulation of phytoplankton and zooplankton in an eutrophication model. The same annual variations in forcing functions are used for all the ten years and the initial values of the state variables are in accordance with measured values from a case study, where the ecosystem (a lake) has found its balance with the prevailing forcing functions. As seen the model output is stable. In this case was ten years simulation selected for the stability test. This was considered sufficiently as the water retention time was 4-6 months. The simulation period should of course be selected sufficiently long to allow the model to demonstrate any possible instability.

The verification may seem very cumbersome, but it is a very necessary step for the model builder to carry out. Through the verification he learns to know his model through its reaction, and the verification is furthermore an important checkpoint in the construction of a workable model. This emphasizes also the importance of good ecological knowledge to the ecosystem, without which the right questions as to the internal logic of the model cannot be posed.

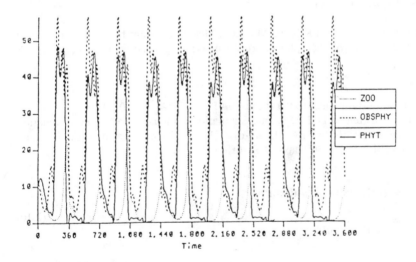

Fig.2.14: Model verification. Ten years simulation of phytoplankton and zooplankton. It can be concluded that the model is long-term stable.

Unfortunately, many models have not been verified properly due to lack of time, but the experience shows that what might seem to be a short cut, will lead to an unreliable model, which at a later stage might require more time consumption to compensate for the lack of verification. It must therefore be strongly recommended to invest enough time in the verification and plan for the necessary allocation of resources in this important phase of the modelling procedure.

## ILLUSTRATION 2.1

Building a model is very time consuming if all steps in the modelling procedure are included, which in fact has to be done to ensure an applicable model. A rather primitive and unrealistic model has therefore been selected to illustrate some of the concepts in this chapter in a few pages.

Fig. 2.15 gives the conceptual diagram of the model, that we want to examine further. The phosphorus cycle in an aquatic ecosystem is modelled. We consider only two state variables: soluble phosphorus, PS, and phosphorus in algae, PA. An input of phosphorus PI takes place and the output of PS and PA follows the outflow of water Q. The volume of the system is V. In addition to these forcing functions, the solar radiation available for photosynthesis can be described as:

$$(2.6)$$

$$S = S_{max} (1 + \sin (0.00\ 8603 * t))$$

where S is the solar radiation, $S_{max}$ is the maximum sunlight equal to 0.5 and t is time. Q/V is equal to 0.01 (time$^{-1}$) PI is 1.0 g P m$^{-3}$.

The uptake of phosphorus by algae (process 1, see fig. 2.15) is described as:

$$(2.7)$$

$$\mu = S \times PA \times S/(PS + K)$$

where $\mu$ is the growth rate and K is the Michaelis-Menten constant here equal to 1.0 g P m$^{-3}$. Processes 2, the loss of algal phosphorus, is described by first order kinetics:

$$(2.8)$$

$$= R * PA$$

where R is a rate constant equal to 0.1 (time $^{-1}$) AT t = 0 PA$_0$ = 1.0 g P m$^{-3}$.

The model is written in DSL-Dynamic Simulation Language/VS, a new simulation language launched recently by IBM. Users of CSMP will recognize some of the statements. The model is run for 365 days. The differential equations are:

$$(2.9)$$

$$dPS/dt = (PIN-PS)Q/V-(\mu-R) \times PA$$

$$(2.10)$$

$$dPA/dt = (\mu-R-Q/V)PA$$

## TABLE 2.8

## A SIMPLE PHOSPHORUS MODEL

PARAMETERS FOLLOWS

```
PARAM  K = 1.0
PARAM  PIN = 1.0
PARAM  Q/V = 0.01
PARAM  R = 0.1
```

PARAM SUNMAX = .5

DIFFERENTIAL EQUATIONS

    DPS = (PIN - PS * QV - ($\mu$ - R) * PA
    DPA = ($\mu$ - R - QV) * PA

INTEGRATORS FOLLOWS

---

    PS = INTGRL (IPS, DPS)
    PA = INTGRL (IPA, DPA)

INITIAL VALUES FOR INTEGRATORS

INCON IPS = 0., IPA = 1.0

ADDITIONALS EQUATIONS FOLLOWS

    PT = PS + PA
    $\mu$ = S * PS/(K + PS)
S = SMAX*(1. + SIN (0.008603 * TIME))

A STATEMENT FOR PLOTTING USE OF VARIABLE

SAVE 5.0, PT, PS, S, $\mu$, PA

GRAPHIC OUTPUT STATEMENTS FOLLOWS

GRAPH (G1, DE = IBM3279) TIME (LE = 10, NI = 5), PA (LI = 71, LE = 8, NI = 5, ...
PS (LI 0 74, LE = 8, NI = 5)
LABEL (G1, DE = IBM3279) A SIMPLE PHOSPHORUS MODEL

CONTROL STATEMENTS FOLLOWS

CONTRL FINTIM = 365.0
END
STOP

The complete computer program is shown in Table 2.8. A comparison between model output and measurements is carried out in illustration 2.3, while illustration 2.2 shows a sensitivity analysis of K in this model.

Figs. 2.16-2.22 give the results of a verification. S is plotted versus the time in fig. 2.16. The behaviour of the state variables is illustrated in fig. 2.17. In figs. 2.18-2.22 the internal logic of the model is tested by recording the response of the model to 1) increased PIN (2.0 g m$^{-3}$) 2) decreased PIN (0.5 and 0.1 g m$^3$ ) 3)increased solar radiation and 4) setting R = O. As seen from the figs. the model response as expected, since 1), 3) and 4) give increased concentration of PA. The mass balance is tested in fig.2.17. As seen the total phosphorus concentration is constant and equal to 1.0 g m$^{-3}$.

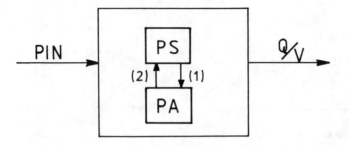

Fig.2.15: Conceptual diagram of the model in illustration 2.1. PS and PA are state variables: soluble phosphorus respectively phosphorus in algae. PIN is input of phosphorus. Q is outflow of water. V is volume of system.

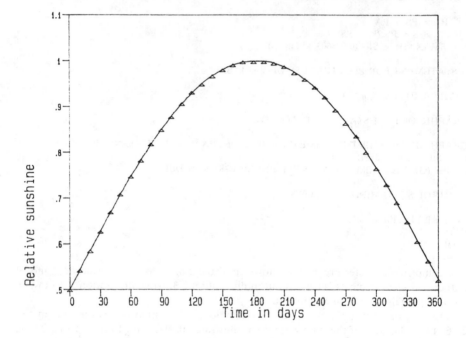

Fig.2.16: S = f(t) is shown.

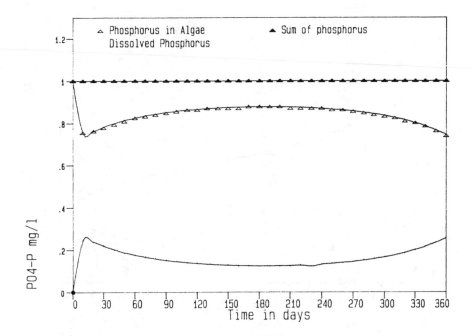

Fig.2.17: PS and PA are plotted versus time, t. Constant total phosphorus is also shown.

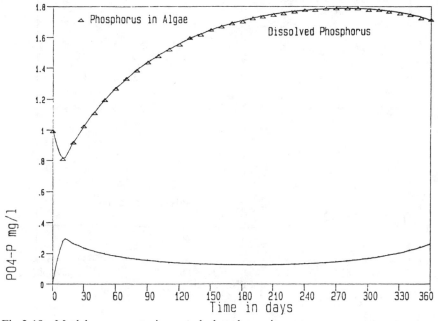

Fig.2.18: Model response to increased phosphorus input.

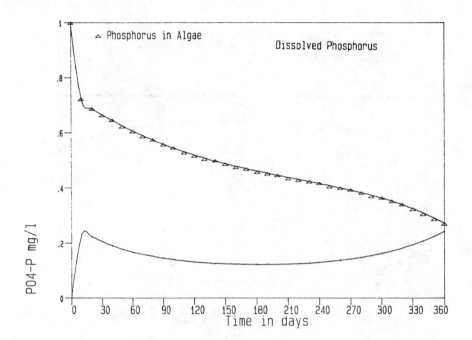

Fig.2.19: Model response to decreased PIN (0.5 g m⁻³), compare with fig. 2.17.

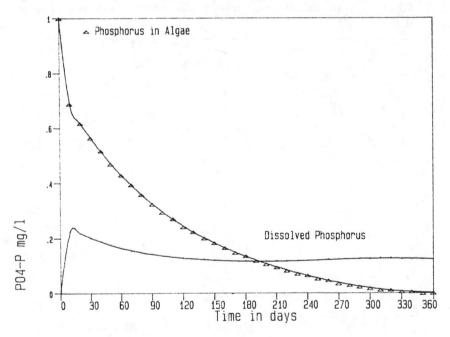

Fig.2.20: Model response to decreased PIN (0.1 g m⁻³), compare with fig.2.17.

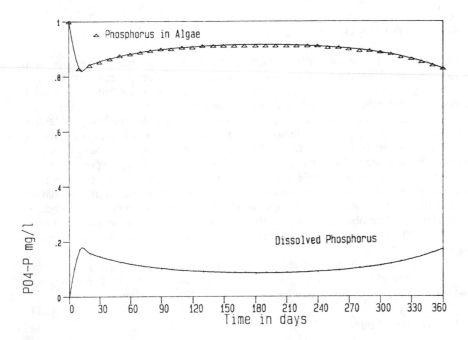

Fig.2.21: Model response to increased solar radiation (S = 0.7), compare with fig.2.17.

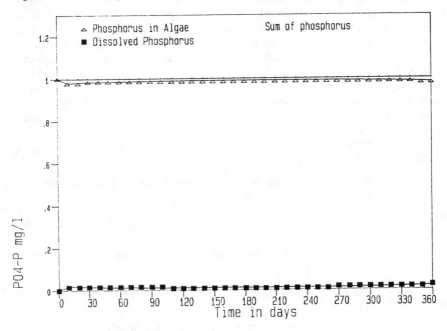

Fig.2.22: Model response to R = O, compare with fig.2.17.

## 2.6. SENSITIVITY ANALYSIS

It is of importance for the modeller to learn the properties of the model. An important step to obtain this knowledge is the above mentioned verification. A sensitivity analysis would be the obvious next step to take. Through this analysis the modeller gets a good overview of the most *sensitive components in the model*.

A sensitivity analysis attempts to provide a measure of the sensitivity of either parameters, forcing functions or submodels to the state variables of greatest interest in the model. If the modeller wants to simulate a response of oxygen concentration in a stream to the discharge of organic matter, he will obviously choose oxygen concentration as the important state variable and he will be interested in the submodels and the parameters, to which the oxygen concentration is most sensitive. If in population dynamics he wants to follow the development of a herbivorous population, the concentration or the total number of this population in a given area will be the important state variable, etc. The first step in the sensitivity analysis is therefore to answer the question: sensitive to what?

In practice the sensitivity analysis is carried out by changing the parameter, forcing function or submodel and observe the corresponding response on the important state variable (x). Thus the sensitivity of a parameter, $S_x$ is defined as follows:

$$S_x = \frac{\partial x/x}{\partial \text{Param}/\text{Param}} \qquad (2.11)$$

where x is the selected state variable and Param the examined parameter.

The relative change in parameters is chosen on basis of our knowledge to the certainty, of the parameters. If the modeller estimates that the parameters are known within $\pm$ 50 % f.inst. he would probably choose a change in the parameters at $\pm$ 10 % and $\pm$ 50 % and record the corresponding change in the state variable (x). It is often necessary to find the sensitivity at two or more levels of parameter changes as the relation between a parameter and a state variable is rarely linear; this implies that it is often crucial to know parameters with the highest possible certainty before the sensitivty analysis is carried out. How this is possible will be discussed in the paragraph below on calibration. It should be added, that the sensitivity might often be higher at the highest level of parameter change.

The interaction between the sensitivity analysis and the calibration could consequently work along the following lines:

1) a sensitivity analysis is carried out at two or more levels of parameter changes. Relatively great changes are applied at this stage.

2) The most sensitive parameters are determined more accurately either by a calibration or by other means (see next paragraph).

3) Under all circumstances great efforts are made to obtain a relatively good calibrated model.

4) A second sensitivity analysis is then carried out using more narrow intervals for the parameter changes.

5) Still further improvements of the parameter certainty is attempted.

6) A second or third calibration is then carried out focusing mainly on the most sensitive parameters.

Table 2.9 shows the result of a partial sensitivity analysis on a eutrophication model. From the results it is evident, that it is of importance to obtain as great certainty as possible for the following parameters: max.growth rate of phytoplankton, max.growth rate of zooplankton, settling rate of phytoplankton, and respiration rate of phytoplankton and zooplankton. Therefore, it would be a great advantage, if these parameters could be determined with great certainty by other means e.g.,by investigations in laboratory aimed for a direct determination of these values.

A sensitivity analysis on submodels (equations) can also be carried out. In this case the change in a state variable is recorded when the equation or submodel is deleted from the model or changed to alternative expressions, e.g.,with more details built into the submodel. Results from such a sensitivity analysis might be used to change the structure of the model, if f.inst. it is found that the submodel has a great impact on the state variable in focus. The selection of the complexity and the structure of the model should therefore work hand in hand with the sensitivity analysis. There is a feed-back from the sensitivity analysis to the conceptual diagram. This idea is in accordance with Halfon's order of model, mentioned in section 2.4.

If it is found that the state variable in focus is very sensitive to a certain submodel, it should be considered which alternative submodels could be used and they should be tested and/or examined further in details either in vitro or in the laboratory.

**It can generally be stated that those submodels, which contain sensitive parameters are also submodels, which are sensitive to the important state variable.** But on the other hand it is not necessary to have a sensitive parameter included in a submodel to obtain a sensitive submodel. A modeller with a certain experience will find these statements intuitively correct, but it is also possible to show that they are correct by analytical methods.

A sensitivity analysis on forcing functions gives an impression of the importance of the various forcing functions and tells us what accuracy is required of the forcing functions data.

## TABLE 2.9:

Analysis of sensitivity (t here = time).
Symbols used: PHYT: phytoplankton, ZOO: zooplankton, NS: soluble nitrogen and PS: soluble phosphorus. Annual average values for sensitivities (S) are shown. t illustrates change in time for occurence of maximum values.

| Definition | Parameter | $S_{PHYT}$ | $S_{ZOO}$ | $S_{NS}$ | $S_{PS}$ | $t_{PHYT}$ | $t_{ZOO}$ | $t_{NS}$ | $t_{PS}$ |
|---|---|---|---|---|---|---|---|---|---|
| Max.growth rate phyt | $CDR_{max}$ | 0.488 | 0.620 | -0.356 | -0.392 | -0.31 | -0.11 | -0.23 | 0.0 |
| Denitrification rate | DENIT | -0.19 | -0.010 | -0.579 | 0.013 | 0.05 | 0.0 | -0.70 | 0.0 |
| Fish concentration | FISH | 0.008 | 0.012 | -0.011 | -0.014 | 0.0 | 0.10 | 0.0 | 0.0 |
| Initial PHYT conc. | PHYT (t=0) | -0.020 | -0.044 | 0.032 | 0.033 | -0.05 | -0.35 | -0.15 | 0.0 |
| Initial ZOO conc. | ZOO (t=0) | -0.169 | -0.223 | 0.252 | 0.282 | 0.0 | -1.58 | -0.43 | 0.0 |
| Rate of mineralization(N) | $KDN_{10}$ ($10^0$ C) | 0.003 | 0.010 | 0.038 | 0.001 | 0.45 | 0.0 | -0.30 | 0.0 |
| - - (P) | $KDP_{10}$ ($10^0$ C) | 0.0 | 0.001 | 0.0 | 0.006 | 0.0 | 0.0 | 0.0 | 0.0 |
| Michaelis—Menten constant (N) | KN | -0.001 | -0.032 | 0.063 | 0.019 | 0.45 | -0.05 | -0.15 | 0.0 |
| Michaelis—Menten constant (P) | KP | -0.003 | -0.014 | 0.021 | 0.034 | 0.05 | -0.05 | -0.25 | 0.0 |
| Max.growth rate ZOO | $MYZ_{max}$ | -2.088 | -4.002 | 2.749 | 4.052 | -1.50 | -25.95 | -17.90 | 0.0 |
| Mortality ZOO | MZ | 2.063 | 1.949 | -3.479 | -3.350 | 1.30 | 21.50 | 8.40 | 0.0 |
| Max.predation rate | $PRED_{max}$ | 0.008 | 0.011 | -0.015 | -0.016 | 0.0 | 0.10 | -0.20 | 0.0 |
| Max.respiration rate PHYT | $RC_{max}$ | -0.243 | -0.201 | 0.139 | 0.153 | 0.45 | 0.05 | -0.35 | 0.0 |
| Max.respiration rate ZOO | $RZ_{max}$ | 0.570 | 0.625 | -0.902 | -0.978 | 0.95 | 5.94 | 1.34 | 0.0 |
| Settling rate detritus | SVD | 0.0 | 0.0 | -0.002 | 0.0 | 0.0 | 0.0 | 0.0 | 0.0 |
| Settling rate PHYT | SVS | -1.042 | -0.823 | 0.321 | 0.388 | -0.05 | 0.20 | 0.15 | 0.0 |
| Max.uptake of C | $UC_{max}$ | 0.629 | 0.636 | -0.428 | -0.481 | 0.05 | 0.10 | -0.25 | 0.0 |
| - - N | $UN_{max}$ | 0.046 | 0.145 | -0.251 | -0.050 | 0.05 | -0.05 | -0.15 | 0.0 |
| - - P | $UP_{max}$ | 0.026 | 0.090 | -0.049 | -0.339 | 0.50 | 0.05 | -0.15 | 0.0 |

**ILLUSTRATION 2.2**

The sensitivity analysis in Table 2.9 shows annual average values of a sensitivity analysis for a rather comprehensive eutrophication model. It would generally be preferable to observe the sensitivity versus the time. In fig. 2.23 is shown PS = f(t) for the model presented in illustration 2.1. The response of three different K-values is shown. Fig.2.24 shows δPS/δK and δPA/δK as function of time. As seen the sensitivity is varying over the year.

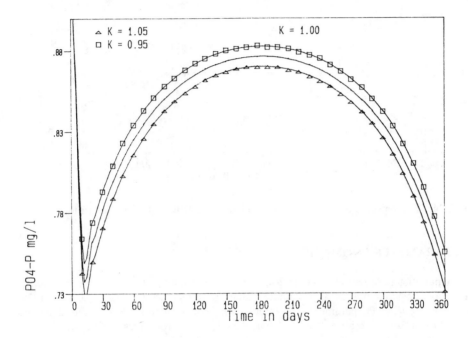

Fig.2.23: PS = f(t) for K = 1.00 +/- 0.05, as seen PS is highly sensitive to changes in K throug out the year.

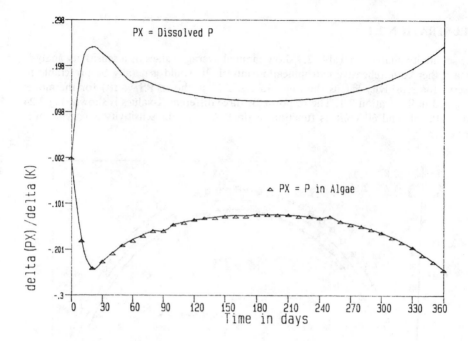

Fig.2.24. The sensitivity of PS and PA to changes in K is plotted versus time.

## 2.7. PARAMETER ESTIMATION

Many parameters in causal ecological models can be found in the literature, not necessarily as constants but as approximate values or intervals. Jørgensen et al., (1979) contains about 40000 parameters of interest for ecological modellers. Table 2.10 and 2.11 illustrate such literature values of parameters, taken from Jørgensen et al., (1979).

However, even all parameters are known in a model from the literature, it is most often required to calibrate the model. Several sets of parameters are tested by the calibration and the various model output of state variables are compared with measured or observed values of the same state variables. The parameter set, that gives the best agreement between model outputs and measured state variables, is chosen.

The need for the calibration can be explained by use of the following characteristics of ecological models and their parameters:

1) As mentioned above most ecological parameters are not known as almost exact values as are many chemical and physical parameters. Therefore all literature values for ecological parameters have a certain uncertainty.

2) All ecological models are *simplifications* of nature. The process descriptions and the system structure do not account for all detail. If the model is selected carefully the model will include all important processes and components for the problem in focus, but still the omitted details might have an influence on the final result. This influence can to a certain extent be taken into account by the calibration. The parameters might be given value slightly different from the real, but unknown value in nature, and this

difference might partly account for the influence due to the omitted details of minor importance for the model problem.

3) By far the most ecological models are lumped models; this implies that one parameter represents the average values of several species. As each species has its own characteristic parameters (see Table 2.10), the variation in the species composition inevitably gives a corresponding variation in the average parameter used in the model. Besides the algebraic average of parameters not necesserily represents the right parameter for the actual species composition. These difficulties do it almost impossible to find a right initial value of a parameter. Here will the calibration phase at least to a certain extent be able to account for the species composition.

A calibration cannot be carried out randomly, if more than a couple of parameters have been selected for calibration. If f.inst. 10 parameters have to be calibrated and the uncertainty justify that 10 different values of the 10 parameters should be tested, it is required to run the model $10^{10}$ times, which is an impossible task. Therefore the modeller will have to learn to understand the behaviour of the model by varying one or two parameters and observing the response to the most crucial state variables. In many cases it is possible to separate the model into a number of submodels, which can be calibrated more or less independently.

In this way the modeller tries to change various parameters one by one to get an acceptable accordance between observed values and model outputs for one or two state variables at the time. In a eutrophication model f.inst. it might be helpful to concentrate on the dynamic of one nutrient at the time and then after the nutrient dynamics is acceptable to go to the phytoplankton dynamics. Before the calibration is satisfactory it might have cost the modeller several hundred of model runs.

Procedures for automatic calibration are available. Table 2.12 gives the characteristics of a software named PSI intended for calibration. However, such procedures do not do the trial and error calibration described above redundant. If the automatic calibration should give acceptable results within a certain time frame, it is necessary to calibrate only 4-8 parameters at the same time and the smaller the uncertainties (it means the intervals used for allowed variations of parameters) are, the more easy will it be to find the optimum parameter set.

# TABLE 2.10

## Max Growth Rate . Symbols: $*$ = x, $**$ = potens.

| ALGAE: MAX GROWTH RATE | | |
|---|---|---|
| ITEM: | VALUE: | CONDITIONS: |
| *ACROSIPHONIA CENTRALIS | 15 MG 02/(GRAM DRY WEIGHT * HOUR) | SPRING, 283 K, LIGHT = 14.4 MW/CM2, BENTHIC, BALTIC SEA |
| *ALGAE | 0.1 1/DAY | MOSSDALE |
| *ALGAE | 0.1 1/DAY | POTOMAC RIVER |
| *ALGAE | 0.58 1/DAY | TEMP. FACTOR IS 1 AT 273K LAKE ONTARIO |
| *ALGAE | 1.-2. 1/DAY | TEMP. FACTOR IS 1 AT 293K |
| *ALGAE | 1.-3. 1/DAY | TEMP. FACTOR IS 1 AT 293K |
| *ALGAE | 0.63 1/DAY | 277 K, NATURAL ASSOCIATION |
| *ALGAE | 0.51 1/DAY | 275.6 K, NATUARAL ASSOCIATION |
| *ALGAE | 0.01-1.2 1/DAY | NONE |
| *CERANIUM STRICTUM | 17 MG 02/(GRAM DRY WEIGHT * HOUR) | AUTUMN, 288 K, LIGHT 6.3 MW/CM2 BENTHIC, ALGAE, BALTIC SEA |
| *CERATAULINA BERGONI | 1.47 DIVISIONS/DAY | VOLUME = 2712 U**3, P CONTENT = 1700*10**(-15) G-ATOM/CELL |
| *CHAETOCEROS CURVISETUS | 1.45 - 2.20 DIVISIONS/DAY | VOLUME = 2170 U**3, P CONTENT = 100-310*10**(-15) G-ATOM/CELL |
| *CHAETOCEROS GRACILIS | 0.107 1/HOUR | N LIMIT.,NH4, 298K, CONT.LIGHT: 0.05LY/MIN. |
| *CHAETOCEROS SOCIALIS | 1.53 - 2.13 DIVISIONS/DAY | VOLUME = 100 U**3, P CONTENT = 3.2 - 7.010**(-15) G-ATOM/CELL |
| *CHAETOMORPHA SP. | 14.0 MICROMOLES C/(MG CHLORO-PHYLL A*HOUR) | 297 K, 1.3*10**17 QUANTA/(S*CM2), BUBBLED WITH AIR |
| *CHLAMYDOMONAS REINHARDTII | 2.64 1/DAY | 298K |
| *CHLAMYDOMONAS REINHARDTIII | 2.64 1/DAY | 298K |
| *CHLORELLA ELLIPSOIDEA | 3.14 1/DAY | 298K |
| *CHLORELLA ELLIPSOIDEA | 1.2 1/DAY | 288K |
| *CHLORELLA ELLIPSOIDEA | 3.14 1/DAY | 298K |
| *CHLORELLA PYRENOIDOSA | 1.36 1/DAY | 292K, NH4 LIMITING, CONTINUOUS CULT., FRESHWATER |
| *CHLORELLA PYRENOIDOSA | 1.451/DAY | 292K, NO3 LIMITING, CONTINUOUS CULT., FRESHWATER |
| *DITYLUM BRIGHTWELLII | 1.20 DIVISIONS/DAY | VOLUME = 175472 UM3**3, P CONTENT = 10000*10**(10-15) G ATOM/CELL |
| *DUMONTIA INCRASSATA | 17 MG 02/(GRAM DRY WEIGHT*HOUR) | SPRING, 283K, LIGHT = 18.4 MW/CM2, BENTHIC ALGAE, BALTIC SEA |
| *DUMONTIA INCRASSATA | 9 MG 02/(GRAM DRY WEIGHT*HOUR) | WINTER, 278K, LIGHT = 5.5 MW/CM2 BENTHIC ALGAE, BALTIC SEA |
| *DUNALIELLA SP. | 1.93 ( + -0.26) DOUBLINGS/DAY | 288K, 95% CONF.LIM., LIGHT =9500LUX BATCH CULTURE |
| *DUNALIELLA TERTIOLECTA | 0.80 1/DAY | 288K, FE LIMITING, CONTINUOUS CULT.,MARINE |
| *DUNALIELLA TERTIOLECTA | 1.83 1/DAY | 298K, NH4 LIMITTING, CONTINUOUS CULT., MARINE |
| *DUNALIELLA TERTIOLECTA | 0.070 1/HOUR | N LIMIT.NH4, 298K, CONT.LIGHT: 5.8CAL/CM**2)*HR, CIM (?048) = 0.1468 |
| *ECTOCARPUS CONFERVOIDES | 14 MG 02/(GRAM DRY WEIGHT*HOUR) | SPRING, 283K, LIGHT = 26.7 MW/CM2 BENTHIC ALGAE, BALTIC SEA |
| *ENTEROMORPHA INTESTINALIS | 30 MG 02/(GRAM DRY WEIGHT*HOUR) | SPRING 283K, LIGHT = 17.5 MW/CM2 BENTHIC ALGAE, BALTIC SEA |
| *ENTEROMORPHA INTESTINALIS | 28 MG 02/(GRAM DRY WEIGHT*HOUR) | WINTER, 278K, LIGHT = 13.5 MW/CM2 BENTHIC ALGAE, BALTIC SEA |
| *ENTEROMORPHA LINZA | 22 MG 02/(GRAM DRY WEIGHT*HOUR) | AUTUMN, 288K, LIGHT = 18.4 MW/CM2 BENTHIC ALGAE, BALTIC SEA |
| *ENTEROMORPHA PROLIFERA | 42 MG 02/(GRAM DRY WEIGHT*HOUR) | AUTUMN, 288K, LIGHT = 18.4 MW(CM2 BENTHIC ALGAE, BALTIC SEA |
| *ENTEROMORPHA PROLIFERA | 13 MG 01/(GRAM DRY WEIGHT*HOUR) | SUMMER, 293K, LIGHT = 5.6 MW/CM2 BENTHIC ALGAE, BALTIC SEA |
| *FUCUS SERRATUS | 5.4 MG 02/(GRAM DRY WEIGHT*HOUR) | WINTER, 278K, LIGHT = 43.6 MW/CM2 BENTHIC ALGAE, BALTIC SEA |
| *FUCUS SERRATUS | 3.1 MG 02/(GRAM DRY WEIGHT*HOUR) | AUTUMN, 288K, LIGHT = 3.1 MW/CM2 BENTHIC ALGAE, BALTIC SEA |
| *FUCUS SERRATUS | 2.4 MG 02/(GRAM DRY WEIGHT*HOUR) | SPRING, 283K, LIGHT =23.6 MW/CM2, BENTHIC ALGAE, BALTIC SEA |
| *FUCUS VESICULOSUS | 6.6 MG 02/(GRAM DRY WEIGHT*HOUR) | AUTUMN, 288K, LIGHT = 23.6 MW/CM2, BENTHIC ALGAE, BALTIC SEA |
| *FUCUS VESICULOSUS | 6.0 MG 02/(GRAM DRY WEIGHT*HOUR) | SPRING, 283K, LIGHT = 28,4 MW/CM2, BENTHIC ALGAE, BALTIC SEA |
| *FUCUS VESICULOSUS | 5.4 MG 02/(GRAM DRY WEIGHT*HOUR) | WINTER, 278K, LIGHT = 30.2 MW/CM2, BENTHIC ALGAE, BALTIC SEA |
| *FURCELLARIA FASTIGIATA | 3.2 MG 02/(GRAM DRY WEIGHT*HOUR) | SPRING, 283K, LIGHT = 10 MW/CM2, BENTHIC ALGAE, BALTIC SEA |
| *FURCELLARIA FASTIGIATA | 2.3 MG 02/(GRAM DRY WEIGHT*HOUR) | WINTER, 278K, LIGHT = 10 MW/CM2, BENTHIC ALGAE, BALTIC SEA |
| *FURCELLARIA FASTIGIATA | 1.8 MG 02/(GRAM DRY WEIGHT*HOUR) | AUTUMN, 288K, LIGHT = 4.3 MW/CM2, BENTHIC ALGAE, BALTIC SEA |
| *GLENODINIUM SP. | 61.7 MICROMOLES C/(MG | 297K, 1.3*10**17 QUANTA/(S*CM2) |

| Species | Rate | Conditions |
|---|---|---|
| | CHLOROPHYLL A * HOUR) | BUBBLED WITH AIR |
| *GYMNODINIUM SPLENDENS | 0.48 (+ -0.06 = DOUBLINGS/DAY | 291K, 95% CONF.LIM., LIGHT = 9500LUX BATCH CULTURE |
| *GYMNODINIUM SPLENDENS | 0.83 (+ -0.42) DOUBLINGS/DAY | 298K, 95% CONF.LIM., LIGHT = 9500LUX BATCH CULTURE |
| *LAMINARIA DIGITATA | 3.9 MG 02/(GRAM DRY WEIGHT * HOUR) | AUTUMN, 288K, LIGHT = 6.3 MW/CM2 BENTHIC ALGAE, BALTIC SEA |
| *LAMINARIA DIGITATA | 3.8 MG 02/(GRAM DRY WEIGHT * HOUR) | SPRING, 283K, LIGHT = 26.7 MW/CM2, BENTHIC ALGAE, BALTIC SEA |
| *LAMINARIA DIGITATA | 1.3 MG 02/(GRAM DRY WEIGHT * HOUR) | WINTER, 278K, LIGHT = 1.3 MW/CM2, BENTHIC ALGAE, BALTIC SEA |
| *LAMINARIA SACCHARINA | 2.0 MG 02/(GRAM DRY WEIGHT * HOUR) | WINTER, 278K, LIGHT = 2.8 MW/CM2, BENTHIC ALGAE, BALTIC SEA |
| *LAMINARIA SACCHARINA | 2.0 02/(GRAM DRY WEIGHT * HOUR) | AUTUMN, 288K, LIGHT = 23.6 MW/CM2 BENTHIC ALGAE, BALTIC SEA |
| *MARINE PHYTOPLANKTON | 3.70 MG C/(M * * 3 * H) | LOW NUTRIENT LEVEL, 15 MILES FROM COAST |
| *CHLORELLA PYRENOIDOSA | 1.95 1/DAY | 298K, P LIMITTING, CONTINUOUS CULT., FRESHWATER |
| *CHLORELLA PYRENOIDOSA | 2.14 1/DAY | 298K, P LIMITTING, CONTINUOUS CULT.,FRESHWATER |
| *CHLORELLA PYRENOIDOSA | 1.84 1/DAY3 | 01,5K, NH4 LIMITTING, CONTINUOUS CULT., FRESHWATER |
| *CHLORELLA PYRENOIDOSA | 2.22 1/DAY | 301.5K, NO3 LIMITTING, CONTINUOUS CULT., FRESHWATER |
| *CHLORELLA PYRENOIDOSA | 3.94 1/DAY | 308K, NH4 LIMITTING, CONTINUOUS CULT., FRESHWATER |
| *CHLORELLA PYRENOIDOSA | 4.32 1/DAY | 308.5K, NO3 LIMITTING, CONTINUOUS CULT., FRESHWATER |
| *CHLORELLA PYRENOIDOSA | 4.26 1/DAY | 312.2K, NH4 LIMITTING, CONTINUOUS CULT., FRESHWATER |
| *CHLORELLA PYRENOIDOSA | 5.65 1/DAY | 312.2K, NO3 LIMITTING, CONTINUOUS CULT., FRESHWATER |
| *CHLORELLA PYRENOIDOSA | 1.96 1/DAY | 298K |
| *CHLORELLA PYRENOIDOSA | 2.15 1/DAY | 298K |
| *CHLORELLA PYRENOIDOSA | 0.2 1/DAY | 283K, HIGH-TEMPERATURE STRAIN |
| *CHLORELLA PYRENOIDOSA | 1.1 1/DAY | 288K, HIGH- TEMPERATURE STRAIN |
| *CHLORELLA PYRENOIDOSA | 2.4 1/DAY | 293K, HIGH-TEMPERATURE STRAIN |
| *CHLORELLA PYRENOIDOSA | 9.9 MICROMOLES C/(MG CHLO-ROPHYLL A * HOUR | 297K, 1,3 * 10 * * 17 QUANTA/(S * CM2), BUBBLED WITH AIR |
| *CHLORELLA PYRENOIDOSA | 0.180 1/HOURS | PECIFIC GROWTH RATE, 308K, SUBSTRATE NO3, LIGHT = 0.05 CAL/(CM2 * MIN) |
| *CHLORELLA PYRENOIDOSA | 1.96 1/DAY | 298K |
| *CHLORELLA PURENOIDOSA | 2.15 1/DAY | 298K |
| *CHLORELLA PYRENOIDOSA | 0.2 1/DAY | 283K |
| *CHLORELLA PYRENOIDOSA | 1.1 1/DAY | 288K |
| *CHLORELLA PYRENOIDOSA | 2.4 1/DAY | 293K |
| *CHLORELLA PYRENOIDOSA | 3.9 1/DAY | 298K |
| *CHLORELLA SPECIES | 1.88 1/DAY | 298K, NO3 LIMITTING, CONTINUOUS CULT., FRESHWATER |
| *CHLORELLA VULGARIS | 1.8 1/DAY | 298K |
| *CHLORELLA VULGARIS | 1.8 1/DAY | 298K |
| *CLADOPHORA GLOMERATA | 27 MG 02/(GRAM DRY WEIGHT * HOUR) | SUMMER, 293K, LIGHT = 18.4 MW/CM2, BENTHIC ALGAE, BALTIC SEA |
| *COCCOCHLORIS SP. | 2.16 1/DAY | 298K, NO3 LIMITTING, CONTINUOUS CULT., MARINE |
| *COCCOCHLORIS STAGNINA | 0.086 1/HOUR | SPECIFIC GROWTH RATE, 298K, SUBSTRATE NO3, LIGHT = 5.8 CAL/(CM2 * HOUR) |
| *COCCOCHLORIS STAGNINA | 0.086 1/HOUR | N LIMIT., NO3, 298K, CONT.LIGHT: 5.8 CAL/(CM * * 2) * HR,CIM (?048) = 0.1848 |
| *COCCOLITHUS HUXLEY | 1.75 DOUBLINGS/DAY | OPTIMAL CONDITIONS |
| *COSCINODISCUS GIGAS | 0.51 - 0.73 DIVISIONS/DAY | VOLUME = 5150805 U * * 3 |
| *COSCINDISCUS GRANII | 0.50 - 0.92 DIVISIONS/DAY | VOLUME = 172292 U * *3, P CONTENT = 1000-4000 * 10 * * (-15) G-ATOM/CELL |
| *CYCLOTELLA NANA | 0.086 1/HOUR | N LIMIT., NO3, 298K, CONT.LIGHT: 5.8 CAL/((CM * * 2) * HR, CIM (?048) = -0.1588 |
| *DELESSERIA SANGUINEA8.8 MG | 02/(GRAM DRY WEIGHT * HOUR) | SPRING, 284K, LIGHT = 10 MW/CM2, BENTHIC ALGAE, BALTIC SEA |
| *DELESSERIA SANGUINEA | 4.0 MG 02/(GRAM DRY WEIGHT * HOUR) | SUMMER, 293K, LIGHT = 4.9 MW/CM2, BENTHIC ALGAE, BALTIC SEA |
| *DELESSERIA SANGUINEA | 3.0 MG 02/(GRAM DRY WEIGHT * HOUR) | AUTUMN, 288K, LIGHT = 8.0 MW/CM2, BENTHIC ALGAE, BALTIC SEA |
| *DELESSERIA SANGUINEA | 2.2 MG 02/(GRAM DRY WEIGHT * HOUR) | WINTER, 278K, LIGHT = 10 MW/CM2, BENTHIC ALGAE, BALTIC SEA |
| *DITYLIUM BRIGHTWELLII | 1.20 DIVISIONS/DAY | VOLUME = 175472 U * *3 P CONTENT = 10000 * 10 * * (-15) G-ATOM/CELL |
| *MARINE PHYTOPLANKTON | 1.20 MG C/(M * * 3 * H)LOW NU-TRIENT LEVEL, | CENTRAL NORTH PACIFIC |
| *MONOCHRYSIS LUTHERI | 0.84 1/DAY | 292K, B12 LIMITTING, CONTINUOUS CULT., MARINE |
| *NANNOCHLORIS ATOMUS | 2.16 1/DAY | 293K |
| *NANNOCHLORIS ATOMUS | 1.54 1/DAY | 283K |
| *NANNOCHLORIS ATOMUS | 2.16 1/DAY | 293K, FLAGGELATE |
| *NANNOCHLORIS ATOMUS | 1.54 1/DAY | 283K, FLAGGELATE |
| *NITZSCHIA ACTINASTROIDES | 2.06 1/DAY | 296K, P LIMITING, CONTINOUS CULT., MARINE |

| Species | Rate | Conditions |
|---|---|---|
| *NITZSCHIA CLOSTERIUM | 1.75 1/DAY | 300K |
| *NITZSCHIA CLOSTERIUM | 1.55 1/DAY | 292K |
| *NITZSCHIA CLOSTERIUM | 1.19 1/DAY | 288.5K |
| *NITZSCHIA CLOSTERIUM | 0.67 1/DAY | 283K |
| *NITZSCHIA CLOSTERIUM | 1.75 1/DAY | 300K |
| *NITZSCHIA CLOSTERIUM | 1.55 1/DAY | 292K |
| *NITZSCHIA CLOSTERIUM | 1.19 1/DAY | 288.5K |
| *NITZSCHIA CLOSTERIUM | 0.67 1/DAY | 283K |
| *OSCILLATORIA AGARDHII GOMOMT | 0.017 1/HOUR | LIGHT = 0.1W/M $\ast\ast$ 2, PH = 7.1-7.6,N LIMIT., NO3, BATCH CULT. |
| *OSCILLATORIA AGARDHII GOMOMT | 0.020 1/HOUR | LIGHT = 3.6W/M $\ast\ast$ 2, PH = 7.3-7.8, N LIMIT., NO3, BATCH CULT. |
| *OSCILLATORIA AGARDHII GOMOMT | 0.20 1/HOUR | LIGHT = 5.5W/M $\ast\ast$ 2, PH = 7.2-7.6, N LIMIT.,NO3, BATCH CULT. |
| *OSCILLATORIA AGARDHII GOMOMT | 0.036 1/HOUR | N LIMIT., NO3, 293K, CONT.SAT.LIGHT, CIM(?048) = 0.092A |
| *PETALONIA FASCIA | 34 MG 02/(GRAM DRY WEIGHT $\ast$ HOUR) | SPRING, 283K, LIGHT = 43.6 MW/CM2, BENTHIC ALGAE, BALTIC SEA |
| *PHYCODRYS RUBENS | 4.8 MG 02/(GRAM DRY WEIGHT $\ast$ HOUR) | SPRING, 283K, LIGHT = 10 MW/CM2, BENTHIC ALGAE, BALTIC SEA |
| *PHYCODRYS RUBENS | 4.1 MG 02/(GRAM DRY WEIGHT $\ast$ HOUR) | AUTUMN, 288K, LIGHT = 4.3 MW/CM2, BENTHIC ALGAE, BALTIC SEA |
| *PHYCODRYS RUBENS | 3.7 MG 02/(GRAM DRY WEIGHT $\ast$ HOUR) | SUMMER, 293K, LIGHT = 4.3 MW/CM2, BENTHIC ALGAE, BALTIC SEA |
| *PHYCODRYS RUBENS | 2.0 MG 02/(GRAM DRY WEIGHT $\ast$ HOUR) | WINTER, 278K, LIGHT = 1.4 MW/CM2, BENTHIC ALGAE, BALTIC SEA |
| *PHYLLOPHORA BRODIAEI | 2.0 MG 02/(GRAM DRY WEIGHT $\ast$ HOUR) | WINTER, 278K, LIGHT = 4.9 MW/CM2, BENTHIC ALGAE, BALTIC SEA |
| *PHYLLOPHORA BRODIAEI | 1.9 MG 02/(GRAM DRY WEIGHT $\ast$ HOUR) | SPRING, 283K, LIGHT = 9.12 MW/CM2, BENTHIC ALGAE, BALTIC SEA |
| | | ALGAE: MAX GROWTH RAGE |
| *PHYLLOPHORA BRODIAEI | 1.2 MG 02/(GRAM DRY WEIGHT $\ast$ HOUR) | AUTUMN, 288K, LIGHT = 6.3 MW/CM2, BENTHIC ALGAE, BALTIC SEA |
| *PHYTOPLANKTON | 2.0 1/DAY | SAN JOAQUIN (RIVER), 293K |
| *PHYTOPLANKTON | 2.5 1/DAY | DELTA(ESTUARY), 293K |
| *PHYTOPLANKTON | 2.0 1/DAY | POTOMAC (ESTUARY), 293K |
| *PHYTOPLANKTON | 1.3 1/DAY | LAKE ERIE, 293K |
| *PHYTOPLANKTON | 2.1 1/DAY | LAKE ONTARIO, 293K |
| *POLYSIPHONIA NIGRESCENS | 5.3 MG 02/(GRAM DRY WEIGHT $\ast$ HOUR) | AUTUMN, 288K, LIGHT = 14.5 MW/CM2, BENTHIC ALGAE, BALTIC SEA |
| *POLYSIPHONIA NIGRESCENS | 4.3 MG 02/(GRAM DRY WEIGHT $\ast$ HOUR) | SPRING, 283K, LIGHT = 18.5 MW/CM2, BENTHIC ALGAE, BALTIC SEA |
| *POLYSIPHONIA NIGRESCENS | 2.3 MG 02/(GRAM DRY WEIGHT $\ast$ HOUR) | WINTER, 278K, LIGHT = 26.7 MW/CM2, BENTHIC ALGAE, BALTIC SEA |
| *PORPHYRA LEUCOSTICTS | 29 MG 02/(GRAM DRY WEIGHT $\ast$ HOUR) | WINTER, 278K, LIGHT = 4.3 MW/CM2, BENTHIC ALGAE, BALTIC SEA |
| *SARGASSUM SP. | 44.0 MICROMOLES C/(MG CHLORO-PHYLL A $\ast$ HOUR) | 297K, 1.3 $\ast$ 10 $\ast\ast$ 17 QUANTA/(S $\ast$ CM2), BUBBLED WITH AIR |
| *SCENEDESMUS OBLIQUUS | 1.52 1/DAY | 298K |
| *SCENEDESMUS OBLIQUUS | 1.52 1/DAY | 298K |
| *SCENEDESMUS QUADRICAUDA | 2.02 1/DAY | 298K |
| *SCENEDESMUS QUADRICAUDA | 2.02 1/DAY | 298K |
| *SCYTOSIPHON LOMENTARIA1 | 2.5 MG 02/(GRAM DRY WEIGHT $\ast$ HOUR) | SPRING, 283K, LIGHT = 26.7 MW/CM2, BENTHIC ALGAE, BALTIC SEA |
| *SELENASTRUM CAPRICORNUTUM | 1.85 1/DAY | BATCH CULTURE, SPECIFIC GROWTH RATE |
| *SELENASTRUM CAPRICORNUTUM | 1.398 1/DAY | 293K, CHEMOST., AMMONIUM-N LIMIT., R = 0.975, MONOD EQ. |
| *SELENASTRUM CAPRICORNUTUM | 1.544 1/DAY | 297K, CHEMOST., AMMONIUM-N LIMIT., R = 0.996, MONOD EQ. |
| *SELENASTRUM CAPRICORNITIUM | 1.540 1/DAY | 300K, CHEMOST., AMMONIUM-N LIMIT., R = 0.989, NONOD EQ. |
| *SELENASTRUM CAPRICORNITIUM | 1.596 1/DAY | 301K, CHEMOST., AMMONIUM-N LIMIT., R = 0.799, MONOD EQ. |
| *SELENASTRUM CAPRICORNITIUM | 1.274 1/DAY | 306K, CHEMOST., AMMONIUM-N LIMIT., R = 0.949. MONOD EQ. |
| *SELENASTRUM CAPRICORNITIUM | 1.365 1/DAY | 293K, CHEMOST., AMMONIUM-N LIMIT. |
| *SELENASTRUM CAPRICORNITIUM | 1.992 1/DAY | 297K, CHEMOST., AMMONIUM-N LIMIT., MONOD EQ., MOST PROB.SPEC.RATE |
| *SELENASTRUM CAPRICORNITIUM | 1.412 1/DAY | 300K, CHEMOST., AMMONIUM-N LIMIT., MONOD EQ., MOST PROB.SPEC.RATE |
| *SELENASTRUM CAPRICORNITIUM | 1.390 1/DAY | 301K, CHEMOST., AMMONIUM-N LIMIT., MONOD EQ., MOST PROB.SPEC.RATE |
| *SELENASTRUM CAPRICORNITIUM | 1.274 1/DAY | 306K, CHEMOST., AMMONIUM-N LIMIT., MONOD EQ., MOST PROB.SPEC.RATE |
| *SELENASTRUM CAPRICORNUTIUM | 1.85 1/DAY | 297K, P LIMITING, CONTINUOUS CULT., FRESHWATER |
| *SKELETONEMA COSTATIUM | 1.27 1/DAY | 292K, B12 LIMITTING, CONTINUOUS CULT., MARINE |
| *SKELETONEMA COSTATIUM | 1.51 - 5.00 DIVISIONS/DAY | VOLUME = 144 U $\ast\ast$ 3 P CONTENT = 8.0-28.0 $\ast$ 10 $\ast$ $\ast$(-15) G ATOM/CELL |
| *THALASSIOSIRA PSEUDONANA | 3.6 DOUBLING/DAY | CLONE 3H |
| *THALASSIOSIRA PSEUDONANA | 2.1 DOUBLING/DAY | CLONE 13-1 |
| *THALASSIOSIRA PSEUDONANA | 1.27 1/DAY | 286.5K, P LIMITTING, CONTINUOUS CULT., MARINE |
| *THALASSIOSIRA PSEUDONANA | 1.14 1/DAY | 291K, P LIMITTING, CONTINUOUS |

| | | CULT., MARINE |
|---|---|---|
| *THALASSIOSIRA PSEUDONANA | 1.46 1/DAY | 297K, P LIMITTING, CONTINUOUS |
| | | CULT., MARINE |
| *THALASSIOSIRA PSEUDONANA | 2.09 1/DAY | 298K, NO3 LIMITTING, CONTINUOUS |
| | | CULT., MARINE |
| *THALASSIOSIRA PSEUDONANA | 2.77 1/DAY | 293K, NO3 LIMITTING, CONTINUOUS |
| | | CULT., MARINE |
| *THALASSIOSIRA PSEUDONANA | 1.34 1/DAY | 289K, B12 LIMITTING, CONTINUOUS |
| | | CULT., MARINE |
| *THALASSIOSIRA PSEUDONANA | 74.6 MICROMOLES C/(MG CHLO-ROPHYLL A * HOUR) | 297K, $1.3*10**17$ QUANTA/S * CM2), BUBBLED WITH AIR |
| *ULOTRIX PSEUDONANA | 35 MG 02/(GRAM DRY WEIGHT * HOUR) | SPRING, 283K, LIGHT = 23.6 MW/CM2, BENTHIC ALGAE, BALTIC SEA |
| *ULVOPSIS GREVILLEI | 47 MG 02/(GRAM DRY WEIGHT * HOUR) | SPRING, 283K, LIGHT = 26.7 MW/CM2, BENTHIC ALGAE, BALTIC SEA |
| *ULVOPSIS GREVILLEI | 13 MG 02/(GRAM DRY WEIGHT * HOUR) | WINTER, 278K, LIGHT = 14.5 MW/CM2, BENTHIC ALGAE, BALTIC SEA |

# TABLE 2.11:

Algae : Ca/biomass Ratio.

| ITEM: | VALUE: | CONDITIONS: |
|---|---|---|
| *ANABAENA SP. | 0.36 % OF DRY WEIGHT | ALGAE FROM PONDS AND LAKES IN SOUTHEASTERN UNITED STATES |
| *APHANIZOMENON SP. | 0.73 % OF DRY WEIGHT | ALGAE FROM PONDS AND LAKES IN SOUTHEASTERN UNITED STATES |
| *CHARA SP. | 8.03 % OF DRY WEIGHT | ALGAE FROM PONDS AND LAKES IN SOUTHEASTERN UNITED STATES |
| *CHLOROPHYTA | 0.157 MG/(GRAM OF ACID-SOLUBLE FRACTIONS) | AVERAGE OF 7 MARINE SPECIES, PUERTO RICO |
| *CHLADOPHORA SP. | 1.69 % OF DRY WEIGHT | ALGAE FROM PONDS AND LAKES IN SOUTHEASTERN UNITED STATES |
| *EUGLENA SP. | 0.05 % OF DRY WEIGHT | ALGAE FROM PONDS AND LAKES IN SOUTHEASTERN UNITED STATES |
| *HYDRODICTYON SP. | 0.69 % OF DRY WEIGHT | ALGAE FROM PONDS AND LAKES IN SOUTHEASTERN UNITED STATES |
| *LYNGBYA SP. | 0.45 % OF DRY WEIGHT | ALGAE FROM PONDS AND LAKES IN SOUTHEASTERN UNITED STATES |
| *MICROCYSTIS SP. | 0.53 % OF DRY WEIGHT | ALGAE FROM PONDS AND LAKES IN SOUTHEASTERN UNITED STATES |
| *MOUGEOTIA SP. | 1.68 OG DRY WEIGHT | ALGAE FROM PONDS AND LAKES IN SOUTHEASTERN UNITED STATES |
| *NITELLA SP. | 1.89 % OF DRY WEIGHT | ALGAE FRO PONDS AND LAKES IN SOUTHEASTERN UNITED STATES |
| *OEDOGONIUM SP. | 0.44 % OF DRY WEIGHT | ALGAE FROM PONDS AND LAKES IN SOUTHEASTERN UNITED STATES |
| *PHAEOPHYCEAE | 0.099 GRAM/(GRAM OF ACID-SOLUBLE FRACTION) | AVERAGE OF 4 MARINE SPECIES, PUERTO RICO |
| *PITHOPHORA SP. | 3.82 % OF DRY WEIGHT | ALGAE FROM PONDS AND LAKES IN SOUTHEASTERN UNITED STATES |
| *RHIZOCLONIUM SP. | 0.52 % OF DRY WEIGHT | ALGAE FROM PONDS AND LAKES IN SOUTHEASTERN UNITED STATES |
| *RHODOPHYTA | 0.106 GRAM/(GRAM OF ACID-SOLUBLE FRACTION) | AVERAGE OF 9 MARINE SPECIES, PUERTO RICO |
| *SPIROGYRA SP. | 0.57 % OF DRY WITGHT | ALGAE FROM PONDS AND LAKES IN SOUTHEASTERN UNITED STATES |
| *SPIROGYRA SP. | 0.84 % OG DRY WEIGHT | ALGAE FROM PONDS AND LAKES IN SOUTHEASTERN UNITED STATES |

In the trial and error calibration the modeller has more or less intuitively set up some calibration criteria. He wants f.inst. to be able in, the first instance, to simulate rather accurately the minimum oxygen concentration for a stream model and/or the time at which this minimum value occurs . When he is satisfied with these model results, he might want to be able to simulate properly the shape of the oxygen concentration versus time curve etc. He calibrates the model to achieve these objectives step by step.

If an automatic calibration procedure is used, it is necessary to formulate objective criteria for the calibration. A possible objective function is, that

$$Y = \left[ \frac{\sum_{i=1}^{i=n} \frac{(x_t^i - x_n^i)}{x_m^i}}{n} \right]^{\frac{1}{2}} \quad \text{is minimum}$$

(2.12)

where $x_t^i$ is the computed value of the $i^{th}$ state variable, $x_n^i$ is the corresponding measured value and $x_m^i$ is the average, observed value of the $i^{th}$ state variable.

However, the modeller is often more interested in a *good accordance* between observations and model output for *one or a few state variables*. In that case he can choose weight for the various state variables. For an eutrophication model f.inst., he might choose the weight 10 for phytoplankton and the weight 5 for the nutrient concentrations, while all other state variables are given the weight 1. He might also be interested in ensuring a very high accuracy of the simulation of the maximum concentration of the phytoplankton and he will therefore give an even higher weight to the phytoplankton concentration at the time, when the spring bloom is expected to occur.

If it is impossible to calibrate a model, this is not necessarily due to an incorrect model, it might also be due to a low quality of the observed data. The quality of the data is crucial to the quality of the calibration. It is furthermore of great importance that **observations reflect the dynamics of the model**. If the objective of the model is to give a description of the  dynamic behaviour of a state variable, which varies from day to day, it is of course not possible to get a good parameter estimation based upon monthly observations. This should be illustrated by an example taken from a eutrophication model.

A eutrophication model is generally calibrated on the basis of annual measurement series with a sampling frequency of once or twice per month. This sampling frequency is, however, not sufficient to describe the dynamics of the lake. If it is the scope of the model to predict maximum values and related data for phytoplankton concentrations and primary production, it is necessary to have a sampling frequency, which is able to give us an estimate of the maximum value in phytoplankton concentration and the primary production.

Fig. 2.25 shows characteristic algae concentrations plotted versus time, April 1-May 15, in a hyper-trophic lake with a sampling frequency of (1) twice per month, and (2) three times per week (denoted as the "intensive" measuring program). As can be seen, the two plots are significantly different and the attempt to get a realistic calibration on the basis of (1) will fail, provided it is the aim to model the day-to-day variation in phytoplankton concentration in accordance with (2). This example illustrates that it is of great importance not only to have data with low uncertainty, but also data sampled with a frequency corresponding to the dynamics of the system.

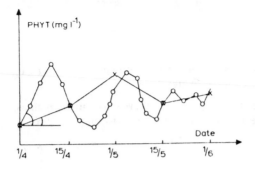

PHYT (mg l⁻¹)

Date

¹/₄        ¹⁵/₄        ¹/₅        ¹⁵/₅        ¹/₆

Fig.2.25: Algae concentration plotted versus time: curve (1) = sampling frequency
twice a month (+) curve (2) = sampling frequency three times a week (0).
Note the difference of d(PHYT)/dt between the two curve

This rule has often been neglected in modelling the eutrophication process, most
probably because limnological lake data, which are not sampled for modelling purposes,
are often collected with a relatively low frequency. On the other hand, the model then
attempts to simulate the annual cycle, and an annual sampling program with a frequency
of three per week will require too many resources. A combination of an annual sampling
program with a frequency of one to three samples per month and an intensive measuring
program placed in periods, where different subsystems show a maximum of changes
gives a good basis for parameter estimations.

The intensive measuring program can, as presented below, be used to estimate state
variables derivatives (for comparison of these estimations by low and high sampling
frequency, see the slopes of curves (1) and (2) in fig. 2.25. These estimates can be used to
set up an overdetermined set of algebraic equations, making the model parameters the
sole unknown.

An outline of the method runs as follows (see fig. 2.26): (For further detail, see
Jørgensen et al., 1981).

Most deterministic lake models are put into a framework of partial or ordinary
differential equations

$$\frac{\partial \bar{\psi}}{\partial t} = \bar{f}\left(\bar{\psi}, \frac{\partial \bar{\psi}}{\partial \bar{r}}, \frac{\partial^2 \bar{\psi}}{\partial \bar{r}^2}, t; \bar{a}\right) \tag{2.14}$$

$$\bar{\psi} = \bar{f}(\bar{\psi}, t; \bar{a}) \tag{2.15}$$

where $\bar{\psi}$ is the vector of state variables,t is the time,r denotes space coordinates and $\bar{a}$ is a
vector of a parameter set.

*Step 1*. **Find cubic spline coefficients, $S_i(t_j)$,** i.e. second order time derivatives at time of
observation $t_j$, of the spline function $s_i(t)$ approximating the observed variable $\psi_i(t)$,
according to the cubic spline method of Mejer et al. (1980).

*Step 2*. **Estimate $\bar{\psi}_i(t_j)/dt$ the spline coefficient found in step 1.**

*Step 3*. Solve eq. (2.15) with $\bar{a}$ (or a subset of the components of $\bar{a}$) regarded as unknown.

*Step 4*. **Evaluate the feasibility of the solution a° found in step 3.** If not feasible, modify the part of the model influenced by āᵒ and go to step 1.

*Step 5.* **Choose a significance level, and perform a statistical test** on constancy of āᵒ. If the test fails, modify appropriate submodels and go to step 1.

*Step 6.* **Use āᵒ as an initial guess in a computerized parameter search algorithm,** such as Marquardt, Powell or steepest descent algorithms, to minimize a performance index

$$(2.16)$$

$$\| \bar{\psi}^{obs} - \bar{\psi}^{model} \|$$

subject to eq. (2.15) and feasibility bounds on ā and $\bar{\psi}^{model}$. Here $\| \ldots \|$ indicates any kind of norm (e.g. a weighted quadratic norm).

Although the model in hand may be highly non-linear regarding the state variables, it usually turns out that this is not the case regarding the parameter set a, or the subset of a that is tuned by calibration. Further, the matrix $\partial \bar{f}/\partial \bar{a}$ usually contains very many zeros.

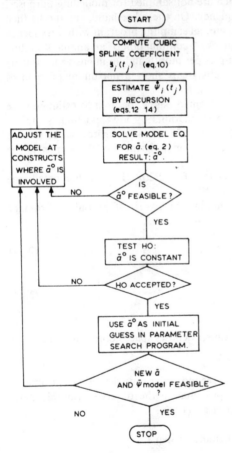

Fig.2.26: Flowchart of the method used.

Since the number of equations of type (2.15) is greater than the number of estimable parameters , eq. (2.15) is overdetermined. It is easy to smooth the solution in some sense, but it is more important to evaluate the constancy of $\bar{a}^\circ$, e.g. variance analysis, test of normality of white noise, etc. Information on standard deviation of $\bar{a}^\circ$ around its average value may eventually be used as point of departure for introducing stochasticity into the model, admitting the fact that parameters in real life may not be as constant as the modellers assume.

As a certain parameter, e.g. $a_k$, seldom appears at more than one or two places in the model equations, an unacceptable value of $a_k$ found as solution to eq. (2.15) quite accurately locates the inappropriate terms and constructs in the model. Experience with the method has shown it valuable as a diagnostic tool to single out unfitted model terms.

Since the method is based on cubic spline approximation, it is essential that observations are dense, i.e. $t_{j+1} - t_j$ should be small in the sense that local third-degree polynomials should approximate observed values well. To test whether this is ful-filled is, in general, difficult (the 'true' $\bar{\psi}_i(t)$ function might have microscopic curls that generate oscillating derivatives $\psi_i(t)/dt$. However, if the method yields basically the same result on a random subset of observations, it may be safe to assume that $\{s_i(t_j)/dt\}$ represent the true rates on a day to day basis. After appropriate adjustment of model equations an acceptable parameter set $a^\circ$ may eventually be obtained.

With $\bar{a}^\circ$ as an initial guess, a better parameter set may be found by systematic perturbation of the set until some norm (performance index)

$$(2.17)$$

$$\| \psi^{obs} - \psi^{model} \|$$

has reached a (local) minimum. At each perturbation, the model equations are solved. Gradients $\{\partial\psi_i/\partial a_k\}$ are hardly ever known analytically. All numerical methods currently in use to solve this kind of problem fail when the number of parameters surpasses four or five unless the initial guess ($\bar{a}^\circ$) is very close to a value that minimizes the performance index. This is why steps 1 and 2 mentioned above are so important. The result of the application of what are called intensive measurements to calibrate the eutrophication model is summarized in Table 2.13. As seen the difference in parameter estimation is pronounced. By use of the parameters before the final calibration took place.

The illustrated use of intensive measurements for a parameter estimation prior to the calibration was as seen based upon determinations of the actual growth of phytoplankton. By determination of the derivatives, it was posssible to fit the parameters to the unknown in the model equations.

Measurements and observations in vitro were used in the referred case to find the derivates. In principle the same basic idea can be used either in the laboratory or by construction of a microcosms. In both cases the measurements are facilitated by use of a smaller unit, where disturbing factors or processes might be kept constant. Current record of important state variables is often possible and provides a high number of data, which decreases the standard deviation.

# TABLE 2.13

## Comparison of parameter values

| Parameter | Parameter (Symbol) | Unit | Application of intensive measurements | Glumsø Lake * | Lyngby Lake * | Litterature ranges |
|---|---|---|---|---|---|---|
| Settling rate | $SVS = D \cdot SA$ | m d$^{-1}$ | 0.30 +/- 0.05 | 0.2 | 0.05 | 0.1-0.6 |
| Max.growth rate [0] | $CDR_{max}$ (reduced) | d$^{-1}$ | 1.33 +/- 0.51 | 2.3 | 1.8 | 1-3 |
| Max.growth rate [0] | $CDR_{max}$ (model) | d$^{-1}$ | 4.71 +/- 1.8 | 4.11 | 3.21 | 2-6 |
| Max.uptake rate P[0] | $UP_{max}$ | d$^{-1}$ | 0.0072 +/- 0.0007 | 0.003 | 0.008 | 0.003-0.01 |
| Min. C:biomass ratio [0] | $FCA_{min}$ | | 0.4 | 0.15 | 0.15 | 0.3-0.7 |
| Min. P: biomass ratio [0] | $FPA_{min}$ | | 0.03 | 0.013 | 0.013 | 0.013-0.035 |
| Min.N: biomass ratio [0] | $FNA_{min}$ | | 0.12 | 0.10 | 0.10 | 0.08-0.12 |
| Max.uptake rate N[0] | $UN_{max}$ | d$^{-1}$ | 0.023 +/- 0.005 | 0.015 | 0.012 | 0.01-0.035 |
| Michaelis-Menten [0] constant N | KN | mg l$^{-1}$ | 0.34 +/- 0.07 | 0.2 | 0.2 | 0.1-0.5 |
| Denitrification rate | DENITX | g m$^{-3}$ d$^{-1}$ | 0.83 +/- 1.05 | | | |
| Respiration rate [0] | RC | d$^{-1}$ | 0.088 | 0.13 | 0.2 | 0.05-0.25 |
| Mineralization rate P | $KDP_{10}$ | d$^{-1}$ | 0.80 +/- 0.47 | 0.40 | 0.25 | 0.2-0.8 |
| Mineralization rate N | $KDN_{10}$ | d$^{-1}$ | 0.21 +/- 0.11 | 0.05 | 0.15 | 0.05-0.3 |
| Max,uptake rate c[0] | $UC_{max}$ | d$^{-1}$ | 1.21 +/- 0.97 | 0.65 | 0.40 | 0.2-1.4 |

Lyngby and Glumsø lakes have approximately same biogeochemical and morphology.
[0] All parameters related to phytoplankton, see also table 3.5 and 7.4.4

An example will be quoted to illustrate this method of parameter estimation. Fish growth can be described by use of the following equation:

$$dw/dt = a \times W^b \tag{2.18}$$

where w is the weight, a and b are constants. It is possible in an aquarium or an aquaculture to follow the weight of the fish versus time. If enough data are available it is easy by static methods to determine a and b in the above shown equation. In this case the feeding is known to be at the optimum level, there is no predator present and the water quality, which influences growth is maintained constant to assure the very best growth conditions for the fish. By variation of these factors it is even possible to find the influence of the water quality, and the available food on the growth parameters. It is often the results of such experiments, which can be found in the literature. However, the modeller might not be able to find the parameter for the species of interest to him, or he cannot find the parameters in the literature under the conditions prevailing in the ecosystem, that he wants to model. Then he might use such experiments to determine parameters of importance to his model. Even if he can find literature values for the crucial parameters he might still want to carry out parameter determinations in the laboratory or in a microcosms, if he estimates that the interval of the parameters in the literature is too wide for the most sensitive parameters.

However, parameters taken from the literature or as resulting from such experiment should be applied precautiously, because the discrepancy between the values in the laboratory or even the microcosms and these in nature is much greater for biological parameters than for chemical or physical parameters. The reasons for this can be summarized in the following points:

1. Biological parameters are generally **more sensitive to environmental factors**. An illustrative example would be: a small concentration of a toxic substance would be able to change growth rates significantly.

2. Biological parameters are **influenced by a great number of environmental factors**, of which some are very variable. For instance, the growth rate of phytoplankton is dependent on the nutrient concentration, but the local nutrient concentration is again very dependent on the water turbulence, which again is dependent on the wind stress, etc.

3. The example in point 2 shows, furthermore, that **the environmental factors influencing biological parameters are interactive**, which makes it almost impossible to predict an exact value for a parameter in nature from measurements in the laboratory, where the environmental factors are all kept constant. On the other hand if the measurements are carried out in situ it is not possible to interpret under which circumstances the measurement is valid, because that would require the determination, at the same time, of too many interactive environmental factors.

4. Often, determinations of biological parameters or variables **cannot be carried out** directly, but it is necessary to measure another quantity which cannot be exactly related to the biological quantity in focus. For instance, the phytoplankton biomass cannot be determined by any direct measurement, but it is possible to obtain an indirect measurement by use of the chlorophyll concentration, the ATP concentration, the dry matter 1-70 $\mu$ etc. However, none of these indirect measurements give an exact value of the phytoplankton concentration, as the ration of chlorophyll or ATP to the biomass is not constant, and the dry matter 1-70 $\mu$ might include other particles (e.g. clay particles). Consequently, it is recommended in practice to apply several of these indirect determinations simultaneously to assure that a reasonable estimate is applied.

   Correspondingly, the growth rate of phytoplankton might be determined by the oxygen method or the C-14-method. Neither method determines the photosynthesis, but the net production of oxygen, respectively the net uptake of carbon - that is, the result of the photosynthesis and the respiration. The results of the two methods are therefore corrected to account for the respiration, but obviously the correction in each individual case which, however, is difficult to do accurately.

5. Biological parameters are finally **influenced by several feedback mechanisms of a biochemical nature.** The past will determine the parameters in the future. For instance, the growth rate of phytoplankton is dependent on the temperature - a relationship which can easily be included in ecological models. The maximum growth rate is obtained by the optimum temperature, but the past temperature pattern determines the optimum temperature. A cold period will decrease the optimum temperature. To a certain extent, this can be taken into account by the introduction of variable parameters (see Straskraba, 1980). In other words, it is an approximation to consider

parameters as constants. An ecosystem is a soft, flexible system, which only with approximations can be described as a rigid system with constant parameters (see Jørgensen, 1981).

It has been pointed out above, that **the calibration is facilitated significantly, if we have good initial guesses of the parameters**. Some might be found in the literature, but it is only a few compared with the number of parameters actually needed, if we want to model all interesting mass flows in all relevant ecosystems. For the nutrient flows the parameters are known from the literature for the most common species only. But if we turn to flows of toxic substances in ecosystems the number of known parameters is even more limited. The earth has million of species and the number of substances of environmental interest is about 50 000. If we want to know 10 parameters for each interaction between substances and species, the number of parameters needed is enormous. If we, f.inst., in the first hand need 5 parameters for the interaction of let us say only 10 000 species with the 50 000 substances of environmental interest, the number of needed parameters is $5 \times 10.000 \times 50.000 = 10^9$ parameters. In Jørgensen et al. (1979) can be found 40 000 parameters and if we estimate that this handbook covers about 20 % of the entire literature, we know only about 10 ppm of the needed parameters.

Physics and chemistry have attempted to solve this problem by setting up some general relationships between the properties of the chemical compounds and their composition. If needed data cannot be found in the literature such relationships are widely used as the second best approach to the problem.

If we draw a parallel to ecology, we need some general relationships which give us some good first estimations of the needed parameters.

The application of such general relationships in chemistry gives a quite acceptable estimation in many cases. In many ecological models used in environmental context the required accuracy is not very high. In many toxic substance models we need, for instance, only to know whether we are far from or close to the toxic levels. However, more experience with the application of such general relationships are needed before a more general use can be recommended.

In this context it should be emphasized that in chemistry such general relationships are used very carefully.

Modern molecular theory provides a sound basis for the predictions of reliable quantitative data on the chemical, physical and thermodynamic properties of pure substances and mixtures. The biological sciences are not based upon a similar comprehensive theory, but it is, to a certain extent, possible to apply basic biochemical mechanisms laws on ecology. Furthermore, the basic biochemical mechanisms are the same for all plants and all animals. The spectrum of biochemical compounds is of course wide, but considering the number of species and the number of possible chemical compounds it is, on the other hand, very limited. The number of different protein molecules is significant, but they are all constructed from only 24 different amino acids.

This explains why the elementary composition of all species is quite similar. All species need, for their fundamental biochemical function, a certain amount of carbohydrates, proteins, fat and other compounds, and as these groups of biochemical substances are constructed from a relatively few simple organic compounds, it is not surprising that the composition of living organisms varies only very little, (see tables in Jørgensen et al. 1979).

The biochemical reaction pathways are also general, which is demonstrated in all textbooks on biochemistry. The utilization of the chemical energy in the food components is basically the same for microorganisms and mammals. It is, therefore, possible to calculate the energy, $E_1$, released by digestion of food, when the composition is known:

$$E_1 = 9 \frac{\text{fat}\%}{100} + 4\frac{(\text{carbohydrates} + \text{proteins})\%}{100}$$

The law of energy conservation is also valid for a biological system, see fig. 2.27. The chemical energy of the food components is used to cover the energy needs for growth, respiration, assimilation, reproduction and losses. As it is possible to set up relations between these needs on the one side, with some fundamental properties of the species on the other, it is possible to put number on the items on fig. 2.27 for different species. This is a rather general but valid approach to parameter estimation in ecological modelling.

The surface area of the species is a fundamental property. The surface area indicates quantitatively, the size of the border to the environments. Loss of heat to the environment must be proportional to this area and to the temperature difference, in accordance with the law of heat transfer. The rate of digestion, the lungs, hunting ground etc. are on the one hand determinant for a number of parameters, on the other hand, they are all dependent on the size of the animal.

It is therefore not surprising that a number of parameters for plants and animals are highly related to the size, which implies that it is possible to get a very good first estimate for a number of parameters only based upon the size. Naturally, the parameters are also dependent on a number of characteristic features of the species, but their influence is minor compared with the size and good estimate, which is valuable in many models, at least as a starting value in the calibration phase.

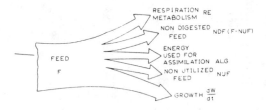

Fig.2.27: The principle of the model of fish growth is shown. The feed is used for respiration, is non-digested, is used for the assimilation processes, or not utilized and is used for growth. Notice that ALG is equal to (F-NUF) (1-NDF) ALC and only (F-NUF) (1-NDF) (1-ALC) is available for respiration and growth. ALC is a parameter (see Jørgensen, 1976).

The conclusion of these considerations must therefore be, that there should be a number of parameters, which might be related to simple properties, such as size of the organisms, and that such relations are based upon fundamental biochemistry and thermodynamics.

There is first of all a strong positive correlation between size and generation time, Tg, ranging from bacteria to the biggest mammals and trees (Bonner, 1965). The relationship is illustrated in fig. 2.28. This relationship can be explained by use of the above mentioned relationship between size (surface) and total metabolic action per unit of body weight. It implies that the smaller the organisms the greater their metabolic activity. The per capita rate of increase, r, defined by the exponential or logistic growth equations:

$$dN/dt = rN \qquad (2.19)$$

respectively:

$$dN/dT = rN(1-N/K) \qquad (2.20)$$

is again inversely proportional to the generation time. This is shown in fig. 2.29 where r and $T_c$ are plotted on a log/log scale. The relationship falls as seen into a narrow straight band with the slope -1.

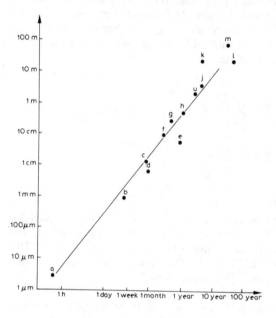

Fig.2.28: Lenght and generation time plotted on log-log scale: (a) pseudomonas, (b) daphnia, (c) bee, (d) house fly, (e) snail, (f) mouse, (g) rat, (h) fox, (i) elk, (j) rhino, (k) whale, (l) birch, (m) fir.

This implies that r is related to the size of the organism, but as shown by Fenchel (1970) falls into three groups: unicellar, poikilotherms and homeotherms, see fig. 2.30. Thus the metabolic rate per unit of weight is related to the size as shown in fig. 2.31. The same basis is expressed in the following equations, giving the respiration, feed consumption and ammonia excretion for fish when the weight, W, is known:

$$\text{Respiration} = \text{constant} \cdot W^{0.80} \tag{2.21}$$

$$\text{Feed Consumption} = \text{constant} \cdot W^{0.65} \tag{2.22}$$

$$\text{Ammonia Excretion} = \text{constant} \cdot W^{0.72} \tag{2.23}$$

This is also expressed in Odum's equation (Odum, 1959, p.56):

$$m = k\, W^{-1/3} \tag{2.24}$$

where K is roughly a constant for all species equal to about 1.4 kcal/$g^{2/3}$ day, and m is the metabolic rate per weight unit.

Similar relationships exist for other animals. The constants in these equations might be slightly different due to differences in shape, but the equations are otherwise the same.

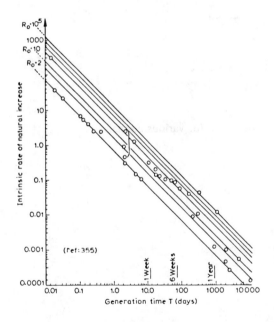

Fig.2.29:  Intrinsic rate of natural increase plotted to generation time with diagonal lines representing net reproduction rate from 2 to 105 for a variety of organisms.

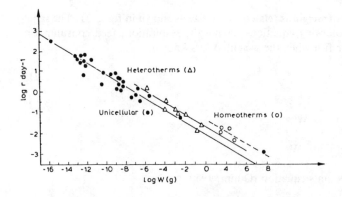

Fig.2.30: Intrinsic rate of natural increase to weight for various animals. Metabolic rate versus weight for various animals.

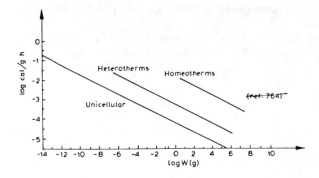

Fig.2.31: Relationship of metabolic rate to weight for various animals.

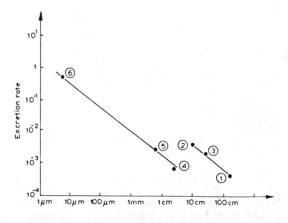

Fig.2.32: Excretion of Cd (24 h)-1 plotted to the length of various animals: (1) Homo sapiens, (2) mice, (3) dogs, (4) oysters, (5) clams (6) phytoplankton.

All these examples illustrate the fundamental relationship between size (surface) and the biochemical activity in the organisms. The surface determines the contact with the environment quantitatively, and thereby the possibility to take up food and excrete waste substances.

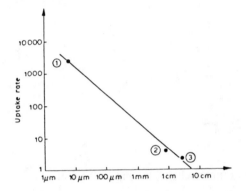

Fig.2.33: Uptake rate ($\mu$g/g 24 h) plotted to the length of various animals (CD): (1) phytoplankton, (2) clams, (3) oysters.

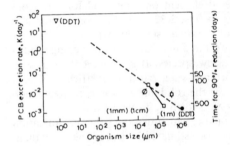

Fig.2.34: Excretion of PBC and DDT versus organism size.

The same relationships are shown in figs. 2.32-2.35, where biochemical processes of toxic substances are involved.

These figures are constructed from literature data, and as seen, the excretion rate and uptake rate (for aquatic organisms) follow the same trends as the metabolic rate (fig. 2.31). This is of course not surprising, as the excretion is strongly dependent on the metabolism and the direct uptake is dependent on the surface.

The concentration factor, which indicates the ratio: concentration in the organism to the concentration in the medium, also follows the same lines, see fig. 2.35. By equilibrium the concentration factor can be expressed as the ratio between the uptake rate and the excretion rate, as shown in Jørgensen (1979). As most concentration factors are determined by equilibrium, the relationship found in fig. 2.35 seems reasonable. Intervals for concentration factors are here indicated for some species, in accordance with the literature (see Jørgensen et al., 1980).

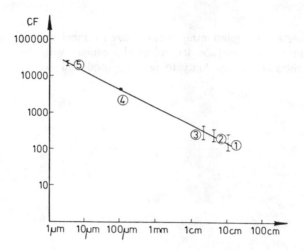

Fig.2.35: CF for Cd versus size: (1) goldfish, (2) mussels, (3) shrimps, (4) zooplankton, (5) algae (brown + green).

The principles illustrated in figs. 2.32 to 2.35 can be applied generally. In other words, it is possible to find the uptake - and excretion rate and concentration factor, provided these parameters are available for the considered element or compound for one - but preferably several species. When a plot similar to figs. 2.32 to 2.35 is constructed it is possible to read the parameters when the size of the organism is known.

These parameters are only known for a very limited number of organic compounds - although this knowledge is required only for a few species. It is, however, possible to estimate CF for organic compounds, if the n-octanol-water partition coefficient, $K_{ow}$, is known, the soil-water sorption coefficient, $K_{oc}$ or the solubility in water, S. Regression equations currently in use for such estimations are shown in Table 2.14.

The biological or ecological magnification factor, which describes the magnification factor from one trophic level to the next can also be estimated by use of solubility data, if they cannot be found in the literature:

$$\log EM = -0.4732 \log S + 4.4806$$
(Metcalf et al., 1975) (mosquito fish, whole)
$$\log EM = -0.3891 \log S + 3.995$$
(Luaand Metcalf, 1975) (mosquito fish, whole)

The partition coefficient can be found in the literature for most organic compounds of environmental interest. But even if this data is missing, it is possible to estimate the partition coefficient from water solubility, as shown in fig. 2.37. Data on water solubility should always be available.

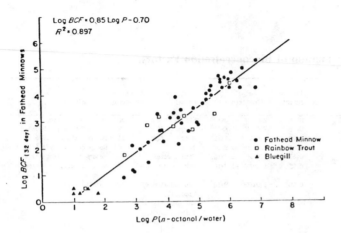

Fig.2.36: Biological concentration factor. (From Veith et al., 1979).

**TABLE 2.14**

## Regression Equations for Estimation of Bioconcentration Factors.

| Indicator | Relationship | Correlation coefficient | Range (Indicator) | Animal | Number of chemicals | References |
|---|---|---|---|---|---|---|
| $K_{ow}$ | $\log CF^b = -0.973 + 0.767 \log K_{ow}$ | 0.76 | $2.0 \times 10^{-2} - 2.0 \times 10^6$ | Fish species[a] | 36 | Kenaga and Goring, 1978 |
| $K_{ow}$ | $\log CF^b = 0.7504 + 1.1587 \log K_{ow}$ | 0.98 | $7.0 \times 10^0 - 1.6 \times 10^4$ | Mosquito fish | 9 | Metcalf et al., 1975 |
| $K_{ow}$ | $\log CF^b = 0.7285 + 0.6335 \log K_{ow}$ | 0.79 | $1.6 \times 10^0 - 1.4 \times 10^4$ | Mosquito fish | 11 | Lu and Metcalf, 1975 |
| $K_{ow}$ | $\log CF^c = 0.124 + 0.542 \log K_{ow}$ | 0.95 | $4.4 \times 10^2 - 4.2 \times 10^7$ | Trout | 8 | Neely et al., 1974 |
| $K_{ow}$ | $\log CF^c = -1.495 + 0.935 \log K_{ow}$ | 0.87 | $1.6 \times 10^2 - 3.7 \times 10^6$ | Fish species[a] | 26 | Kenaga and Goring, 1978 |
| $K_{ow}$ | $\log CF^c = -0.70 + 0.85 \log K_{ow}$ | 0.95 | $1.0 \times 10^0 - 1.0 \times 10^7$ | Fathead minnow | 59 | Veith et al., 1979 (see fig.2. ) |
| $K_{ow}$ | $\log CF^b = 0.124 + 0.542 \log K_{ow}$ | 0.90 | $1.0 \times 10^2 - 5.0 \times 10^7$ | Fathead minnow, bluegill, mosquito fish, rainbow trout, green sunfish | 59 | Lassiter, 1975 |
| $K_{oc}$ | $\log BCF^b = 2.024 + 1.225 \log K_{oc}$ | 0.91 | $0.4 \times 10^0 - 4.3 \times 10^4$ | Fish species[a] | 22 | Kenaga and Goring, 1978 |
| $K_{oc}$ | $\log BCF^c = 1.579 + 1.119 \log K_{oc}$ | 0.87 | $3.2 \times 10^0 - 1.2 \times 10^6$ | Fish species[a] | 13 | Kenaga and Goring, 1978 |
| $S(mg\ l^{-1})$ | $\log BCF^b = 2,183 - 0,629 \log S$ | -0.66 | $1.7 \times 10^{-3} - 6.5 \times 10^5$ | Fish species[a] | 50 | Kenaga and Goring, 1978 |
| $S(ug\ l^{-1})$ | $\log BCF^b = 3.9950 - 0.3891 \log S$ | -0.92 | $1.2 \times 10^0 - 3.7 \times 10^7$ | Mosquito fish | 11 | Lu and Metcalf, 1975 |
| $S(ug\ l^{-1})$ | $\log BCF^b = 4.4806 - 0.4732 \log S$ | -0.97 | $1.3 \times 10^0 - 4.0 \times 10^7$ | Mosquito fish | 9 | Metcalf et al., 1975 |
| $S(mg\ l^{-1})$ | $\log BCF^c = 2.791 - 0.564 \log S$ | -0.72 | $1.7 \times 10^{-3} - 6.5 \times 10^5$ | Fish species[a] | 36 | Kenaga and Goring, 1978 |
| $S(umol/l)$ | $\log BCF^c = 3.41 - 0.508 \log S$ | -0.96 | $2.0 \times 10^{-2} - 5.0 \times 10^3$ | Trout | 7 | Chiou et al., 1977 |

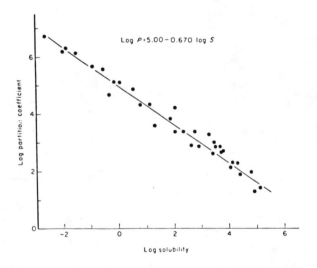

Fig.2.37: The linear regression of log P on log S, for a range of organic compounds that includes aliphatic and aromatic hydrocarbons, aromatic acids, organochlorine and organophosphorus insectides, and polychlorinated biphenyls. P is the n-octanol: water partition coefficient, determined at room temperature, and S is the solubility in water (umol/l), determined at 10-30 C. (From Chiou et al., 1977).

It is obvious that it would be impossible to measure all the parameters needed for modelling the environmental distribution of all toxic substances of interest. It is, therefore, necessary to use the presented rules, which can be applied when needed data is not available. Such rules will, of course, give estimations, which have a higher standard deviation than corresponding measured data, but in most cases such estimations give a satisfactory accuracy.

However, more experience with such estimations and rules is needed before a clear conclusion on the applicability can be taken. The results presented demonstrate how it is possible to obtain several parameters on the basis of the size of the organisms and the partition coefficient.

From these considerations recommendations can now be extracted as to how it is feasible to carry out a calibration of an ecological model:

1. In the first place find **as many parameters as possible from the literature.** *Even a wide range* for the parameters should be considered very valuable, as approximate initial guesses for all parameters are urgently needed.

2. If some of the parameters cannot be found in the literature one of the **general relationships** mentioned above **could be used** or a **determination in the laboratory or in a microcosms** could be carried out.

3. Having intervals for all parameters a **sensitivity analysis must be carried out** by use of the best initial guesses for the parameters. The median value of the interval could f.inst. be used as initial guess.

4. By use of **the most sensitive parameter for calibration the modeller attempts to fit model output to observed data.** He tries as far as possible to consider subsystems or a few processes at a time, if possible keeping some of the state variables constant.

5. If this phase fails the modeller should consider **to use intensive measurements or more accurate determinations of crucial parameters** in the laboratory or by use of microcosms. If the model aims at description of a certain dynamics of the system, the data should under all circumstances reflect the dynamics. In other words intensive measurements are essential, if the available data are not dense enough to follow the dynamics of the state variables modelled, see fig. 2.25.

6. The results from point 5 are used **in a repeated calibration** using the trial and error procedure mentioned in point 4.

7. **A second sensitivity analysis** is then carried out now using the parameters found in point 6 as starting values for the parameters.

8. The most sensitive parameters are finally calibrated **by use of an automatic calibration procedure**, weight the state variables in accordance with their importance for the model result. An objective function must be defined at this stage, see f.inst. equation (2.12).

The automatic calibration is preferably carried out for the selected parameters for the interval determined by average value +/- standard deviation. The standard deviation is either estimated from previous calibrations found in the literature or determined by use of microcosms, intensive measurements or laboratory investigations.

A calibration along this line is rather cumbersome, but it is crucial for the further application of the model to be able to trust the parameter and not question the parameters, when the model results are evaluated.

## ILLUSTRATION 2.3

The following measurements were obtained for the model presented in illustration 2.1:

**TABLE 2.15**

| | Measurements | |
|---|---|---|
| t | PS | PA |
| 0 | 0 | 1.0 |
| 80 | 0.160 | 0.80 |
| 160 | 0.140 | 0.85 |
| 240 | 0.180 | 0.80 |
| 365 | 0.120 | 0.88 |

K and R, the two most essential parameters in the model, must be calibrated to give a possible better agreement between model output and measured values. As mentioned in illustration 2.1 the initial guesses for the two parameters are: K = 1.0 and R = 0.1.

The DSL program has facilitied optimizing the calibration. The model was run 102 times. The objective function, which in this case is defined as $\Sigma$(measured value - model output)$^2$, was improved from 29.68 to 27.99 by changing the parameters to K = 1.759 and R = 0.0573.

## 2.8. VALIDATION

When the modeller has satisfactorily terminated the calibration, his next obvious question would be:

Do the parameters found by the calibration represent the real values in the system? It might be possible, even in a data rich situation, to force a wrong model by selection of the paramters to give outputs, that fit well with the data. It is therefore crucial for the modeller **to test selected parameters with an independent set of data, this is termed validation.** It must, however, be emphasized that the validation only confirms the model behaviour under the range of conditions represented by the available data. Consequently, it is preferably to validate the model by use of data obtained from a period, in which other conditions prevail than from the period of data collection for the calibration. F.inst. if a eutrophication model is used, the ideal situation would be to have observations from the modelled ecosystem over a wide range of nutrient inputs, as the model is used to predict the ecosystem response to changed nutrient loadings. This is often difficult as it must correspond to a complex validation of the prognosis, but then the data can be used directly and the model prognosis is redundant. However, it might be possible and helpful to obtain data from a certain range of nutrient loadings f.inst. from a humid and a dry summer. Alternatively, it might also be possible to get data from a similar ecosystem with approximately the same morphology, geology and water chemistry as the ecosystem modelled in the first place.

Similarly, a BOD/DO model should be validated under a wide range of BOD-loadings, a toxic substance model under a wide range of concentrations of toxic substances and a population model by different levels of the populations etc.

If an ideal validation cannot be obtained, it does not imply that the construction of models are useless. As mentioned in chapter 1 models are multi purpose tools, and if the best validation cannot be carried out, it is still of importance to validate the model. Furthermore, the model can always be used as a management tool, provided that the modeller presents all the open questions of the model to the manager and as we gain more experience in use of models the number of open questions will be reduced.

The method of validation is dependent on the objectives of the model. A comparison between measured data and model output by use of the objective function shown equation (2.12) in section 2.7 is an obvious test. This is , however, most often insufficient, as it does not focus in the main objectives of the model, but only on the general ability of the model to describe the state variables of the ecosystem, correctly. It is therefore required to translate the main objectives of the model into a few validation criteria. They cannot be formulated generally, but are individual for the model and the modeller. If f.inst. a BOD/DO model is used to predict the water quality of a stream, it will be useful to compare the minimum concentration of oxygen predicted by the model with the corresponding measured data. For an eutrophication model the maximum phytoplankton concentration and the maximum production could be used for validation. For a population model the modeller might be interested in the minimum or maximum level of certain species etc.

In a data-poor situation it might be impossible to meet such validation criteria, but it could then be useful to compare average situations, as the model due to the quality of data available does not describe the dynamics of the system very well, but only can give information of the general level or average of important variables.

The discussion on the validation, can be summarized into the following issues:

1. **Validation is always required.**

2. Attempt should be made to get data for the validation which are **entirely different from those used in calibration.** It is important to have data from a **wide range of the forcing functions,** that are defined by the objectives of the model.

3. Validation criteria are formulated **on basis of the objectives of model and the quality of the data.**


## 2.9. CONSTRAINTS ON MODELS

Most ecological models at present are of the form presented in equation (2.14): (Among the exceptions are discrete models, stochastic models, self-modifying algorithmic models, etc.).

A naive question is: Which subset of all possible equations of type (2.14) are ecological models? Which constraints should be put on eq. (2.14) to make it an ecomodel?

The answers to this question are numerous, and they cover a wide range from solid facts to tentative hypotheses and down to pure guesswork. Model constraints might be grouped as follows:

1) Conservation of mass and elements.
2) Conservation of momentum.

3) Conservation of energy.
4) Boundary conditions.
5) Initial conditions.
6) Narrow bands of biomass composition.
7) The second law of thermodynamics.
8) Law of chemical and biochemical processes.
9) Laws of biological processes (and other biological restrictions).
10) Other thermodynamic restrictions.
11) Topological restrictions.

The constraints should be considered an instrument to find workable equations in accordance with nature for the model. Models must reflect our knowledge of natural science, which is formulated as laws or constraints. It is therefore advantageous for the modeller to apply as many of the above mentioned constraints as possible. In most cases it is not possible to use them all, as it depends on the model. If f.inst. the model is a geobiochemical model, a description of energy relations are not included and the principle of energy conservation cannot be formulated in the model. However, in most modelling cases several of the constraints can be used simultaneously.

1) Conservation of mass.
In a general geobiochemical system mass balances may be formulated as:

$$\frac{\partial \psi_c}{\partial t} = -\nabla \cdot (\overline{J}_c + \psi_c \overline{v}) + \sum_r \nu_{rc} \dot{\xi}_r \qquad (2.25)$$

here $\psi_c$ is concentration of compound c in a volume element moving with velocity $\overline{v}$. $\overline{J}_c$ is the netflux of c out of the volume element and $\nu_{rc}$ are signed stoichiometric coefficients for all reactions r (with reaction rates $\dot{\xi}_r$) in which c participate. Although many excellent models handle transformations and transport phenomena by choosing other state variables than concentration of compounds (e.g. number of individuals in prey-predator models), it is recommended, wherever possible to choose $\psi$ as molar concentrations (or equivalent) taking advantage of the extra check that the sums of equations of type (2.26) over all components containing an element e should yield:

$$\frac{\partial \psi_e}{\partial t} = -\nabla \cdot (\overline{J}_e + \psi_e \cdot \overline{v}) \qquad (2.26)$$

where $\psi_e = \Sigma \psi_c$ and $\overline{J}_e = \Sigma \overline{J}_c$.

2) Conservation of momentum.
This physical principle is in essence based on Newton's law. In hydrodynamics Newton's law is the point of departure for setting up the well-known Navier-Stokes equation and the (less general) Reynold's equation. (See e.g. TRACOR (1971), which also contains a multitude of examples of ecological interest).

3) Conservation of energy.
Energy conservation in a volume element may be expressed by vector calcus:

$$\frac{\partial u}{\partial t} = -\nabla \cdot \overline{J}_u \quad, \quad \overline{J}_u = \overline{J}_q + u\overline{v} + \overline{\overline{P}} \cdot \overline{v} + \sum_c \phi_c \cdot \overline{J}_c \qquad (2.27)$$

Symbols used are: u (energy density), $\overline{J}_q$ (heat flux), $\overline{\overline{P}}$ (stress tensor), $\phi_c$ (specific potential energy belonging to the external force per unit mass,). Terms on the right hand side of (2.27) describe: heat flow, advection, mechanical work and the work performed by external forces.

For a detailed example of energy conservation applied to wind mixing and temperature profile prediction in stratified lakes, see Harlemann (1977).

4) Boundary conditions.

The number and type of boundary conditions depend strongly on the ecosystem, on the specific problem of interest and on the type of model, that we intend to construct. An ecosystem is an open system and at the surface separating the ecosystem from its surroundings, we must prescribe the boundary values of: the fluxes of energy and matter across the surface and/or the temperature and concentrations of the constituents of the ecosystem at the surface. If f.inst. thermal effects are included in the model, a boundary condition on the temperature is required. Either the temperature or the heat flux or both are specified.

Boundary conditions reflect the constraints of the surroundings acting upon the components of the ecosystem.

5) Initial conditions.

Most dynamic models are dependent on the initial conditions. The state of the ecosystem might even after a long period of time be different, when different initial conditions are used. The initial condition represents the time (history) of the system prior to the time, that the model is intended to describe. As the system is dependent on its history (see f.inst. Patten (1985), it is important to define the initial stage properly.

The initial stage might be defined by the objectives of the model f.inst., when the model is intended to describe the response of a ecosystem under specific conditions to various patterns of forcing functions.

Another possibility is to measure the state variables at the initial stage and use these values as initial condition. When it is required that the model should be long term stable, the initial conditions can be found, which are those initial values of the state variables, which with the same seasonal of diurnal fluctuations of forcing functions bring the state variables again and again back to the same values or fluctuations.

It is recommendable under all circumstances to carry out a **sensitivity analysis of the initial values** to observe, which ones are required to be known with the highest accuracy. Furthermore, simulations with a wide range of initial values of the state variables are often carried out in the verification phase in order to observe the stability of the model.

Steady state models often include an initial situation, which is either an instant or average description of the ecosystem. The model will then be used to observe changes in the ecosystem, when the initial forcing functions are changed. The response of the ecosystem to changed conditions is found by a comparison of the initial state with the state under changed circumstances.

6) Narrow bands of biomass composition.

Tables in Jørgensen et al. (1979) illustrate typical compositions of biomass. They might vary but within relatively narrow bands. If for example, the phosphorus cycle is very well understood including rates and rate coefficient, we know that the corresponding rates for nitrogen will be f.inst. 2.5 - 15 times those for phosphorus. Such information is of course of great value to the modeller. The constraints on chemical composition enable him to construct the model with much fewer data.

7) The second law of thermodynamics.
The 2. law of thermodynamics extended to open systems far from equilibrium, may be formulated as the inequality (Prigogine, 1955)

$$\sigma > 0 \tag{2.28}$$

where $\sigma$ is the local entropy production per volume calculated by

$$T\sigma = -\bar{J}_s \cdot \nabla t - (\bar{\bar{P}} - P\bar{\bar{I}}):\nabla\bar{v} - \sum_c \bar{J}_c \cdot (\nabla\mu_c - \bar{f}_c) + \sum_r \xi_r A_r \tag{2.29}$$

Here T denotes temperature, $\bar{\bar{P}}$ is the stress tensor, P the hydrostatic pressure, $\bar{\bar{I}}$ an identity matrix, $\bar{v}$ the velocity of the volume element, $\bar{J}_c$ the flux of component c, $\mu_c$ its chemical potential, $\bar{f}_c$ the force per unit mass acting on c, $\xi_c$ is the reaction rate for reaction r, $A_r$ its affinity (calculated from chemical potentials and stoichiometry), and finally

$$\bar{J}_s = \frac{\bar{J}_q}{T} + \frac{1}{T}\sum_c \mu_c \bar{J}_c \tag{2.30}$$

is the entropy flux out of the volume element. (2.29) shows how dissipation (T$\sigma$) of free energy is due to: entropy influx (-$\bar{J}_s$) induced by temperature gradients, velocity gradients impelled by shear stresses, transport of matter caused by diffusion and external forces, and (bio-)chemical reactions enforced by chemical affinities. Entropy balance is the expressed by

$$\frac{\partial s}{\partial t} = -\nabla \cdot (\bar{J}_s + s\bar{v}) + \sigma \tag{2.31}$$

where s is entropy per volume. The change of total system entropy (S) is found from (2.31) by integration:

$$\frac{dS}{dt} = \frac{d_eS}{dt} + \frac{d_iS}{dt} \tag{2.32}$$

$$\tag{2.33}$$
where $\frac{d_eS}{dt} = -\int_{surface}(\bar{J}_s + s\bar{v}) \cdot d\bar{a}$ and $\frac{d_iS}{dt} = \int_{volume} \sigma dV > 0$.

This equation - among other things - shows that systems only can maintain a non-equilibrium steady state (ds/dt = 0) by compensating the internal entropy production ($d_iS/dt > 0$) by negative entropy influx ($d_eS/dt < 0$). Such an influx induce order into the

78

system. In ecosystems the ultimate negative entropy influx comes from solar radiation, and the order induced es e.g. biochemical molecular order.

The dissipation function (2.32) may be used to identify generalized system forces $(-\nabla T, (\overline{\overline{P}}-\overline{\overline{PI}}), -(\nabla \mu_c \cdot \overline{f}_c), A_r)$. Ecomodels have mainly been concerned with flows and processes, maybe the forces causing these flows and processes have received too little attention. This may partly be due to relics from thermostatics (equilibrium thermodynamics - the queen of science for over a century) in which local forces vanish.

8) Law of chemical and biochemical processes.
Chemistry and biochemistry set up laws for processes. These processes occur also in ecosystems and should be described by use of underlying chemical and biochemical knowledge. Potentially, significant chemical processes in ecosystem can be classified broadly into:
acid-base reactions
coordination reactions of metal ions and ligands
precipitation and dissolution of solid phases
adsorption and desorption
redox processes
gas solution and outgassing processes.
Decomposition reactions are often well described by use of first order kinetics, f.inst. the sink term for BOD:

$$\frac{d\,\mathrm{BOD}}{dt} = -k_1 \cdot \mathrm{BOD} \tag{2.34}$$

The same equation in integrated form is:

$$\tag{2.35}$$
$$\mathrm{BOD}_t = \mathrm{BOD}_0 \cdot e^{-k_1 \cdot t}$$

Enzymatic reactions or a process with one limiting factor uses the so called Michaelis-Mentens expression:

$$\tag{2.36}$$
$$V = K_2 \cdot \frac{S}{k_m + S}$$

where V is the reaction rate $(= \frac{dS}{dt})$, $K_2$ is a rate constant, $k_m$ the so called Michaelis-Menten constant and the concentration of substrate for the enzymatic reaction.

In chapter 3 physical, chemical and biochemical processes are reviewed together with their mathematical description used in ecological modelling. The review is comprehensive but not complete. Many textbooks in physics chemistry and biochemistry will supply the missing information.

9) Laws of biological processes.
These are still hard to quantify. Many of the proposed "evolution criteria" (evolution is here not used in the Darwinistic sense) takes the form of an extremum principle, e.g. H.T. Odum's "maximum power principle" Odum, 1955).

The Glansdorff-Progogine principle on minimum dissipation rate has only been proved to be strictly valid near equilibrium. Minimizing niche overlap or the

"not-putting-all-eggs-in-one-basket"-principle (May 1973), may in the future be quantified as an extremum principle. It is related to the problem of quantifying adaptability. It
is certainly true that change in the environment induces latent machinery in the system.
This "requsite variety" (as Ashby (1967) puts it) is seldom considered in out models.
Even if total information in the genetic pool is treated as a constant, it should be possible
to simulate various degrees of realization of this information activated by changed
environment.

In pure trophic models cybernetic refinements are easily overlooked. To quote
Morowitz (1968): "A misplaced methyl group may eventually kill a whale".

It has been shown mathematically (Smale, 1966) that structural stable systems are
rare in more than 3 dimensions. Evolutionary processes have filtered out those tiny and
mathematically atypical regions of parameter space, which furnish the system with long
term stability.

The role of ecosystem history is sometimes remarkable: for example, just compare
the fauna of old Lake Malawi and young Lake Victoria.

Productive and predictable environment will often favour greater specialization;
cold regions and unpredictable environments may favour generalists. These few remarks
suggest that one single principle covering all kinds of ecosystem development in detail is
Utopian.

For restricted classes of systems, without delicate cybernetic elegance practical
measures of structure-response relations may be given. Ecological buffer capacity is
used to explain why hypereutrophic systems are simple, but still stable (Jørgensen and
Mejer, 1979). Furthermore tentative guesses as to the form of the functional to be
optimized is suggested.

If the biological constraints show up to take the form of an extremum principle,
attention should be paid to the Pontryagin maximum principle, which (in one of its
versions) states:

To maximize $\int_{t1}^{t2} Q(\psi, C) dt$ varying $C(t)$

subject to $\dot{\psi} = f(\psi, C)$

you may instead maximize the Hamiltonian:

$$H = Q(\overline{\psi}, C) + \sum_c f(\overline{\psi}, C) \tag{2.37}$$

This is just mentioned to point out that mathematically there is hope, while the real
problem is: what should be optimized, and for the well-being of who? For the individual
organism, for the evolutionary development of the species, for the well-being of the
whole ecosystem or even the total biosphere?

Chapter 3 reviews biological process equations, which can be used as constraints in
the model, that contain a description of these processes. Modelling would be much
simpler if only one expression was valid for one biological process, but as it will be
demonstrated in chapter 3, for a number of processes, there are several alternative
equations available. Which equation to select for a given problem of ecological modelling depends on:

1) **the quality and quantity of data.** The better the data the more accurately it is possible to describe the processes and the more complex equations can be selected. They will contain more parameters, but the data should be able to cope with the calibration.

2) **the ecosystem.** Some biological processes do not have a general, site independent description. The site differences are often caused by different foodwebs or other biological details, which are not considered in the model, but which might influence the selected state variables.

10) Other thermodynamic restrictions.

Many scientists believe that further thermodynamic constraints might be the answer of the future to models, which should contain the characteristic properties of the ecosystems on system level. This will be further discussed in chapter 9 and it would be going into too much detail in this context to discuss at this stage all the possible thermodynamic approaches to this problem.

However, two concepts should be mentioned: exergy and ecological buffer capacity.

**Exergy:**

If the ecosystems were in thermodynamic equilibrium, the entropy seq would be higher than in non-equilibrium. The excess entropy is the thermodynamic information

$$(2.38)$$

$$I = S^{eq} - S$$

It is an old fact that I also equals the Kullbach measure of information (Brillouin, 1956):

$$(2.39)$$

$$I = k \sum_j p_j^* \cdot \ln \frac{p_j^*}{p_j}$$

where $p_j^*$ and $p$ are probability distributions, a posteriori and a priori to an observation of the molecular details of the system, and $k$ is Boltzmann's constant.

It may be shown (see e.g. Evans, 1966) that (2.38) leads to

$$(2.40)$$

$$I = \frac{U + PV - TS - \sum X_j n_j}{T}$$

where $P$, $T$ and $X_j$ are intensive properties of reservoirs that the system is assumed to interact with. nj are mole numbers.

R.B. Evans (op.cit.) defined the exergy as

$$(2.41)$$

$$E = T \cdot I$$

so from (2.40):

$$E = U + PV - TS - \sum_j X_j n_j \tag{2.42}$$

It is easily seen that E degenerates to well-known thermodynamic potentials at special circumstances. If e.g. a chemical inert system interacts with a heat reservoir only (2.42) becomes E = U-TS (2.43), i.e. the Helmholtz free energy, since V and $n_j$ remain constant. Similarly, E = U + PV-TS (2.44) equates with the Gibb's potentential, if the system is coupled to heat and work reservoirs only. E will equate with the enthalpy U + PV, if only volume can be exchanged with the surroundings.

One of the main postulates in equilibrium thermodynamics is, that entropy may be expressed by U, V and $n_j$ in the form

$$S = \frac{U}{T} + \frac{P}{T}V - \frac{1}{T}\sum_j X_j n_j \quad \text{at equilibrium} \tag{2.45}$$

It then follows from (2.42) that $\quad E = 0$

$$\tag{2.46}$$

This also follows from (2.41) and (2.43) which further shows that

$$\tag{2.47}$$

$$E \gtrless 0 \text{ in general}$$

It should be stressed once more that the intensive properties (P, T, $X_j$) in (2.42) are reservoir properties. So exergy depends on the surroundings. An evacuated vessel normally have a positive exergy, but E = 0 if it is brought to outer space. Exergy - as a free energy concept - is a measure of the maximum useful (entropy free) work that may be extracted from the system on its way to that thermodynamic equilibrium state, which is compatible with reservoir properties.

The exergy balance is found by differentiation of (2.42)

$$\frac{dE}{dt} = - \int_{\text{surface}} \overline{J}_E \cdot d\overline{a} - T \int_{\text{volume}} \varrho \, dV \tag{2.48}$$

where $\overline{J}_E$ is the exergy outflux:

$$\overline{J}_E = \left(1 - \frac{T^R}{T}\right)\overline{J}_u + T^R \sum_c \left(\frac{X_c}{T} - \frac{X_c^R}{T^R}\right)\overline{J}_c \tag{2.49}$$

(R denotes reservoir properties).
(E cannot be defined as a local property, since it depends on its surroundings, i.e. sum of subsystem exergy will not equal the exergy of the combined system).

### Ecological buffer capacity.

Assuming relaxation times for processes inside the system to be shorter than time constants of driving variables, we may speak of (slowly moving) steady states.

Near steady state generalized "surplus" forces $X_c$ may be derived from an "elastic potential function" L as

$$X_c = \frac{\partial L}{\partial \psi_c} \qquad \qquad (2.50)$$

where $\psi_c$ are state variable displacements from steady state. (compare with equation (2.5).

The phrase "ecological buffer capacity" is coined for $\beta$ (and related concepts such as inverse driving variable sensitivities). $\beta$ may be conceived as stiffness coefficients or elastic moduli for the systems.

The development of ecological modelling goes toward inclusion of more and more constraints. The scope is to limit the number of possible models to facilitate the selection of an ecological sound model and to build as much ecological realism into the models as possible. Further treatment of this important subject will take place in chapter 9.

## 2.10. COMPUTERS AND ECOLOGICAL MODELLING

The conversation shown in table 2.16 - deliberately made naive - is supposed to illustrate the idea of an expert system operating in a lake management situation.

TABLE 2.16

**A hypothetical conversation between the computer ("C") and you ("Y").**

C:Is the lake eutrophic?
Y:YES
C:Which nutrient is a limiting factor?
Y:P
C:Is loading of P from tributaries > 1 g/m2/y?
Y:NO
C:Could eutrophication be due to release of P from sediment?
Y:YES
C:Try to fix the problem by one of the following measures:
- sediment removal
- hypolimnic water exchange
- add alumina oxide
C:Let's diagnose next problem...

To extend the idea further, let's assume that the answer to the last question was NO and that the computer replies:
C:Sorry. I can't diagnose the problem.
If you have a suggestion, type it here...
If you type:
Internal recycling

the computer may prompt you to type in a question, the answer of which is YES for "internal recycling" and NO for "release of P from sediment".

Next time this conversation takes place you and the expert system, the computer may give up less easily: the system learns by adding information to its knowledge base.

Knowledge in this context means not only facts, but relationships between these facts as well. Software apt to deal with facts and logical relationships and rules governing interrelations between the facts has been emerging during the last decade. Programming languages like Prolog, LISP and even Logo have quite different structure compared to classical procedural languages like Pascal, FORTRAN and PL/I; they are more adaptive to the problem in hand, they may extend themselves and they are based on formal logic rather than algorithms. The central component of expert systems is the data base. In an ecological expert system the data base holds information on numeric and symbolic facts about the ecosystem and knowledge about the biological, chemical and physical (e.g. hydrological) structure and interrelations in the system.

Around the data base revolves software systems like:

- data base management systems (DBMS), including e.g. a query language, updating procedures, spreadsheet programs and editors;
- presentation packages, e.g. for graphics and report generation;
- statistical packages;
- a simulation language, including facilities for parameter estimation, system identification and optimization.

Together this may be named the "simulation environment" (fig. 2.38)

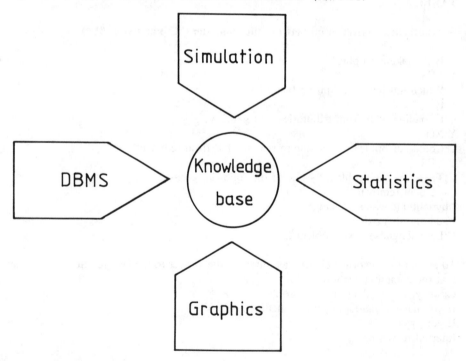

Fig.2.38:Simulation environment.

Traditionally data bases, simulation - and to a less degree statistics and graphics - have been implemented on main frames (big computers), but since ab. 1980 the revolution of microcomputer technology made it feasible to use these much cheaper - and still more powerful - tools in a simulation environment. Either the smaller systems may be used isolated as personal computers (PC's), as interconnected units such as small and local area networks (SAN and LAN), or as intelligent workstations linked to a host or a network of hosts (metropolitan and wide area networks, MAN and WAN). At present external and internal storage in microcomputers put some limits on the size of the data base and their speed makes them less suited for large scale simulations. By linking microcomputers to mainframes these problems may be solved as well as data and models may be shared by the community of ecological experts. Although expert systems as described in the previous section are still rare among ecologists, they are emerging in other areas, such as medicine, natural sciences, law, psychology etc. Despite of this, the concept of the "simulation environment" (fig. 2.38) still applies, if you substitute "Knowledge base" with "Ecologist".

Among the components of simulation software, probably simulation languages and statistics are the most widely implemented on computers used by modellers of ecological systems. The purpose of a simulation language for ecologists is to facilitate computer implementation of more or less well structured conceptions about ecosystems. A simulation language should lighten the burden of the modeller by supervising routine operations irrelevant to the modelling process. If f.inst. an ecosystem may be conceived as a continuous system with time as the only independent variable, then the simulation language should take care of procedures such as integration by time, multiple run administration, sorting of model equations into a possibly non-contradictory order, performing interpolation or perhaps smoothing on discrete time series data and provide suitable output such as tables and graphs.

An example of a lake model program is shown in table 2.17. The model itself is further described in illustration 2.1.

**TABLE 2.17**

**A CSMP program describing the standard model example. The order of the statements (except for END, STOP, ENDJOB) could have been changed without any consequences.**

DPS = (PIN - PS ) * Q/V - (MU - R - Q/V) * PA
DPA = (MU - R - Q/V) * PA
MU = PS/(K + PS)
 S = Smax (1 + sin (0.008603 . t)) PARAM PIN = 1., V = 1.E5, Smax = 0.5, K = 1.0, R = 0.1, ℚ = 0.01
PS = intgrl (0.0 DPS)
PA = INTGRL )1.0, DPA)
TIMER FINTIM = 365
PRPLOT PA(PS)
END
STOP ENDJOB

Table 2.18 summarizes some features for a few of the common available simulation languages.

# TABLE 2.18

**Simulation language features. A " + " indicates that the feature is well developed, a "-" that it is absent. "(+)" and "(-)" usually means that some programming effort is necessary or that not all versions of the language provide the feature.**

| Language / Feature | CSMP | DYNAMO | Simscript | GPSS | ACSL | SOMNON | COSY | GASPDSL |
|---|---|---|---|---|---|---|---|---|
| Continuous systems | + | + | (+) | - | ++ | + | (+) | + |
| Discrete systems (Event handling) | - | - | + | + | (+) | - | + | (+) |
| Stochasticity (Monte Carlo) | + | (+) | + | + | (+) | - | + | + |
| Other statistics | - | - | + | + | - | - | + | + |
| Data base interface | (-) | - | - | (+) | - | - | (+) | (+) |
| User selectable numerical methods | + | (-) | (-) | (-) | + | + | + | + |
| PDE | (+) | - | - | - | (+) | - | + | (+) |
| Interpolation | + | (+) | - | - | + | - | + | + |
| Parallelism (sorting, networks) | + | (+) | - | + | + | + | - | + |
| Modularity (macros etc) | + | - | - | - | + | - | + | + |
| Graphics | (+) | - | - | + | (+) | + | + | + |
| Interactiveness | - | - | - | - | (+) | + | - | - |
| Sensitivity an. Parameter est. | (-) | - | - | - | (+) | (+) | + | (+) |
| Portability | (-) | - | - | - | - | (-) | (+) | + |

Broadly speaking, simulation languages divide into tools for continuous and discrete systems modelling. Continuous systems are described as sets of ordinary or partial differential equations, so the basic component of a continuous simulation language is an integration routine. Discrete systems programs provide not only for a difference equations solver (which is almost identical to algorithms applied to continuous systems), but are event oriented programs, including transaction flow, activity scanning, entities and attributes.

Typical early continuous systems languages are CSMP (from IBM) and DYNAMO (developed at MIT, Boston), and most modern programs are actually descendants of these; discrete systems simulation language ancestors like GPSS and Simscript (from various sources) have modern offspring that combine discrete and continuous aspects (an example being COSY from ETH, Z∗rich).

Almost all simulation languages provide a random number generator for one or more distribution functions, but only a few allow to fit the statistical parameters of these functions to measured or calculated data and to analyse many replicates of a stochastic model (Monte Carlo simulation).

Statistical report generation is mainly developed for discrete simulation packages. Instead of using integrated software model output and model input, may well be analysed more conveniently by dedicated and professional statistical packages providing options for e.g. variance analysis, time series techniques, correlation and regression. The only requirement to do this is a clearcut and flexible interface between the simulation language and the statistical software to facilitate exchange of data files.

One exception to this separation of simulation and statistics is on-line parameter estimation in noisy systems by methods like Kalman filters, extended Kalman filters and other recursive techniques. In this case deviations between observed and predicted values are fed back to the model during runs, not between runs. Ready-to-use software seems not available at present for this purpose.

Data base interface deals with retrieval and storing of model input and output. The ideal simulation language would automatically search for input time series and parameters in an allocated data base and store output in the base for merging results from different runs. Also submodels and other structural information should be transferable between data base and model. This is not a very developed feature for present day simulation languages. With some programming effort it would be possible, although, to interface a 4.th generation statistical/data base system like SAS (from SAS Institute, N. Carolina) with some of the existing simulation software.

Numerical methods, like an integration method in a continuous simulation language, are selectable by the user in some systems. CSMP offers 5 built-in integration methods plus one provided by the user. IBM's new DSL - the modern descendant of CSMP - provides p integration methods, two of which handle stiff systems. "Stiff" means roughly that the model has time constants of widely different magnitude, or more formally that the largest absolute value of the real part of the eigen-value is much greater than the smallest value.

In large ecological networks or spatial distributed systems the number of actual interconnected nodes is usually much smaller than the theoretical maximum number of connections. The connectivity matrix is "sparse". Sparse matrix techniques to solve the model equations may economize computations dramatic. Present day simulation languages do not use this method, although it is very easy to implement.

Partial differential equations (PDE's) are normally treated in simulation languages by discretizing all other independent variables but time (usually only one other variable is allowed). The PDE's are converted to sets of ordinary differential equations (compartment models). This works fine for parabolic equations, such as the diffussion equation, but hyperbolic and elliptic problems are still a challenge. Finite element methods will probably find their ways into general purpose simulation languages in the future.

Table look-up, interpolation and smoothing is another critical issue. All continuous simulation languages offer linear interpolation in one variable, some deal with more than one variable and expansion by base functions (e.g. splines). If table values are sufficiently dense, gradients may be estimated, which is an attractive method for estimating parameters (see Mejer and Jørgensen, 1981).

A common feature in continuous systems simulation languages is internal sorting of statements into a possibly non-contradictionary sequence. Violation of this possibility is known as algebraic loops, such as:

A = B
B = C
C = A

The counterpart of statement sorting in discrete systems is the parallellism in networks.

Modularity, which is a common feature in most modern high-level languages, are found with varying sophistication in simulation languages: include files, procedures, macros and submodels with or without library maintenance and hierarchial nesting.

The amount of output from ecological models is often tremendous and graphical visualization rather than tables is a must. Not necessarily as an integrated part of the simulation language, but like statistics, graphics should be available through a clean interface between the modelling system and its environment. Future development will probably include graphical input as well as output: for the occasional user or modellers with little to no programming experience the conceptual model - whether formulated as boxes and arrows or as a more formalized ikonic language like Odum's or Forrester's symbols - may serve as model specifications possibly in an interactive mode between man and machine.

Another help for the inexperienced user is menus providing selctions and data entry. There is no doubt that future simulation languages will learn from the user-friendliness of the many "plastic" packages available for microcomputers, such as spreadsheet programs like Lotus 1-2-3.

Sensitivity analysis and parameter estimation (adjustment of model parameters until some measure of mismatch between modelled and observed data eventually reach an acceptable low value) involves usually more than one run of the model (exceptions being the above mentioned recursive techniques). Although multiple run administration is common in simulation languages, only a few offer easy-to-use estimation techniques. The crucial point here is to implement the strategy to move around in parameter space based on previously calculated residuals.

The final feature mentioned in table 2.18 is portability. Fully portable modelling packages are usually just a library of FORTRAN subroutines that are to be modified by the user. The easy-to-use languages are mostly rather machine dependent. What is lacking is a revision of the old standards from the sixties. Hopefully, standards for simulation languages will emerge - and be accepted - in the near future.

A few hints - and hopefully wishes - for future developments have already been mentioned: recursive parameter estimation, sparse matrix techniques, finite element methods, graphical input, high degree of interactiveness, enhanced portability, agreement upon standards and data base handling.

Computer technology advances points in future directions: speech synthesis and recognition is coming out of the research laboratories and it may influence the interactiveness of the modelling exercise. More important is the shift from procedural high-level languages towards logic languages, artificial intelligence and expert systems. Ecological simulation will not be isolated from knowledge bases, statistics, advanced graphics (CAD), spreadsheets, optimization, decision support, on-line sampling, control systems, computer networks etc.

The boom in microcomputers will certainly influence the future ecological simulation environment. Systems will be distributed, providing more adaptive man-machine interaction, but at the same time be more integrated. Integration means here both physical links between workstations and mainframes and aggregation of software components around a data base system. Despite the progress in hardware, software development is a very labor intensive process and an evolution rather than revolution is to be expected.

## PROBLEMS CHAPTER 2.

1. Estimate the concentration factor and the biomagnification factor of bluegills for a phenolic compound with a solubility in water of 100 mg 1-1. What would the concentration factor be for 5 $\mu$ phytoplankton?

2. Explain the importance of verification, calibration and validation.

# 3. ECOLOGICAL MODELLING

This chapter reviews mathematical expressions of processes relevant to ecological modelling. Some readers might not insist in many of the details in this chapter and might therefore use the chapter as a "handbook in mathematical formulations of ecological processes".

Physical processes are reviewed in section 3.2 i.e. transport processes, sorption, temperature dependence and evaporation. Transport processes, based upon Fick's first and second laws, are widely used in models of auqatic ecosystems and transport of pollutants in air and soil, see sections 7.2, 7.3, 7.8 and 7.9. Sorption accumulates pollutants in sediment and soil and will be a component in many models of eutrophication, distribution of toxic substances and soil pollution, see sections 7.4, 7.6 and 7.9. All processes are dependent on the temperature, which is often considered a forcing function. Many of the models presented in the following chapters will therefore include a dependence of the temperature. The transport process of evaporation is mainly used in models of soil pollution and crop production, see section 7.9.

Chemical processes take place in all ecosystems and are widely used in models of toxic substances, see section 7.6.

Biological processes are important components in most ecological models. Models of photosynthesis are used in eutrophication models, wetland models and models of crop production, see sections 7.4, 7.5 and 7.9. It is characteristic that biological processes in most cases can be described by a wide range of mathematical expressions, see f.inst. table 3.7. Which one to select is a matter of considerations in each modelling case. Growth in general is an important component throughout chapter 6 and 7, while zooplankton grazing is mainly used in eutrophication models, see 7.4. Adaption will be further mentioned in chapter 9.

## 3.1. APPLICATION OF UNIT PROCESSES IN ECOLOGICAL MODELING

It is characteristic of models based on natural sciences, as has been stressed in chapter 1, that they contain a description of the significant processes that have actually taken place in the ecosystem modelled. This implies that a quantitative description of the processes is needed. **The flow rate is determined by the driving force or tension, and the time integrals of these two variables can be considered as state variables, determining the state of the system.** In most cases only the time integrals of the flows are of ecological interest.

We can use Ohm's Law as an illustration of these ideas:

$$i = dQ/dt = R \times \Delta V$$

Here V is tension and i the flow or current. In accordance with Ohm's Law they are proportional. The state variable in this case will be the electrical charge, since:

$$Q = \int_1^2 i \, dt$$

Chemical and biological processes can in principle also be included in this framework, but it is most obvious to use it for physical processes. For chemical processes the tension would be the chemical affinity, the current would be the reaction rate and the state variable would be the concentration.

Pollution problems are caused by an anthropogenic transfer of mass and/or energy to the ecosystems. **The transfer and transport processes are of a physical nature** and their quantitative description must therefore be based on physics.

The relation between the concentration of a component in an ecosystem and the effect on the biota will often be described by means of biological processes.

Many components undergo chemical reactions in the environments such as *oxidation, photolysis, hydrolysis, reduction, acid-base reaction etc.*, which implies that a description of chemical processes must be included in an ecological model.

Biological processes are obvious elements in ecological models, such as *growth, predation, mortality, immigration, emigration and the biochemical effect of various substances on biota.*

The use of process equations to construct ecological models can be illustrated by menas of the conceptual diagram for a simple model in fig. 3.1.

A and B are state variables, which are expressed in the units of concentrations in bio-geochemical models f.inst. mg/1 or in the units of numbers pr. km$^2$ or m$^2$ in population models. The arrows (1) - (6) indicate processes. A dynamic model of the conceptual model in fig. 3.1 might be formulated by use of the mass conservation principle as follows:

Accumulation = input - output   i.e.

$$dA/dt = \text{processes } (1) + (2) + (3) - \text{process } (4) \tag{3.1}$$

$$dB/dt = \text{processes } (4) + (6) - \text{process } (5) \tag{3.2}$$

As seen, it is crucial to have good formulation for the processes,as the mathematical formulation contains the process equations. A static equation is constructed by setting:

$$dA/dt = dB/dt = 0$$

These equations can be used to find A and B in the steady state situation, as some of the processes normally will be dependent on A and B. The steady state model can of course not account for diurnal or seasonal variations, but represent a static situation or an average situation, see also section 2.3.

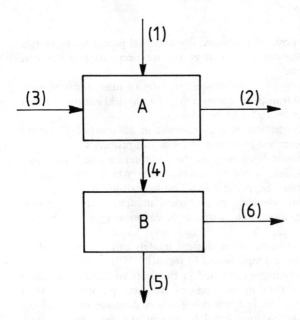

Fig.3.1:  Conceptual diagram of a simple model. It consists of 2 state variables A and B.

In principle there are no differences between models of ecosystems and models of ecological processes, which are used as elements in ecological models. Ecological processes might also be described by use of more or less details. The latter will require as for ecosystem models less data than the former. Therefore the survey of various physical, chemical and biological processes given in this chapter contain alternative equations for some processes. This is, not surprisingly, most pronounced for the biological processes, while for physical and chemical processes it is often a question, as to how many processes should be included at all?

The construction of an ecological model is therefore not just a question of collocating relevant process equations, but a selection of the equations fitted to the problem, the ecosystem and the data must take place.

**TABLE 3.1**

**Transport Processes.**

| Process | Tension | Current or Flow | State Variable | Equation |
|---------|---------|-----------------|----------------|----------|
| Hydraulic or Pneumatic | Pressure (P) $N/m^2$ | Flow (Q) $m^3/s$ | Volume (V) $m^3$ | $Q = \frac{dV}{dt} = K \cdot \Delta P$ (Poiseville's law) |
| Thermal Condition | Temperature (T) $C^O$ | Heat Flow (q) $\frac{J}{m^2 \cdot s}$ | Energy U | $q = -\lambda \nabla T$ (Fourier's law) |
| Diffusion | Concentration gradient g $m^{-4}$ * | Mass Flow (N) $\frac{g}{m^2 \cdot s}$ | Concentration $\left(\frac{g}{m^3}\right)$ | $N = -D_m \cdot \frac{\partial c}{\partial x}$ (Fick's law) |
| Water Flow in Soil | Hydraulic gradient $\left(\frac{dh}{dx}\right) m^2$ | Water Flow (Q) $\frac{m^3}{s}$ | Concentration $\left(\frac{m^3}{m^3}\right)$ | $Q = K \cdot \frac{dh}{dx}$ (Darcy's law) |
| Gas Transfer | Concentration gradient g $m^{-4}$ * | Mass Flow (N) $\frac{g}{m^2 \cdot s}$ | Concentration | $N = K_v \Delta c$ |

*) mol $m^{-4}$ might also be used

## 3.2. PHYSICAL PROCESSES

### 3.2.1. Transport processes

Table 3.1 lists a number of physical transport processes of ecological interest.

*Diffusion* is one of the most basic processes in nature and accounts at the molecular level for most of the transport that takes place. The diffusing phase can be a gas, solid or liquid, while the dispersing phase most often is a liquid or gas.

The basic relationships governing the diffusion are called *Fick's first* law, (see also Table 3.1).

If we consider the mass balance of a dispersed phase diffusing along the x-coordinate through a volumetric element of fluid, we have Accumulation = mass in - mass out

or
$$\frac{1}{dA}\frac{\partial m}{\partial t} = \left(-D_m \frac{\partial c}{\partial x}\right)_1 - \left(-D_m \frac{\partial c}{\partial x}\right)_2 \tag{3.3}$$

where m is the mass of the considered component, dA the area = dy × dz, $D_m$ the molecular diffusion coefficient and c the concentration.

Since the gradients at the two planes are related, the following equation is valid:

$$\left(-D_m \frac{\partial c}{\partial x}\right)_2 = -\left(D_m \frac{\partial c}{\partial x}\right)_1 + \left(-D_m \frac{\partial}{\partial x}\left(\frac{\partial c}{\partial x}\right)dx\right) \tag{3.4}$$

Substitution of (3.4) for (3.3) yields:

$$\frac{1}{dA} \cdot \frac{1}{dx}\frac{\partial m}{\partial t} = \frac{\partial c}{\partial t} = D_m \frac{\partial}{\partial x}\left(\frac{\partial c}{\partial x}\right) = D_m \frac{\partial^2 c}{\partial x^2} \qquad (3.5)$$

This equation is called *Fick's second law*.

For the general three-dimensional case the equation can be written:

$$\frac{\partial c}{\partial t} = D_m\left(\frac{\partial^2 c}{\partial x^2} + \frac{\partial^2 c}{\partial y^2} + \frac{\partial^2 c}{\partial z^2}\right) = D_m \nabla^2 c \qquad (3.6)$$

When both advection and diffusion occur, the effect of the two phenomena are additive, and we get:

$$\frac{\partial c}{\partial t} + U_x\frac{\partial c}{\partial x} = D_m \nabla^2 C \qquad (3.7)$$

where $U_k$ = the flow rate

$\nabla$ = an operator, see equation (3.6).
    This general form of relationship has validity even when turbulence exists: (one dimensional case)

$$\frac{\partial c}{\partial t} + (\overline{U} + U_t)\frac{\partial c}{\partial x} = D_m\frac{\partial^2 c}{\partial x^2} \qquad (3.8)$$

where $\overline{U}$ = temporal mean velocity $U_t$ = turbulent velocity fluctuations.

For typical environmental situations equation (3.7) is modified, so that the advective term is replaced by cross-sectional average of the velocity U, and the possible effects, due to the lateral distribution of velocity, are included in the dispersion term:

$$\frac{\partial c}{\partial t} + U\frac{\partial c}{\partial x} = E\frac{\partial^2 c}{\partial x^2} \qquad (3.9)$$

where c = cross-sectional average concentration, U = cross-sectional average of velocity, E = longitudinal dispersion coefficient, including turbulent diffusion coefficient.

(3.9) is the basic equation used in many hydrodynamic models, see section 7.3.
    The molecular diffusion coefficient is proportional to the absolute temperature and inversely proportional to the molecular weight of the diffusing phase and the viscosity of the dispersing phase.
    Diffusion and dispersion coefficients characteristic of various environments are indicated in Table 3.2.
    Equation (3.9) can be expanded to include chemical reactions and sources and sinks, see chapter 7).

**TABLE 3.2:**

**Diffusion and Dispersion Coefficients, characteristics of various Environments.-**

| Environment | Diffusion Coefficient cm²s⁻¹ |
|---|---|
| Ionic solutes in sediments and soils | $7.10^{-9} - 7.10^{-6}$ |
| Ions in water (Thermal diffusion) | $5.10^{-9} - 8.10^{-7}$ |
| High molecular weight components in water | $10^{-7} - 2.10^{-6}$ |
| Molecular diffusion of salt and gases in water | $8.10^{-6} - 10^{-4}$ |
| Turbulent diffusion (Vertical) | $10^{-2} - 2$ |

Transfer processes, describing the flow from one phase to another are also of major environmental interest. Volatilization can be described by use of the so-called two-film theory.

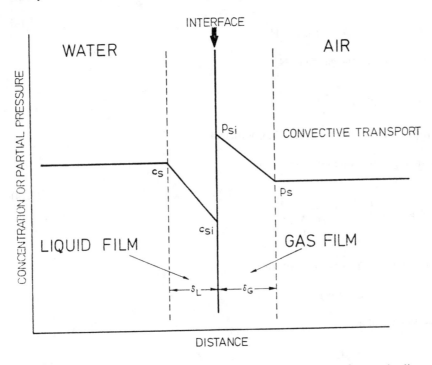

Fig.3.2: The two-film model of volatilization from the surface of water bodies.

Fig. 3.2 illustrates the major features of a *two-film model* of mass transfer, which is generally applied in chemical engineering. The water phase is assumed to be well mixed, so that any volatile compound is at a uniform concentration $C_s$ except in the vicinity of the interface. A stagnant liquid film of thickness δL separates the bulk of the water phase from the interface. Movement of a volatile component through this film is due to

95

diffusion. The concentration decreases across the film from $C_s$ to $C_{si}$, and the rate at which the Si component is transported across the film, $N_S$, is

$$N_s = K_L^S (C_s - C_{Si})$$

(3.10)

where $K_L^S$ is the liquid film mass transfer coefficient (m h⁻¹).

A stagnant gas film having a thickness of $\delta_G$ is on the air side. The partial pressure $P_{Si}$ on the air side is related to $C_{Si}$ (molar concentration on the water side of the interface) in accordance with Henry's law: ($x_{Si}$ mole fraction)

$$P_{Si} = H_c C_{Si} = H \cdot x_{Si}$$

(3.11)

where $H_c$ and $H$ are Henry's law's constants expressed in M or mole fraction unit. The relation between $H_c$ and $H$ is:

$$H_c = H \times 18/100 = 1.8.10^{-3} H$$

(3.12)

The transport across the gas film, $N_s$, is

$$N_S = \frac{K_G^S}{RT}(P_{Si} - P_s)$$

(3.13)

where $K_G^S$ is the gas film mass transfer coefficient (m h⁻¹).

By continuing these equations, we obtain

$$K_V^S = \frac{A}{V}\left(\frac{1}{K_L^S} + \frac{RT}{H_C^S K_G^S}\right)^{-1}$$

(3.14)

where

$K_V^S$ is the overall transfer coefficient (h⁻¹)

A is the interfacial area (m²)

V is the liquid volume

T is the absolute temperature.

We assume that

$$K_L = \frac{D}{\delta_L}$$

(3.15)

where D is the molecular diffusion coefficient, and similarly:

$$K_G = \frac{D}{\delta_G} \qquad (3.16)$$

It has been shown that if molecules are spherical, molecular diffusion coefficients in solution are inversely proportional to molecular diameters, d, so that

$$\frac{K_v^S}{K_v^O} = \frac{D^S}{D^O} = \frac{d^O}{d^S} \qquad (3.17)$$

where S indicates the considered component and O is oxygen.

$d^O$ is 2.98 A.

If data on the diffusion coefficients or molecular diameter for the

component are not obtainable, the molecular diameters can be estimated from the critical volume, $V_c$, since

$$\frac{\pi d^3}{6} = \frac{V_c}{2N} \quad \text{or} \quad \frac{V_c}{3N} \qquad (3.18)$$

where N is Avogadro's number.

$H_c^S$ can be estimated from solubility and vapor pressure:

$$H_c^S = \frac{p^S}{S_{wo}} \qquad (3.19)$$

where $p^S$ is the vapor pressure of S in pure form and $S_{wo}$ is the solubility in water. When data for the considered component are not available, data for a related component can be used.

The theory of this process has been developed by several authors (Liss et al., 1974; MacKay et al., 1976; and Smith et al., 1977).

The same considerations are used when re-aeration is described, (see 7.2). The flow is conveniently expressed in concentration units, but in this case it is only the liquid film resistance that is of significance, which implies that we can set up the following equations:

$$\frac{dC}{dt} = K_L \cdot \frac{A}{V}(C_s - C) \qquad (3.20)$$

where

C  = oxygen concentration in water

$C_s$ = oxygen concentration in water at saturation

A  = surface or interfacial area

V = liquid volume.

Since A/V is the depth, H, and $C_s$ - C is the oxygen deficit -D, the equation (3.20) yields:

$$\frac{dC}{dt} = -\frac{K_L}{H} \cdot D \qquad (3.21)$$

$\frac{K_L}{H}$ is called the re-aeration coefficient, $K_a$, and Table 7.2 lists a comprehensive review of the available empirical methods to estimate $K_a$.

### 3.3. Sorption

Sorption of components on to suspended matter, sediment and biota is a significant process in the environmental context.

Available data might fit either Langmuir's or Freundlich's adsorption isotherms:

$$S_s = \frac{DC_w}{K_m + C_w} \quad \text{(Langmuir's adsorption isoterm)} \qquad (3.22)$$

$$ \qquad (3.23)$$

$S_s = K.C_w^{1/n}$ (Freundlich's adsorption isotherm)

where $S_s$ is the weight of the considered component sorbed per g of sorbent, $C_w$ is the weight in solution per litre or ml of solution and D, K, $K_m$, and n are constants. At low substrate concentrations, n is often close to 1, and K becomes a partition coefficient.

Smith et al. (1977) have shown, in a limited number of case studies, that for a given sorbent the logarithm of the partition coefficient and the logarithm of the solubility linear are related (see fig. 3.3). Although this relationship seems to be generally valid, compounds that interact via an ion exchange process probably would not fit this plot.

Adsorption processes are of importance for modelling the pollution from non-point sources, for modelling the distribution of toxic substances in an aquatic ecosystem for a quantitative description of the exchange of nutrient or toxic substances between the sediment and water, and for modelling soil processes. All the models mentioned are treated in chapter 7.

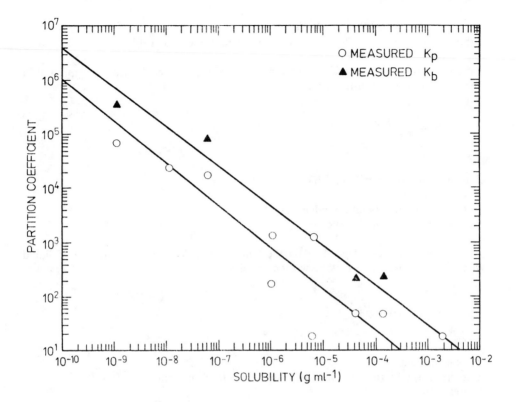

Fig.3.3: Solubility versus partition coefficient on Coyte Creek sediments
(K$_p$) and on a mixed population of bacteria (K$_b$).

In most cases the adsorption process is very fast and the equilibrium is attained in minutes, or hours at the most. This implies that in models using days as the time steps, it is sufficient to describe the process by application of the above mentioned equilibrium terms. If it requires to include the rate of the adsorption process, this can be done by using a first order expression:

$$\frac{dS_S}{dt} - K(S_e - S_s)$$  (3.24)

where

S$_e$ is S$_s$ at equilibrium

Sorption includes not only the physical process of adsorption, but also a number of chemical processes named chemosorption. Here the mechanisms are more complicated, but in many cases it is possible to use the same basic description.

99

### 3.3.1. Temperature dependence

Flow rates are dependent on the temperature in the same way as the constants in the expressions, relating the flow rate with the driving force, (see Table 5.1), are dependent on temperature.

The relation between the temperature and the diffusion coefficients is expressed in the so-called Stokes-Einstein equation:

$$D = \frac{\varkappa \cdot T}{6\pi\eta r} \tag{3.25}$$

where

T is the absolute temperature (K)
$\eta$ is the viscosity
$\varkappa$ is a constant (temperature independent)
r is the radii of the molecules

The viscosity is again dependent on the temperature. A typical dependence of liquid viscosity on temperature may be written as follows:

$$\eta = A \exp\left(\frac{\Delta E}{RT}\right) \tag{3.26}$$

where

A is a constant
E is the energy barrier that must be overcome before the elementary flow process can occur.

*The reaeration coefficient $K_a$* should contain the diffusion coefficient and therefore have the same temperature dependence as expressed in equations (3.25) and (3.26). In practice, however, an empirical expression is used, (see section 7.2), which might have the following form:

$$K_{a,t} = K_{a,20} \cdot V^{(t-20)} \tag{3.27}$$

where V is a constant

Chemical reaction rates, R, are also dependent on the temperature:

$$R = A \exp\left(\frac{-E_a}{RT}\right) \tag{3.28}$$

where

A is a constant
$E_a$ is the activation energy.

However, for rather narrow temperature ranges the following approximation can be used (compare with (3.27):

$$R_t = R_{t_0} \cdot \theta^{(t-t_o)} \tag{3.29}$$

$\theta$ is a constant.

Another approximation uses the so-called $Q_{10}$ value, which is defined as:

$$R_t = R_{t-10} \cdot Q_{10} \tag{3.30}$$

since for many chemical and biochemical reactions $Q_{10}$ is in the order of 2.0, see the discussion on $Q_{10}$ for biological processes later in this chapter.

## 3.3.2. Evaporation.

It is often significant to include equations for the water balance in ecological models of aquatic ecosystems. This implies that equations for evapora tion are needed. Evaporation from a free water surface is dependent on the climatic conditions and tables of its function of time and/or temperature can be found in the literature. It is reduced by the presence of vegetation, which is discussed in section 7.9.

## 3.4. CHEMICAL PROCESSES

The reaction rate of chemical processes can generally be described by a n. order differential equation, where the order m of the reaction is dependent on the nature of the chemical reaction.

Reaction order is determined as the sum of exponents appearing in the equation of the reaction rate. A simple equation generally applied for the process:

$$nA + mB \rightarrow pC + qD \text{ is:} \tag{3.31}$$

$$\text{process rate} = K.[A]^n .[B]^m \tag{3.32}$$

where k is a constant. As seen the reaction order is n + m. A and B indicate concentrations in moles/l.

Many reactions can, however, be described as first-order reactions, including settling of detritus or phytoplankton, photolysis (see below), and even the biochemical degradation of organic matter. First order reactions are widely used in the models presented in chapter 7.

Many chemical processes are extremely rapid, compared with the time step of interest in the ecological models. For such processes equilibrium expressions should be used for inclusion in the ecological model. For the process

$$aA + bB = cC + dD \qquad (3.33)$$

the equilibrium can be described by use of the following equation:

$$\frac{[C]^c [D]^d}{[A]^a [D]^b} = K_m \qquad (3.34)$$

where $K_m$ is the equilibrium constant

A survey of the chemical processes of interest for ecological models is given below, and it is indicated, which type of approach is generally applicable to models of environmental management.

### 3.4.1. Chemical Oxidation.

Chemical oxidation of organic or even inorganic compounds may be important under some environmental conditions. In most cases a first-order reaction scheme seems to give an acceptably accurate description of the process.

### 3.4.2. Photolysis.

Photochemical transformation is a significant process for many toxic components (see Wolfe et al., 1975 and Zepp et al., 1977). The rate of absorption of light, $I_A$, by a chemical, is determined by:

$$I_A = \epsilon I_\lambda [S] = K_a [S] \qquad (3.35)$$

where S is the concentration of the chemical, $\epsilon$ the molar extinction coefficient, $I_\lambda$ the intensity of the incident light, and $K_a = \epsilon . I_A$. By multiplying $I_A$ by $\phi$ the quantum yield, which is the efficiency for converting the adsorbed light into chemical energy, we get the rate of direct photolysis:

$$-\frac{d[S]}{dt} = k_a \cdot \phi \cdot [S] = k_p \cdot [S] \qquad (3.36)$$

where $k_p = k_a . \phi$.

As shown, the photochemical transformation is a first-order reaction, where $k_p$ is dependent on the intensity of the incident light. Zepp et al., 1977 have demonstrated how the half-lives for photolysis vary with the season due to variation in $I_\lambda$. It is suggested by Wolfe et al., 1976 that   and   are measured in laboratory experiments and from these values $k_p = f(I_\lambda)$ can be calculated as a function of time, day, season and latitude.

### 3.4.3. Hydrolysis.

Hydrolysis of organic compounds usually results in the introduction of an HO-group:

$$(3.37)$$

$$RX + H_2O \rightarrow ROH + RHX$$

The rate of hydrolysis can be expressed as

$$(3.38)$$

$$R_h = k_h[S] = k_B[OH^-][S] + k_A[H^+][S] + k_N.[H_2O][S]$$

where $k_h$, $k_B$, $k_A$ and $k_N$ are rate constants.

Only a few data on $k_h$, as a function of pH, are available. Wolfe et al., 1977 give, however, kinetic data for methoxychlor and DDT, and other data can be found in Wolfe et al., 1976.


### 3.4.4. Ionization, Complexation and Precipitation.

These processes are all rapid and can therefore be included in a model by application of the equilibrium expression. The mass equation constant can often be found in one of the many handbooks containing these data.

Precipitation of heavy metals all have a low solubility in water due to the formation of hydroxides. Table 3.3 gives the solubility product for some important heavy metals. However, while the precipitation process can be described in most ecological models by means of the equilibrium expression, the settling of the suspended matter formed by means of the equilibrium expression, can be described as a first-order reaction.

The formation of heavy metal complexes is of equal environmental interest, as the toxic effect of the heavy metal in most cases is related to the free ions. The stability constants of some heavy metal complexes of environmental interest are given in table 3.4.

103

**TABLE 3.3**

**Solubility of Metals (ppb) Brooks et al., 1968.**

| Sulphide | $K_{sp}$ (solubility product) | Solubility of metals (ppb) in $S^{2-}$ -free water |
|---|---|---|
| CdS | $3.6 \times 10^{-29}$ | $8 \times 10^{-4}$ |
| CoS | $3.0 \times 10^{-26}$ | 0.1 |
| CuS | $8.5 \times 10^{-45}$ | $6 \times 10^{-12}$ |
| FeS | $3.7 \times 10^{-19}$ | 39 |
| NiS | $1.4 \times 10^{-24}$ | 0.08 |
| ZnS | $1.2 \times 10^{-22}$ | 0.3 |

**TABLE 3.4**

**Stability Constants (-pK) of complexes present in water (Zitko et al., 1976).**

| Cation | Glycine | ATPG | Lutathione | Acetic acid |
|---|---|---|---|---|
| $Ca^{2+}$ | 1.31 | 3.60 | | 0.39 |
| $Mg^{2+}$ | 3.44 | 4.00 | | 0.28 |
| $Zn^{2+}$ | 5.52 | 4.85 | 8.30 | 1.57 |
| $Cd^{2+}$ | 4.80 | | 10.50 | 1.70 |
| $Cu^{2+}$ | 8.62 | 6.13 | | 2.24 |

## 3.5. PHOTOSYNTHESIS

The photosynthetic process may be divided into independent reaction systems, the light absorption-energy producing system (known as the light reaction), and the reductive system of carbon dioxide fixation (known as the dark reaction).

The light reaction transmutes the energy of sunlight into the two biochemical energy sources ATP and $NADPH_2$ via the two main photochemical pathways. Absorption of photons excite chlorophyll electrons to higher energy levels, which are then utilized via either cyclic photophosphorylation or non-cyclic photophosphorylation.

The dark reaction uses the biochemical energy sources ATP and $NADPH_2$ to reduce carbon dioxide to organic carbon. $CO_2$ is combined with ribulose 1.5-diphosphate and forms two molecules of 3-phosphoglyceric acid, which form one molecule of fructose 1.6-diphosphate. The over all reaction can be simplified as:

$$(C_5)-P \xrightarrow[ATP \quad ADP]{} (C_5)-2P \xrightarrow[CO_2]{} (C_6)-2P \xrightarrow[\substack{ATP \quad ADP \\ NADPH_2 \quad NADP}]{H_2O} (C_3)-P \longrightarrow$$

$$\downarrow (C_6)$$

or summarized as

$$CO_2 + H_2O \xrightarrow[\substack{radiant \\ energy}]{} (CH_2O) + O_2.$$

Obviously, then photosynthesis involves two sets of external limiting factors, availability of energy and inorganic elements ($CO_2$), and these two elements govern the rates of the light and dark reactions. Additionally, internal limiting factors are involved since transport mechanisms are involved in providing the elements essential for synthesis of organic matter. Besides this, organisms need time to adapt to fluctuations in environment conditions, for instance a change in radiant intensity, and so both internal pools of essential elements (C, P, $H_2O$, S, etc.) and the "reaction tools" (enzymes, transport mechanisms, respiration level, leaf index, reproductive stage, etc.) may limit the rate of photosynthesis, fig. 3.4.

The common mathematical description of photosynthesis normally involves a coupling of light and essential element dependency, and may consequently be catagorized as an empiric model. If no changes in adaption occurs, then photosynthesis may be quoted as

$$(3.39)$$

PHOTO = k . f (max. requirement of limiting factors)

where PHOTO is the photosynthesis measured as uptake of $CO_2$, $O_2$ produced, increased organic energy, or similar units, and f(x) represents the optimal yield of the maximal available elements, external as well internal. Fig. 3.4 gives some basic experimental results to illustrate different types of limiting factors and adaptation cases.

The use of photosynthetic equations in models must be related to the total system or subsystems. In aquatic models normal demographic equations are used, e.g. equations treating an average exposed biomass of one or more species. In terrestial systems, where the number of single species is low, intra-specific adaption plays an important role.

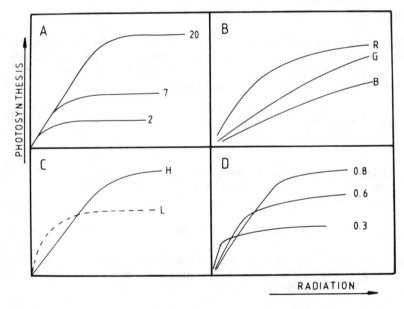

Fig.3.4: Rate of photosynthesis as a function of radiation energy A: at different temperatures (in °C). B: at different chromatic intervals, R = red, B = blue, G = green. C: of two light-intensity adapted stages, H = high, L = low. D: at

different leaf index stage.

In algal populations, where the photosynthetic description probably is the best-attempted simulation of nature, the models have been extensively developed (see also 7.4). Thus only a few examples will be quoted here. Chen et al., 1975 considered four external factors, nitrogen (N), phosphate (P), radiant intensity (I) and temperature (Y), given by a product of Michäelis-Menten expressions (q.v.) and an optimal temperature dependency (f(T)),

$$\mu = \mu_{max} \cdot f(T) \cdot I\frac{1}{I + k_i} \cdot N\frac{1}{1 + k_n} \cdot P\frac{1}{1 + k_p} , \tag{3.40}$$

where $\mu$ is the actual specific photosynthesis and $\mu_{max}$ is the maximal specific photosynthesis.

   If several limiting factors regulate the growth simultaneously, there are description methods referred to in the literature for the total effect of interacting factors. The most important description methods are summarized as follows:

1) The minimum value of the limiting factors is used, see 1) Table 3.7. This is in accordance with Liebig's minimum law.

2) The product of the limiting factors is used, see f.inst. Table 3.5. 5)

3) It is presumed that the factors work in parallel, see 13) Table 3.5.

4) The average of the limiting factors is used, see 8) Table 3.7. A review of photosynthesis equations is given in Table 3.5.

Nyholm (1976) and Jørgensen (1976) also considered internal factors. These describe pools of phosphorus and nitrogen, which are readily available during periods of expotential growth (q.v.) This principle is shown here by phosphate-limited photosynthesis, but may be used for each essential element vital to the process

$$\mu = \mu_{max} \cdot \frac{k + CM - CA}{CM - CA} \cdot \frac{CI - CA}{k + CI - CA} , \tag{3.41}$$

where
  CM is maximal internal concentration of phosphorus
  CA is minimal internal concentration of phosphorus
  CI is actual internal concentration of phosphorus.

The use of this particular expression involves a separate description of the nutrient uptake from the exterior.
   Temperature dependency f(T) has been described in different ways. Chen et al., 1975 gives a simple equation, which does not take temperature optimization into account

$$\tag{3.42}$$

$$f(T) = k_{20} \cdot V^{(T-20)},$$

where $V$ normally is recorded between 1.0 and 1.2.

Many processes in biological systems have a temperature optimum and some models take this into consideration, as e.g. Jørgensen (1976)

$$f(T) = K_{opt} \cdot \exp\left(-2.3 \cdot \frac{T - T_{opt}}{15}\right) \tag{3.43}$$

where $K_{opt}$ is the optimal process rate and $T_{opt}$ the corresponding temperature (see Table 4.4) The equation assumes that the temperature dependency is symmetrical around the optimal value.

Different types of light functions have been used including or excluding light inhibition due to photo-oxidation in the chlorplasts.

In a homogeneous dispersed biomass (e.g. a phytoplankton community) Vollenweider (1965) has formulated the radiant dependency as

$$f(I) = \frac{I}{k_m \sqrt{1 + (I/k_m)^2}} \cdot \frac{1}{\left(\sqrt{1 + (aI)^2}\right)^{k_2}}, \tag{3.44}$$

where $k_m$ is the light saturation value, I is the actual light intensity, and a and $k_2$ are constants.

# TABLE 3.5

## Photosynthesis Equations
(see list of abreviations)

1) Broqvist (1971)

$$PHOTO = MY(T) \cdot PHYT \cdot min \left( \frac{I}{I(0)}, \frac{PS}{PS_0}, \frac{NS}{NS_0} \right)$$

2) Chen (1970) and Chen et al. (1975)

$$PHOTO = MY(T) \cdot \frac{I}{I+IK} \cdot \frac{NS}{KN+NS} \cdot \frac{PS}{KP+PS} \cdot PHYT$$

$$MY(T) = MYMAX \cdot f(T)$$

3) Parker (1972)

$$PHOTO = MY(T) \cdot I \cdot PS \cdot NS \cdot PHYT$$

4) Anderson (1973)

$$PHOTO = MYMAX \cdot PHYT \cdot (PS+NS)$$

5) Dahl-Madsen et al. (1974)

$$PHOTO = MY(T) \cdot f(I) \cdot \frac{PS}{KP+PS} \cdot \frac{NS}{KN+NS} \cdot \frac{CS}{KC+CS}$$

$$MY(T) = MYMAX \cdot g(T)$$

6) Jansson (1972)

$$PHOTO = MY \cdot PHYT \cdot PS \cdot I \cdot T$$

7) Lassen et al. (1972)

$$PHOTO = MY \cdot PHYT \cdot f(PS) \cdot FD \cdot f(I)$$

8) Patten et al. (1975)

$$PHOTO = MY \cdot PHYT \cdot f(I) \cdot \frac{f(NS) + f(PS) + f(CS)}{3}$$

9) Larsen et al. (1974)

$$PHOTO = MY \cdot PHYT \cdot min(f(I), f(PS), f(NS))$$

$$f(I) = \frac{F_{max} \cdot R(t)}{Ik + R(t)}$$

corresponding equations for f(PS) and f(NS)

10) Bierman et al. (1974)

$$UP = UPMAX \left( \frac{1}{1-PKI \cdot PA} - \frac{1}{1-PKI \cdot PS} \right)$$

11) Gargas (1976)

$$MY = MYMAX \cdot f(I) \cdot f(PS) \cdot f(NS) \cdot f(T) \cdot FD \cdot FAC$$

12) Cloern (1978)

$$PHOTO = MY(T) \cdot \frac{I}{I_{op}(T)} \cdot exp(1 - \frac{I}{I_{op}(T)}) \cdot f(PA) \cdot f(NA)$$

$$MY(T) = 0.02 \cdot exp(0.17\ T)$$

$$I_{op}(T) = 0.06 \cdot exp(0.22\ T)$$

$$f(PA) = \frac{PAP}{PAP-PAMIN} \ , \quad PAP = \frac{PA}{PHYT}$$

$$f(NA) = \frac{NAP}{NAP-NAMIN} \ , \quad NAP = \frac{NA}{PHYT}$$

$$UP = UPMAX \cdot \frac{PS}{KP+PS}$$

$$UN = UNMAX \cdot \frac{NS}{KN+NS}$$

13) Nyholm (1978)

$$PHOTO = MY \cdot f(I) \cdot f(NA,PA)$$

$$f(NA,PA) = \frac{2}{\frac{1}{f(PA)} + \frac{1}{f(NA)}} \quad \text{(see also Bloomfield et al. (1974))}$$

$$f(PA) = \frac{KPA + PAMAX - PAMIN}{PAMAX - PAMIN} \cdot \frac{\frac{PA}{PHYT} - PAMIN}{KPA + \frac{PA}{PHYT} - PAMIN}$$

$$f(PN) = \frac{\frac{NA}{PHYT} - NAMIN}{NAMAX - NAMIN}$$

excess supply of nutrients

$$\frac{dPS}{dt} = PAMAX \cdot \frac{dPHYT}{dT}$$

$$\frac{dNS}{dt} = NAMAX \cdot \frac{dPHYT}{dT}$$

$$UP = MY \cdot PAMAX$$

$$UN = MY \cdot NAMAX$$

limiting conditions

UP and UN equal to supply

14)  Jørgensen (1976)

$$PHOTO = MYMAX \cdot f(T) \cdot f(PA) \cdot f(NA) \cdot f(CA)$$

$$f(PA) = \frac{PA - PAMIN \cdot PHYT}{PA}$$

$$f(NA) = \frac{NA - NAMIN \cdot PHYT}{NA}$$

$$f(CA) = \frac{CA - CAMIN \cdot PHYT}{CA}$$

$$UC = f(I) \cdot UCMAX \cdot \frac{CAMAX \cdot PHYT - CA}{CAMAX \cdot PHYT - CAMIN \cdot PHYT} \cdot PHYT \cdot \frac{CS}{KC+CS}$$

$$UP = UPMAX \cdot \frac{PAMAX \cdot PHYT - PA}{PAMAX \cdot PHYT - PAMIN \cdot PHYT} \cdot PHYT \cdot \frac{PS}{KP+PS}$$

UN analogous

___

Besides light intensity, the day length influences photosynthesis in different species. Many terrestrial plants contain day-length "receptors", which govern photosynthesis efficiency or more commonly the minimum threshold for starting photosynthesis. Data from studies on phytoplankton indicate that day length may be mainly responsible for population succession. These mechanisms will not be treated further in this chapter, fig. 3.5.

Most plants contain mechanisms to adapt to the actual light regime by changing the orientation of the chloroplasts in leaves, or the leaves themselves, or by altering enzymatic concentrations. The processes have not yet been modelled satisfactorily.

In many higher plants there is a well-developed stomata-controlling system, which regulates the $CO_2/O_2$ exchange between leaf and atmosphere. Accordingly, very complex models must be constructed to simulate actual photosynthesis in most terrestrial plants.

The number of external factors controlling photosynthesis and primary production in higher plants, or terrestrial ecosystems, is very high. Some of the most potent environmental factors are the intensity of solar and diffuse radiation; humidity and wind; and infrared radiation and day length. In addition to environmental factors, the infra-organization of the terrestrial subsystem is a most important element. The determination of reflection and transmission in the leaf system, leaf-area index, orientation, and the $CO_2/O_2$-gas-exchange system are the preliminary processes for the development of a mathematical description of production.

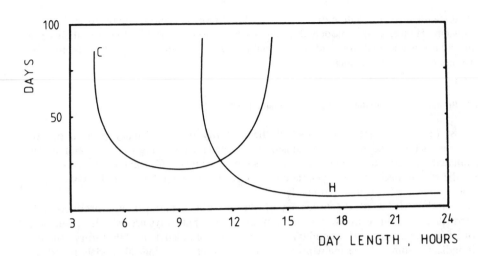

Fig.3.5: Time of flower development (maturity) as a function of day length.
C = Chrysanthemum sp. "short day plant", H = Hyascomus sp. "long day plant".

Obviously, two basic types of process description can be developed, namely, I) by the use of single compartments (leaves, leaf galleries, individuals or species)
or
II) by using total community production, fig. 3.6

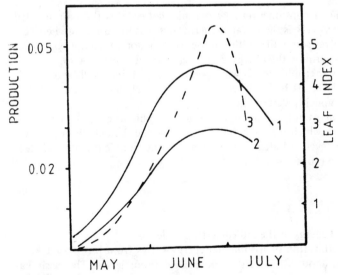

Fig.3.6: Production, respiration and leaf index during the growth period for Hordeum sp. (barley). (1) and (2) are gross primary production, and net primary production, (g dry matter m$^{-2}$ $^{-1}$) and (3) is leaf index.

Most difficult is the modelling of the effects of soil moisture and temperature profiles in the system. Hence, simple models describing production in terrestrial environments use solar radiation and leaf-area index as variables, (see 7.9), and let all other factors be constants or stochastic elements.

### 3.5.1. Production at Secondary and Higher Trophic Levels.

In most of the ecosystems investigated the majority of net primary production (or energy) is metabolized by microorganisms (e.g. bacteria and fungi) and the remainder assimilated by herbivores. This ratio becomes increasingly important to the ecosystem since the rate in decrease in available energy governs the complexity of the system and the development of different trophic levels.

If our knowledge of the processes of photosynthesis and primary production is limited, then data on secondary production and energy pathways are even less. Further, most of our knowledge of secondary production is compiled from laboratory studies. This indicates some of the main problems in constructing ecological models based on present information, but efforts continue because eventually they will produce a more holistic case study.

One of the main problems in understanding the energy transfer through a trophic level is the measurement of energy utilization in searching for food, and in maintaining a certain temperature level in homeothermal animals. As a simplified system, secondary production may be quoted as:

$$P_s = C - F - E - R = A - R, \tag{3.45}$$

where C denotes the ration of consumed energy/organic matter, F is the non-ingested energy, E is the excreted energy, R is the respired energy and A is the assimilated energy.

This simplification involves some difficulties in the construction of energy systems as it is nearly impossible to determine the energy cost at each trophic level. Also, it is not possible to determine in lower trophic levels what percentage of F may still be available for that particular trophic level. At higher levels the percentage of caught food, which is non-ingestigable, may be implicitly determinable.

Thus, the amount of energy utilized at organism and trophic levels is known from many recent holistic studies, which makes it possible to describe the energy in progressing up through the trophic levels, Lindeman's efficiency or the trophic transfer efficiency, seems to vary within the range 10-20 %:

$$E_{tl} = \frac{I_n}{I_{n+1}}, \tag{3.46}$$

where $I_n/I_{n+1}$ is the ingested energy at the indicated trophic levels.

At different levels the utilization of assimilated energy varies from less than 1 % in photosynthetic processes to about 90 % at the upper carnivorous levels. This assimilation efficiency is given by

$$E_a = \frac{A_n}{I_n} \cdot \frac{GPP}{SR}, \tag{3.47}$$

where $A_n$ is the assimilated energy, GPP is the gross primary production, and SR is the incident solar radiation.

A third useful term in discussing energy transfer in an ecosystem concerns the organism's ability to use the ingested energy for catabolism, i.e. that part of the energy ingested at level n, which would be available to level n+1 (called "ecoligal efficiency" by many authors). This ratio had long been argued to be as high as 10-15 % for homeotherms, but the basic assumptions used for these calculations have recently been discredited. It is now the general opinion that ecological efficiency never exceeds 2-4 % for homeotherms and 5-20 % for poikilotherms. Moreover, the scientific material which is our background for making these general assumptions, is based on less than 50 holistic investigations. Thus, from a theoretical point of view, the term ecological efficiency is quoted as

$$E_e = \frac{P_n}{I_n}, \qquad (3.48)$$

where $P_n$ is the net production at level n.

During the last decades much effort has been devoted to modelling secondary production as an interaction between compartments in the ecosystem. The majority of the works presented concern homeostatic systems, describing the energy transfer processes as first-order differential equations. It is obvious from the large amount of work involved in the description of the complexity of the secondary production, that the models may only provide guidelines to the reaction patterns of the variation in external factors.

### 3.5.2. Energy Flow in Secondary Production

The assimilation and loss of energy in a food web is governed by the trophic structure of the ecosystem. This structure and the corresponding functions are determined by the amount of species and their abundance, the ratio between species and biomass, and the complexity in the food relations

In complex ecosystems the different populations have the option of food selection corresponding to the state of their development. Consequently, a simple way of describing energy transport is not always possible. Thus, at any given time the energy flow may be categorized into as few levels (trophic levels) as

| | |
|---|---|
| 1. Primary producers | autotrophic level |
| 2. Herbivores | |
| 3. Carnivores 1. order | heterotrophic levels |
| 4. Carnivores 2.-n orders | |
| 5. Top carnivores | |
| 6. Decomposers | saprotrophic level |

Dividing into levels thus indicates, that populations within the same trophic level may prey or be preyed upon by different populations from another trophic level. Further, a species may, during its development. change trophic levels corresponding to a change in its food selection.

The thermodynamic concept of energy conservation can be determined by examination of the energy flow in the ecosystems. In each energy transfer the loss of heat energy must be calculated. To illustrate energy flow an example is shown in fig. 2.8.

A simple description of energy flow through a population at a trophic level may be given as

$$I = P + R + F$$
$$A = P + R$$
$$P = G + (E + S) + N$$

(3.49)

where
I is the ingested energy,
P is the produced biomass energy,
R is the respired energy (heat loss)
F is the energy loss by faeces,
A is the assimilated energy,
G is the growth energy,
E is the excretion energy,
S is the storage energy, and
N is the energy needed for reproduction.

No distinct differentiation between heterotrophs and saprotrophs can be established, and in most cases, with the exception of bacteria, herbivores and carnivores may include stages of saprotrophic assimilation.

The energy transfers in secondary production are dependent in most cases on the age of the individuals. If the population is a cohort (q.v.) this age dependency follows the population.

The individual increase in weight or biomass has been described for juvenile individuals as

$$dW/dt = k.W,$$

(3.50)

where W is the organism weight or biomass.

In more general approximation of individual weight increase the formula has been quoted as

$$dW/dt = k.(W_{max} - W),$$

(3.51)

where $V_{max}$ is the maximal weight or biomass of an individual, and k is a species-specific constant. A more comprehensive review on growth equations is given in chapter 6.

In a more simple rearrangement the average secondary population production is given as

$$\frac{dW_p}{dt} = W_p \frac{k'}{n} \sum_{0}^{i=n} \frac{(W_{max} - W_i)}{W_i} = W_p \cdot k' \cdot \frac{(\overline{W}_{max} - \overline{W})}{\overline{W}},$$

(3.52)

where $\overline{W}$ is average individual weight, $W_p$ is the total population weight or biomass and n is the number of individuals. $W_i$ is the weight of the $i^{th}$ individuals.

114

**TABLE 3.6:**

**Zooplankton Grazing Models.**
(see list of abriviations)

1) Dodson (1975)
$$GRZ = K \cdot PHYT \cdot ZOO$$

2) Steele (1974)
$$GRZ = MYZ \cdot \frac{PHYT - KTR}{KZ + PHYT} \cdot ZOO$$
$$MYZ = MYZMAX \cdot f(T)$$

3) Walsh et al. (1971)
$$GRZ = g(ZOO)(PHYT > KTR)$$
$$e.g. \quad g(ZOO) = MYZ \cdot ZOO$$

4) O'Brien et al. (1972)
$$GRZ = ZOO \cdot MYZ(1 - \exp(D_p(PHYT > KTR))$$
$$MYZ = MYZMAX \cdot f(T)$$

5) Lotka (1924)
$$GRZ = MYZ \cdot (1 - \frac{ZOO}{CK}) \cdot ZOO$$

6) Odum (1972)
$$GRZ = MYZ \cdot AV \cdot (1 - \frac{ZOO}{CK}) \cdot ZOO$$
AV represents a variable related to the availability of food
$$AV = f(PHYT)f(OX)f(T)f(TOX)$$

7) Gargas (1976)
$$GRZ = \begin{cases} MYZ \cdot \frac{PHYT - KTR}{KZ + PHYT} ZOO & PHYT > KTR \\[2mm] MYZ \cdot \frac{PHYT}{ZOO} & PHYT < KTR \end{cases} \qquad MYZ = MYZMAX \cdot f(T)$$

8) Canale (1976); Chen et al. (1975)
$$GRZ = MYZ \cdot \frac{PHYT}{KZ + PHYT} \cdot ZOO$$

9) Canale (1976)
$$GRZ = MYZ \cdot \frac{KMFM \cdot PHYT + KFLM}{PHYT + KFLM} \qquad PREF = \frac{\alpha_K^Z \cdot PHYT}{\Sigma \alpha_K^Z \cdot PHYT}$$

10) Jost et al. (1973)
$$GRZ = MYZ \cdot \frac{PHYT^2}{(KZ1 + PHYT)(KZ2 + PHYT)} \cdot ZOO$$

11) Gause (1934)
$$GRZ = \frac{MYZ \cdot PHYT^{\frac{1}{2}} \cdot ZOO}{(KZ + PHYT^{\frac{1}{2}})}$$

12) Dugdale (1975)
$$GRZ = MYZ \cdot ZOO(1 - \exp(-KZ \cdot (PHYT - KTR)))$$

13) Richey (1977)
$$GRZ = AK \cdot (BL)^2 \cdot T$$

14) Jørgensen (1976)
$$GRZ = MYZ \cdot \frac{PHYT - KTR \cdot ZOO}{PHYT + KZ \cdot ZOO} \cdot ZOO$$

In Table 3.7. some examples of the temperature dependency of biological processes are given.

**TABLE 3.7:**

**Models of Temperature Dependency.**
(see list of abreviations)

---

1) Chen et al. (1975)

$$K(T) = K_{20} \cdot KOT^{T-20}$$

2) Lassiter et al. (1974)

$$K(T) = K_{opt} \cdot e^{a(T-T_{opt})} \cdot \left( \frac{T_{max} - T}{T_{max} - T_{opt}} \right)^{a(T_{max} - T_{opt})}$$

3) Lehman et al. (1976)

$$K(T) = K_{opt} \cdot exp(-2.3(T-T_{opt})^2/(T_{max}-T_{opt})^2) \quad \text{for } T > T_{opt}$$

$$K(T) = K_{opt} \cdot exp(-2.3(T_{opt}-T)^2/(T_{opt}-T_{min})^2) \quad \text{for } T \leqslant T_{opt}$$

4) Jørgensen (1976)

$$K(T) = K_{opt} \cdot exp\left(-2.3 \left| \frac{T-T_{opt}}{15} \right| \right)$$

$$K(T) = K_{opt} \cdot exp(K \cdot T)$$

5) Lammana et al. (1965)

$$K(T) = K_{opt}(\frac{T}{T_{opt}})^n \cdot exp(1 - (\frac{T}{T_{opt}})^n \quad 0 < T < T_{opt}$$

$$K(T) = K_{opt}(1 - (\frac{T - T_{opt}}{T_{max} - T_{opt}})^m) \quad T_{opt} < T < T_{max}$$

6) Park et al. (1979)

$$PHOTO\ (T) = exp\left( K(K_1 T^2 - K_2 T^{K_3} - 1) \right)$$

$$K = -ln\left( PHOTO\ at\ 0^oC \right)$$

$$K_1 = K_2 (T_{max})^{K_3 - 2}$$

$$K_2 = \frac{1 + ln\left( PHOTO\ at\ T_{opt} \right)/K}{(T_{max})^{K_3 - 2} \cdot (T_{opt})^{K_1} - (T_{opt})^{K_3}}$$

$$\frac{2^{\frac{1}{K_3-2}}}{k_3} = \frac{T_{opt}}{T_{max}}$$

7) Straskraba (1976)

$$T_{opt} = T + 28 \exp(-0.115 \cdot T)$$

$$PHOTO\ (T) = PHOTO\ (T_{opt}) \cdot \exp(-(K(T_{opt}-T)^2))$$

### 3.5.3. Decomposition.

The decomposition of organic matter is the most essential process in a natural ecosystem to maintain the biogeochemical cycles. In such systems the decomposition will be closely related to the production by several feed-backs and feed forward mechanisms. However, this balance is very sensitive to external factors, as an anthropogenic influence causes increasing input of organic matter or nutrients, which results in an unsteady state between production and decomposition.

The major part of the total decomposition of dead organic matter, detritus, particulate or dissolved, is performed by bacteria and fungi. However, in most systems several animal species may involved in the break up of larger particles, fig. 3.7.

The principle of decomposition is to convert exogenic organic bound energy into endogenic metabolic energy, through a series of controlled redox processes. The dominant part of these redox processes takes place in the presence of oxygen and can be simplified as

$$C_{106} H_{263} O_{110} N_{16} P_1 S_1 + R(0) + decomposers \longrightarrow$$

$$a\ CO_2 + b\ NH_4^+ + c\ HPO_4 + d\ HS^- + f\ H^+$$
('new decomposers' energy)

$$+\ e\ H_2O,$$

where $\cong$ R 228

and a, b, c, d, e, f are values depending on decomposition efficiency. The degree of decomposition depends on different factors, for instance the type of decomposers present, the absence or presence of $O_2$, $NO_3^-$, $SO_4^{2-}$, and temperature.

117

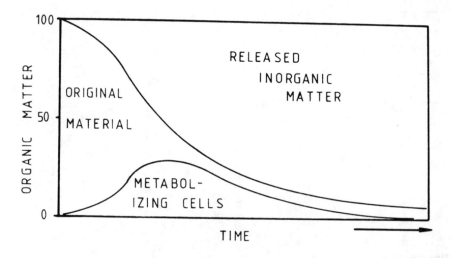

Fig.3.7: General trend in decomposition of dead organic material (detritus) showing the internal ratio between initial organic matter, decomposer biomass and metabolized inorganic matter.

In the presence of $O_2$ the decomposition may be mainly aerobic. Hence, the rate of consumption of oxygen is a measure of the metabolic rate. Assuming for the moment that all organic matter has the same susceptibility to degradability, then decomposition may be expressed as

$$\frac{ds}{dt} = -\frac{1}{Y}\frac{dx}{dt} = -\frac{1}{Y}\mu\frac{S}{K_s + S}X ,$$
(3.53)

where S is the concentration of the organic substrate, Y is the average growth yield, X the concentration of microorganisms expressed as biomass per volume, $\mu$ the maximum growth rate and $K_s$ a constant.

Since the specific growth rate is generally much larger than the specific decay rate, a portion of the organic material in the water is rapidly incorporated into the total biomass of the standing crop of microorganisms, followed by a longer period in which the biomass wastes away in decay.

In the measurement of organic pollution in water, it is common practice to express concentrations in terms of the rate at which oxygen is utilized in the biological decomposition of organic material, i.e. biochemical oxygen demand (BOD).

Assuming first order kinetic, we get

$$dL/dt = = k \times L$$
(3.54)

where L is the BOD and K the rate constant for biochemical oxidation. Upon integration between the limits of t = 0 and t = t, the equation becomes:

$$L_t = L_0 \cdot e^{-k \cdot t}$$
(3.55)

where $L_0$ is the initial BOD, and $L_t$ is the BOD at time t.

This equation can also be expressed as

$$Y = L_0 \left(1 - e^{-k \cdot t}\right)$$
(3.56)

where Y is the BOD that has been exerted by time t. For an elaboration and application of these equations, see section 7.2.

Besides decomposition of organic matter, the release of energy may be generated by the oxygenation of inorganic compounds in the presence of oxygen. These compounds may for instance be $NH_4^+$, S or $H_2S$. The generalized processes are given as:

$NH_4^+\ 1\frac{1}{2}\ O_2 = 2H^+ + H_2O$ ($F^o$ = -276 k /mole)
$NO_2^- + \frac{1}{2}\ O_2 = NO_3^-$ ($F^o$ = -75 k /mole)
$NH_4^+\ 2\ O_2 = NO_3^- + H_2O + 2H^+$ ($F^o$ = -351 k /mole)
$H_2S + \frac{1}{2}\ O_2 = S + H_2O$ ($F^o$ = 172 k /mole)
$S + 1\frac{1}{2}\ O_2 + H_2O = H_2SO_4$ ($F^o$ = -495 k /mole)
$H_2S + 2\ O_2 = H_2SO_4$ ($F^o$ = -667 k /mole)

As the processes are normally generated by a single population of bacteria or perhaps a few populations, the kinetics may fit a Monod expression or a first order reaction, as in the case of nitrification:

$$dA/dt = k.A, \quad or$$
(3.57)

$$dA/dt = \max \begin{cases} K \cdot A_1 \\ 0 \end{cases}$$
(3.58)

where A is the ammonia concentration, $A_1$ is the threshold concentration for initiating nitrification, and k is the reaction constant. This expression is often used in eutrophication models, see section 7.3.

In the absence of free oxygen, the decomposition of organic material may be effected by the use of oxygen from nitrate, nitrite or sulphate.

In aqueous environments denitrification and sulphate reduction are some of the most dominant metabolic pathways under anaerobic conditions:

$C_6H_{12}O_6 + 12\ NO_3^- = 12\ NO_2^- + 6\ CO_2 + 6\ H_2O$ ($F^o$ = -1927 k /mole)

$C_6H_{12}O_6 + 8\ NO_2^- = 4\ N_2 + 2\ CO_2 + 4\ CO_3^- + 6\ H_2O$ ($F^o$ = -3017 k /mole)

The energy released is nearly the same as that from the aerobic decomposition of glucose ( $F^o$ = -2929 k /mole). Moreover, denitrification may start soon after the development of anaerobic conditions, because many bacteria involved are facultative anaerobes.

Denitrification (or nitrate reduction) has been described as a simple first order process:

$$(3.59)$$

$$dN/dt = -k \times N,$$

Similar equations may be established for sulphate reduction.

### 3.5.4. Adaptation.

Many ecological processes are subject to adaption to input changes. The spectrum of adaptation processes is large and covers all levels from cellular to ecosystem level. However, only few adaption processes have been modelled so far, but it might be essential to include them in ecological models, as the stability of the ecosystem often is dependent on adaptation.

These aspects will be further discussed in sections 9.1 and 9.2.

**TABLE 3.8.**

**List of Abbreviations.**

Many previously mentioned process formulations include adaptation e.g. the two-steps phytoplankton growth model, where a regulation of the growth occurs in accordance with the internal nutrient concentration.

| | | |
|---|---|---|
| $\alpha_K^2$ | = | preference coefficient |
| AK | = | Constant |
| AV | = | Availability of food |
| BL | = | Body length mm |
| CA | = | Concentration of phytoplankton C g m$^{-3}$ |
| CAMAX and CAMIN | = | Max. and min. concentration of C in phytoplankton g C per g dry matter |
| CK | = | Carrying capacity g m$^{-3}$ |
| CS | = | Concentration of inorganic soluble carbon g m$^{-3}$ |
| f (x) | = | Function of x |
| FAC | = | Correction factor for the biochemical growth activities during dark periods |
| FD | = | Relative day-length |
| $\overline{F}_{max}$ | = | Maximal fractional reduction in daily specific growth rate over eutrophic depth |
| g (y) | = | Function of y |
| GRZ | = | Grazing rate g m$^{-3}$ 24 h$^{-1}$ |
| I | = | Irradiance |
| I (0) | = | Irradiance, surface intensity |
| $I_{op}$ | = | Optimum irradiance |
| IK | = | Light saturation parameter |
| K | = | Constant, unspecified |
| $K_{20}$ | = | Constant or coefficient at 20$^{o}$C |
| K (T) | = | Constant or coefficient at T$^{o}$C |
| $K_{opt}$ | = | Constant at optimum temperature |
| KC | = | Half saturation constant of CS-uptake g m$^{-3}$ |
| KFLM | = | Food level, where multiplier is $\frac{1}{2}$(1 + KMFM) |
| KLYS | = | Constant for light inhibition |
| KMFM | = | Minimum filtering rate multiplier |
| KN | = | Half saturation constant of NS-uptake g m$^{-3}$ |
| KP | = | Half saturation constant of PS-uptake g m$^{-3}$ |
| KPA | = | Saturation constant for intracellular phosphorus |
| KTR | = | Threshold concentration for grazing g m$^{-3}$ |
| KZ | = | Half saturation concentration for grazing g m$^{-3}$ |
| MY | = | Growth rate, phytoplankton 24 h$^{-1}$ |
| MYMAX | = | Max. growth rate, phytoplankton 24 h$^{-1}$ |
| MYZ | = | Growth rate, zooplankton |
| n | = | Constant, a number |
| NA | = | Concentration of nitrogen in phytoplankton g m$^{-3}$ |

NAMAX
and    = Max. and min. concentration of N in phytoplankton g N per
NAMIN    g dry matter

NS     = Concentration of soluble inorganic nitrogen g m$^{-3}$

$NS_0$   = Constant

OX     = Oxygen concentration g m$^{-3}$

PA     = Concentration of phosphorus in phytoplankton g m$^{-3}$

PAMAX
and    = Max. and min. P-concentration in phytoplankton g P per
PAMIN    g dry matter

PHOTO  = Photosynthetic rate g m$^{-3}$ 24 h$^{-1}$

PHYT   = Phytoplankton concentration g m$^{-3}$

PKI    = Equilibrium constant for reaction brtween phosphorus and
         carrier in lit/mole

PRE F  = Preference ratio

PS     = Concentration of soluble inorganic phosphorus g m$^{-3}$

$PS_0$   = Constant

R (t)  = Total daily radiation (unit as I)

T      = Temperature, $^{\circ}$C $T_j$ = T for j in element, $T_i$ and $T_o$ = T
         in inflow and outflow

$T_{max}$,
$T_{min}$  = Temperature, maximum, minimum. Rate coefficient = 0 at
           $T_{max}$ or $T_{min}$

$T_{op}$   = Temperature, optimum

TOX    = Concentration of toxic material g$^{-3}$

UC     = Uptake rate of inorganic carbon g m$^{-3}$ 24 h$^{-1}$

UCMAX  = Max. uptake rate of inorganic carbon 24 h$^{-1}$

UN     = Nitrogen uptake rate g m$^{-3}$ 24 h$^{-1}$

UNMAX  = Max. nitrogen uptake rate 24 h$^{-1}$

UP     = Phosphorus uptake rate g m$^{-3}$ 24 h$^{-1}$

UPMAX  = Max. phosphorus uptake rate 24 h$^{-1}$

ZOO    = Concentration of zooplankton g m$^{-3}$

# 4. CONCEPTUAL MODELS

Ten different methods of conceptualization, with their advantages and disadvantages, are presented in this chapter. A general recommendation at to which method to use is not given, as it is hardly possible. The problem, the ecosystem, the application of the model and the habits of the modeller will determine the preference of the conceptualization method.

A conceptual model has a function of its own. If flows and storages are given by numbers, the diagram gives an exellent survey of a steady state situation. It can be applied to get a picture of the changes in flows and storages if one or more forcing functions are changed and another steady state situation emerges. If first order reactions are assumed it is even easy to compute other steady state situations, which might prevail under other combinations of forcing functions, (see also chapter 5). Three illustrations and one example of this application of conceptual models are included in section 4.3 to give the reader an idea of these possibilities.

## 4.1. APPLICATION OF CONCEPTUAL MODELS.

Conceptualization is one of the early steps in the modelling procedure, see section 2.2, but it might also have a function of its own, as it will be illustrated in this chapter.

**A conceptual model can be considered as a list of state variables and forcing functions of importance to the ecosystem and the problem in focus, but will also show how these components are connected by means of the processes.** It is employed as a tool to create abstractions of reality in ecosystems and to delineate the level of organization that best meets the objectives of the model. A *wide spectrum* of conceptualization approaches is available and will be presented here. Some give only the components and the connections, other imply mathematical descriptions. It is hardly possible to give general recommendations as to which one to apply. It will be dependent on the problem, the ecosystem, the class of model and to a certain extent also on the habits of the modeller.

It is hardly possible to model without a conceptual diagram to visualize the modellers concepts of the system. The modeller will usually play with the idea of constructing various models of different complexity at this stage in the modelling procedure, making the first assumptions and selecting the complexity of the initial model or alternative models. It will require intuition to extract the applicable parts of the knowledge about the ecosystem and the problem involved. It is therefore not possible to give general lines on how a conceptual diagram is constructed, except that it is often better at this stage to use a slightly too complex model than a too simple approach. In the later stage of modelling it will be possible to exclude redundant components and processes. On the other hand it will do the modelling too cumbersome, if a too complex model is used even at this initial stage.

Generally, will good knowledge about the system and the problem facilitate the conceptualization step and increase the chance to find close to the right complexity for the initial model. The questions to be answered are: *What components and processes of the real system are essential to the model and the problem? Why? How?* In this process a suitable balances sought is between elegant simplicity and realistic detail.

Identification of the level of organization and selection of the needed complexity of the model are not trivial problems. Miller (1978) indicates 19 hierarchical levels in living systems, but to include all of them in an ecological model is of course an impossible task, mainly due to lack of data and general understanding of nature.

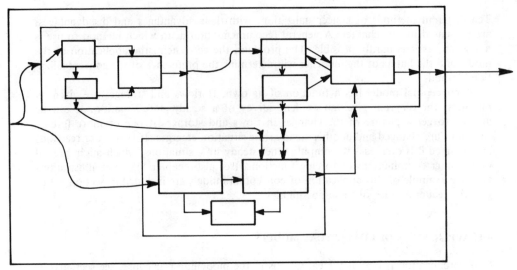

Fig.4.1:  Conceptualization showing three levels of hierarchical organization.

It is, however, in most cases not necessary to include more *than a few or even only one hierarchical level* to understand a particular behaviour of an ecosystem at a particular level, see Pattee (1973), Weinberg (1975), Miller (1978) and Allen and Star (1982). Fig. 4.1 illustrates a model with three hierarchical levels, which might be needed if a multi goals model is constructed. The first level could f.inst. be a hydrological model, the next level a eutrophication model and the third level a model of phytoplankton growth, considering the intracellular nutrients concentrations.

   Fig. 4.2 illustrates an actual case study, where the water quality of the Upper Nile Lake System has been constructed. The figure shows how models of the next hierarchical level are connected to form the total model.

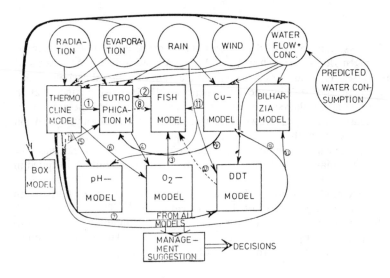

Fig.4.2: Connections of models to form a total model of the Upper Nile Lake System.

Each of the submodels shown has its own conceptual diagram, see f.inst. the conceptual diagram of the phosphorus flowing in a eutrophication model fig. 2.9. In this latter submodel there is a subsubmodel considering the above mentioned growth of phytoplankton by use of intracellular nutrients concentrations, which is conceptualized in fig. 4.3, symbols used in this figure, see fig. 4.4.

Models, which considers as well the distribution and effects of toxic substances, might often require three hierarchical levels: one for the hydrodynamics or aerodynamics to account for the distribution, one for the chemical and biochemical processes of the toxic substances in the environment and the third and last for the effect on the organism level.

### 4.2. TYPES OF CONCEPTUAL DIAGRAMS

Ten types of conceptual diagrams are presented and reviewed. Table 4.1 gives a summary of the characteristics of the various types of diagrams. In the table is also indicated, where each diagram example can be found with reference to a figure number.

**Word models** use a verbal description of model component and structure. Language is the tool of conceptualization in this case. Sentences can be used to describe a model briefly and precisely. However, word models of large complex ecosystems quickly become unwieldy and therefore they are only used for very simple models. The proverb "One picture is worth thousand words" explains, why the modeller needs to use other types of conceptual diagrams to visualize the model.

**Picture models** use components seen in nature and place them within a framework of spatial relationships. Fig.7.18 shows such a picture model of a cypres dome. It indicates the components that must be included in the model. Another example is shown in fig. 4.5 taken from Seip (1983). The latter example illustrates the direction of interactions between the elements in a food web representation.

**Box models** are simple and commonly used conceptual design for ecosystem models. Each box represents a component in the model and arrows between boxes indicate processes. Fig. 2.8 shows an example of the P-flows in a eutorphication model. A similar diagram for the nitrogen flows is shown in fig. 2.1. The arrows indicate mass flows of mass caused by processes. Fig. 4.6 gives a conceptual diagram of a global carbon model, used as basis for predictions of the climatic consequences of the increasing concentration of carbon dioxide in the atmosphere. The numbers in the boxes indicate the amount of carbon on a global basis, while the arrows give information on the amount of carbon transferred from one box to another per annum.

A model to predict the carbon dioxide concentration in the atmosphere can easily be developed on basis of the mass conservation principle by use of the numbers included in the diagram.

The term **black box models** is used, when the equations are set up on basis of an analysis of input and output relations f.inst. by statistical methods. The modeller is not concerned with the causality of these relations. Such a model might be very useful, provided that the input and output data are of sufficient quality. However, the model can only be applied on the case study, for which it has been developed.

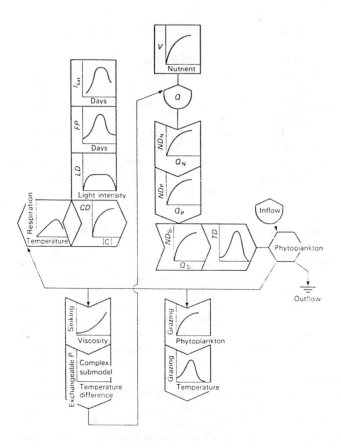

Fig.4.3: Flow chart of the phytoplankton model of Jørgensen (1976) and Jørgensen et al. (1978).

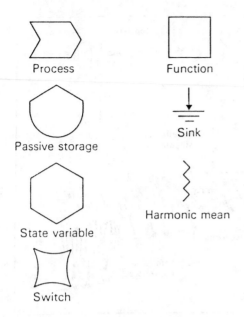

Process Function

Passive storage Sink

State variable Harmonic mean

Switch

Fig.4.4: Flow chart of the phytoplankton model of Chen and Orlob (1975). LD, ND, TD: light nutrient and temperature dependence.

New case studies will require new data, a new analysis of the data and consequently new relations.

**White box** models are constructed on the basis of causality for all processes. This does not imply that they can be applied on all similar case studies, because, as discussed in section 2.4, a model always reflects ecosystem characteristics. But in general a white box model will be applicable to other case studies with some modification.

In practice **most models are grey,** as they contain some causalities but also apply empirical expressions to account for some of the processes. Some modellers prefer other geometric shapes, for example, Wheeler et al. (1978) prefer circles to boxes in their conceptualization of a lead model. This leads to no principal difference in the construction and use of the diagram.

**Input/output models** differ only slightly from box models, as they can be considered as box models with indications of in- and outputs.The global carbon model,see fig.4.6 can be considered to be an input/output model as all in- and outputs of the boxes are indicated with numbers. Another example is shown in fig. 4.7. It is an oyster model, developed by Patten (1983).

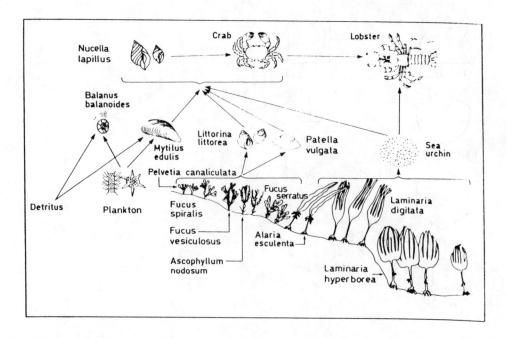

Fig.4.5: Generalized food webs of the Hardangerfjord rocky shores.

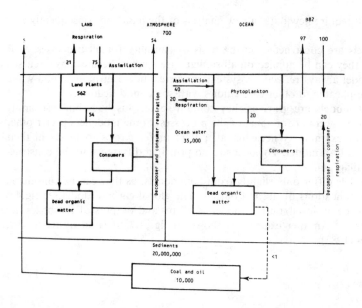

Fig.4.6: Carbon cycle, global. Values in compartments are in tons and in fluxes $10^9$ tons/year.

The same model is illustrated by use of **matrix conceptualization** in fig. 4.8. The first upper matrix is a so-called adjacency matrix, which indicates the connectivity of the system. This matrix has $a_{ji} = 1$ if a direct causal flow (or interaction) exists from compartment j (column) to compartment i (row), and $a_{ji} = 0$ otherwise. The lower matrix, called a flow or in/output matrix, represents the direct effects of compartment j on compartment i. The number expresses the probability that a substance in j will be transferred to i in one unit of time. P is a one step transition matrix in Marko chain theory and can be computed readily from storage and flow information. Notice that fig. 4.7 uses the units kcal/m² and kcal/m² day, while the flow matrix in fig. 4.8 uses six hours as unit. The number for $a_{12}$ is therefore found as $15.7915/(4.2000) = 0.1974 \ 10^{-2}$ indicated in the matrix as 1.974-3.

The two matrices provide a survey of the possible interactions and their quantative role.

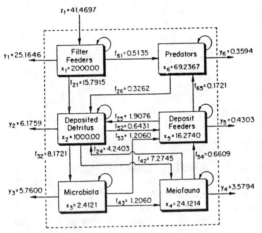

Fig. 4.7: Input/output model for energy flow (kcal m⁻²d⁻¹) and storage (kcal m⁻²) in an oyster reef community.

The feedback dynamics diagrams use a symbolic language introduces by Forrester (1961), see fig. 4.9. Rectangles represent state variables. Parameters or constants are small cirles. Sinks and sources are cloudlike symbols, flows are arrows and rate equations are the pyramides that connect state variables to the flows.

A modification is developed by Park et al.(1979) the symbols in fig. 4.4 and the phytoplankton model fig. 4.3, which uses these symbols. It differs from the Forrester diagrams mainly by giving more information on the processes, which are shown by a graphic representation.

A **computer flow chart** might be used as a conceptual model. The sequence of events shown in the flow chart can be considered a conceptualization of the ordering of important ecological processes. An example is given in fig. 4.10, which is a swamp model developed by Phipps (1979). The model subjects each of the three species in the swamp to the same sequence of events with specific parameters as function of species. Trees are born, grow and die off due to old age (KILL), lumbering (CUT) or environmental forces (FLOOD). Birth dependent on al other processes. This type of model is very usefull in setting up computer programs, but does not give information on the interactions. F.inst. it is not possible to read on fig. 4.10 that GROW is a subroutine, which takes into account the interactions between water table and crowding on the individual tree species.

A subcategory of computer flow charts is analog computer diagrams. An example is shown in fig. 4.11. Analog symbols are used to represent storages and flows. An amplifier is used to sum and invert one or more inputs. By adding a capacitor to an amplifier we get an integrator. Analog computers have found only a limited use in ecological modelling. For descriptions see Patten (1971).

**Compartments**

(a)

| from / to | 1 | 2 | 3 | 4 | 5 | 6 | Row Sum |
|---|---|---|---|---|---|---|---|
| 1 | 1 | 0 | 0 | 0 | 0 | 0 | 1 |
| 2 | 1 | 1 | 0 | 1 | 1 | 1 | 5 |
| 3 | 0 | 1 | 1 | 0 | 0 | 0 | 2 |
| 4 | 0 | 1 | 1 | 1 | 1 | 0 | 3 |
| 5 | 0 | 1 | 1 | 1 | 1 | 0 | 4 |
| 6 | 1 | 0 | 0 | 0 | 1 | 1 | 3 |
| Column Sum | 3 | 4 | 3 | 3 | 3 | 2 | 18 |

(b)

| from / to | 1 | 2 | 3 | 4 | 5 | 6 | Row Sum |
|---|---|---|---|---|---|---|---|
| 1 | 9.948-1 | 0 | 0 | 0 | 0 | 0 | 9.948-1 |
| 2 | 1.974-3 | 9.944-1 | 0 | 4.395-2 | 2.930-2 | 1.178-3 | 1.071 |
| 3 | 0 | 2.043-3 | 1.530-1 | 0 | 0 | 0 | 1.551-1 |
| 4 | 0 | 1.818-3 | 1.250-1 | 9.121-0 | 0 | 0 | 1.039 |
| 5 | 0 | 1.608-4 | 1.250-1 | 6.850-1 | 9.614-1 | 0 | 1.093 |
| 6 | 6.419-5 | 0 | 0 | 0 | 2.644-3 | 8.975-1 | 1.000 |
| Column Sum | 9.969-1 | 9.985-1 | 4.030-1 | 9.629-1 | 9.934-1 | 9.987-1 | 5.353 |

Fig.4.8: Oyster reef model first order matrices (a) A for paths, and (b) P for causality. Example entry in P: $9.948^{-1} = 9.948 \times 10^{-1}$.

**Signed digraph models** extend the adjacency concept. Plus and minus signs are used to denote positive and negative interactions between the system components in the matrix and the same information is given a box diagram, see fig. 4.12, where a general benthic model is shown (Puccia, 1983). Lines connecting the components represent the causal effects. Positive effects are indicated with arrows and lines with a small circle head indicate a negative effect.

**Energy circuit diagrams**, developed by Odum (see Odum 1971, 1972 and 1983) are designed to give information on thermodynamic constraints, feed-back mechanisms and energy flows. The most commonly used symbols in this language are shown fig. 4.13. As the symbols have an implicit mathematical meaning, it gives many informations about the mathematics of the model. It is, furthermore, rich in conceptual information and hierarchical levels can easily be displayed, as demonstrated in fig. 4.14 and 8.9. Numerous other examples can be found in the literature, see f.inst. Odum (1983). A review of these examples will reveal, that energy circuit diagrams are very informative, but they are difficult to read and survey, when the models are a little more complicated.

**TABLE 4.1**

**Types of conceptual Diagrams**

| Conceptual design | Characteristics, advantages and disadvantages | Example see figure |
|---|---|---|
| Word models | Sentences describe model Simple to use. Cannot be used for complex models. | |
| Picture models | Picture of ecosystem components Very illustrative. Difficult to transfer to mathematical formulation. | Figs. 4.5 and 7.18 |
| Box models equations | Components are boxes, processes are arrows, simple to use. Relatively easy to transfer to mathematical formulation, but give little information on process | Fig. 2.1, 2.8 and 4.6 |
| quations. | | |
| Black box models | Based upon statistical analysis. Relate input and output without causality | |
| Input/output models | Box models with indication of input and output as rates. Assume often linearity and lack temporal dynamics. | Fig. 4.6 and 4.7 |
| Matrix model | Matrix notation used to indicate connectivity and flow rates. Assume linearity and lack temporal dynamics. | Fig. 4.8 |
| Forrester diagrams (with modifications) | Include feed-backs. Give more information by use of symbolic language. | Fig. 4.3, 4.4 and 4.9 |
| Computer flow charts | Easy to set up computer program. Difficult to give information on processes and interactions. | Fig. 4.10 and 4.11 |
| Signed digraph models | Contain logic gates and qualitative interactions. Matrix notation easy to use. Assume linearity and lack temporal dynamics. | Fig.4.12 |
| Energy circuit diagrams | Give detailed information on thermodynamic constraints. Feed-back mechanisms and energy flow. Relatively difficult to survey. | Fig. 4.14 and 8.9 |

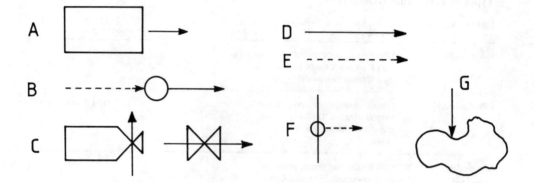

Fig.4.9: Symbolic language introduced by Forrester (Jeffers, 1978). A state variable, B Auxiliary variable, C Rate equations. D Mass flow, E Information, F Parameter. G Sink.

### 4.3. THE CONCEPTUAL DIAGRAM AS MODELLING TOOL

The Word models, Picture models and Box models give all a description of the relation between the problem and the ecosystem. They are very useful as a first step in modelling, but their application as a modelling tool of its own is rather limited. Additional information is needed to be able to answer even semiquantitative questions. It is, however, possible by use of many of the other conceptual approaches, which will be demonstrated in this paragraph.

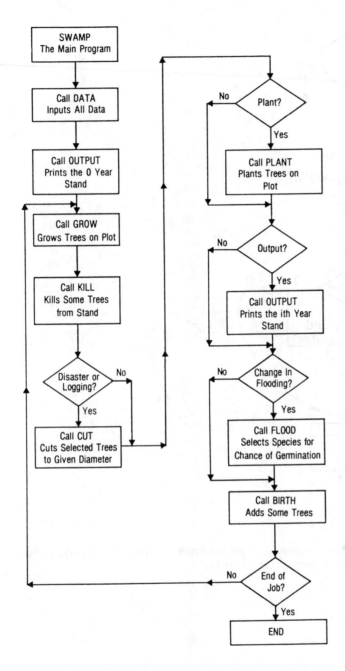

Fig.4.10:  Flow chart of SWAMP (modified from Phipps, 1979).

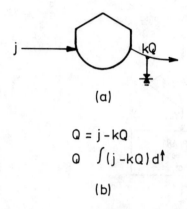

(a)

$$Q = j - kQ$$
$$Q \quad \int (j - kQ) \, d^t$$

(b)

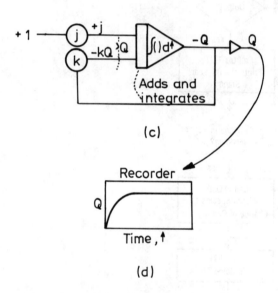

(c)

Recorder

Q

Time, ↑

(d)

Fig.4.11: Example of an analog computer conceptualization; (a) storage; (b) equations, (c) analog model; (d) output (Odum, 1983).

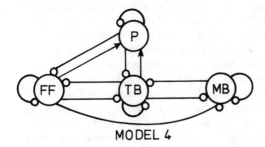

MODEL 4

| Parameter | Change in Equilibrium | | | |
|-----------|-----|-----|-----|-----|
| Input to  | FF  | TB  | MB  | P   |
| FF        | +   | −   | +   | −   |
| TB        | −   | +   | −   | −   |
| MB        | +   | −   | +   | +   |
| P         | +   | +   | +   | +   |

Fig.4.12: A general signed digraph model for east coast (USA). Benthic organisms from a sandy environment (from Puccia, 1983).

**ILLUSTRATION 4.1**

In fig. 4.6 the global C-cycle is shown. It is seen that the input of carbon dioxide due to the use of fossil fuels increases the atmospheric carbon dioxide concentration by (5/700) per annum or (5/7%). If the amount of carbon dioxide dissolved in the sea is deducted the increase will only amount to (2/7%). As the carbon dioxide concentration at present is 0,032% on a volume-volume basis it is easy to see that at the present rate of fossil fuel combustion, the concentration will reach 0,040% in about 87 years. It is, of course, also possible to compute, the concentration at year x, if a certain trend in the use of fossil fuels is given, or the time, it will take with a certain global energy policy to reach a given threshold concentration. These computation assume that the percentage of carbon dioxide transferred to the sea is constant or at least as given. A far more complex computation is, of course, required to find the carbon dioxide concentration in the atmosphere if we want to incorporate the actual mechanisms for these transfer processes from the atmosphere to the sea, but as seen by this illustration it is possible to get some first approximations by use of a conceptual diagram with indication of storages, in - and output flows.

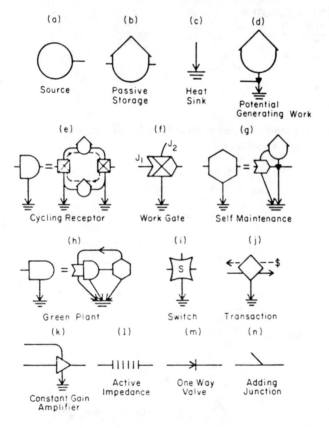

Fig.4.13: Diagrammatic energy circuit language of Odum (1971, 1972, 1983) developed for ecological conceptualization and simulation applications.

## ILLUSTRATION 4.2

Patten (1983) uses the matrix representation directly to compute, what he calls the indirect effects. If the adjacency matrix is multiplied by itself, the product A2 indicates the number of indirect paths of the length 2 from one compartment to another. In general the product of the matrix $A^n$ will represent the number of length n paths from compartment j to compartment i. Fig. 4.15 shows the tenth order matrix. As seen the number of paths of length 10 is incredibly high. There are more than 500 000 length 10 paths in the model. The reason is that the length of a cyclic path is infinite. Matter, energy and information may pass around such a path until it either dissipates or exists from the cycle.

In fig. 4.16 is shown the $P^{10}$ for influences. Smaller values than corresponding nondiagonal entries in $P^3$ i are underlined. Thus indirect effects are generally still tending to grow at the 10th order level due to the enormous number of paths. Patten demonstrates with this simple analysis the importance of indirect effects. These aspects will be further discussed in section 5.2.

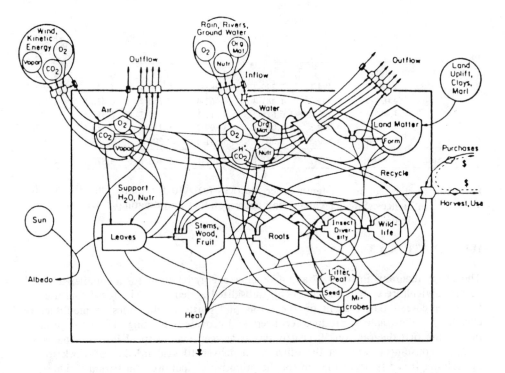

Fig.4.14: General conceptual energy circuit model of a floodplain swamp (from Odum 1983).

Odum (1983) uses widely energy circuit diagrams not only as a basis for mathematical models, but also directly as management tool. As the diagrams contain the essential elements of the system, the processes and the quantitative relations, they provide the modeller with a rather detailed picture of the system.

**Compartments**

(a)

| from / to | 1 | 2 | 3 | 4 | 5 | 6 | Row Sum |
|---|---|---|---|---|---|---|---|
| 1 | 1 | 0 | 0 | 0 | 0 | 0 | 1 |
| 2 | 23696 | 34729 | 23697 | 27201 | 23696 | 16168 | 149187 |
| 3 | 11033 | 16168 | 11032 | 12664 | 11033 | 7528 | 69458 |
| 4 | 16169 | 23696 | 16168 | 18560 | 16169 | 11033 | 101795 |
| 5 | 23695 | 34729 | 23696 | 27201 | 23696 | 16169 | 149186 |
| 6 | 11032 | 16169 | 11033 | 12664 | 11032 | 7527 | 69457 |
| Column Sum | 85626 | 125491 | 85626 | 98290 | 85626 | 58425 | 539084 |

Fig.4.15

(b)

| from / to | 1 | 2 | 3 | 4 | 5 | 6 | Row Sum |
|---|---|---|---|---|---|---|---|
| 1 | 9.491-1 | 0 | 0 | 0 | 0 | 0 | 4.494-1 |
| 2 | 1.883-2 | 9.494-1 | 7.290-2 | 2.988-1 | 2.410-1 | 1.137-2 | 1.592 |
| 3 | 4.029-5 | 2.303-3 | 1.581-4 | 6.662-4 | 5.254-4 | 2.430-5 | 3.718-3 |
| 4 | 1.416-4 | 1.396-2 | 6.616-2 | 4.011-1 | 1.915-3 | 8.512-5 | 4.833-1 |
| 5 | 3.353-5 | 4.009-3 | 1.089-1 | 3.899-2 | 6.753-1 | 2.014-5 | 8.282-1 |
| 6 | 6.203-4 | 4.520-5 | 3.003-2 | 5.831-4 | 2.203-2 | 9.755-1 | 1.002 |
| Column Sum | 9.690-1 | 9.697-1 | 2.521-1 | 7.401-1 | 9.408-1 | 9.870-1 | 4.859 |

Fig.4.16: Oyster reef model tenth order matrices (a) A10 for paths, and (b) p10 for influences. Smaller values than corresponding non-diagonal entries in p3 are underlined.

## ILLUSTRATION 4.3

The energy diagrams in fig. 4.17 and 4.18 show energy relations for a complex adapted natural agriculture system and an energy subsidized agriculture. The fossil fuel used in the latter case is partly spent on the farms and partly in the cities to manufacture chemicals, build tractors, produce fertilizers and provide marketing. The basic production in the first diagram on 50 kcal/m²day is as seen mainly spent in maintenance of complex consumers, who on the other hand deliver 10 kcal/m² day of work to the agriculture. It is a typical situation for the primitive agriculture: most food produced is used as energy supply to the agriculture family and only a minor amount of food can be sold.

By comparison of possible alternatives in energy input it is feasible to see the consequences. It is shown in fig. 4.18 that by the use of 60 kcal/m² day it is possible to increase the primary production from 50 kcal/m² to 100 kcal/m² day. It is not possible to have a very high production without the energy subsidies.

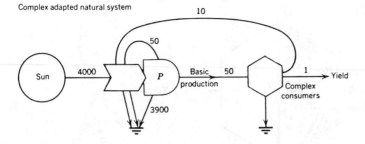

Complex adapted natural system

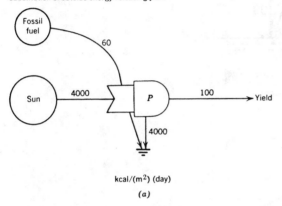

Substitution of outside energy releasing yield

kcal/(m²) (day)

*(a)*

Fig.4.17 and 4.18: Comparison of a complex natural system adapted to maximize its basic production through the works of its diversified organization of consumer species with the same system after fossil-fuel supported works of man have eliminated the natural species and substituted industrial services for the services of those natural species, releasing the same basic production to yield. (a) Plant production and yield; (b) animal production and yield.

## ILLUSTRATION 4.4

Figs. 4.19 and 4.20 compare animal husbandry with and without fossil fuel. The same picture as above is seen. In what is called ecological agriculture it is possible to couple the systems in fig. 4.17 and 4.19, which of course does it possible to save some energy, but a reduction in production is unavoidable when less fossil fuel is used. The figures 4.18 and 4.20 show agriculture systems close to an economical optimum, but an economical-ecological optimum will most probably correspond to a smaller energy use than in fig. 4.18 and 4.20. To find this optimum requires, however, the use of a more complex model, which takes into account the effects of applied pesticides, fertilizers and other pollutants (f.inst. from manufacturing of these chemicals, smoke from the tractors etc.).

Fig. 4.21 shows the situation of man in a system of industrialized high yield agriculture.

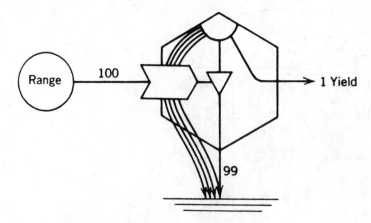

Fig.4.19   Animal husbandry without fossil fuel.

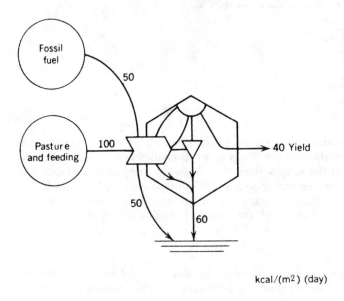

kcal/(m²) (day)

Fig.4.20   Animal husbandry with fossil fuel.

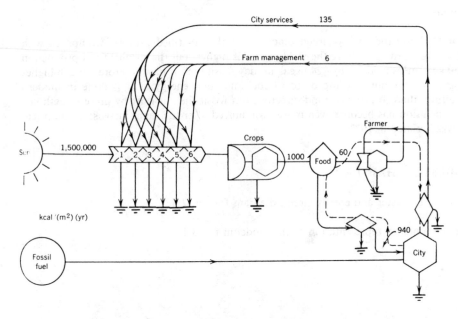

Fig.4.21 System of industrialized, high-yield agriculture. Energetic input include flows of fossil fuels which replace the work formerly done by man and his animals.

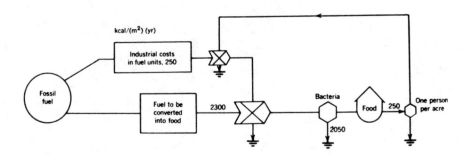

Fig.4.22 System converting methane to food using industrial-microbiological proces-ses.

**Example 4.1**

It is possible to convert fossil fuel (methan and/or petroleum) into bacterial tissue and hence food. In accordance with information from such a plant it is possible to produce $10^4$ kg food with 50% proteins from $4.6 \cdot 10^8$ kcal of fossil fuel. Proteins have 4 kcal/g and the other 50% of the food 5 kcal/g. Set up an energy circuit diagram for this production when it is known that a plant with a production of $10^4$ kg per day requires an area of 100 000 m$^2$. Compare this diagram with the figures 4.17-4.21 and comment on the energy economy of the process.

**Solution**

Fig. 4.22 shows the energy circuit diagram for the described process. Compared with energy subsidized agriculture the yield per m² is higher, but the surplus food production of 150 kcal/m² is paid for by 2240 kcal/m² day fossil fuel. It has therefore a much higher efficiency to combine the use of fossil fuel and solar energy as it is done in modern agriculture, than to produce food directly from fossil fuel, at least by present methods. This conclusion has become even more pronounced after the rapid increase of oil prices that was initiated in 1973.

**PROBLEMS, CHAPTER 4.**

1. Draw a Forrester and energy circuit diagram for fig. 2.8.

2. Set up a matrix representation of the model in fig. 3.1.

# 5. STATIC MODELS

Static models give important information on flows and storages by steady state. Their advantages are that they give good pictures of average situations and it is easy to compare different steady states from prevailing forcing functions. If first order reactions are assumed, it is possible by use of matrix computations to carry out these comparisons, as demonstrated by the input/output environ analysis presented in section 5.2.

The last part of this chapter is devoted to another type of static models: response models. They are based on the simplification, that there exists a relation between one or more prevailing forcing function(s) and a selected sensitive state variable. A few examples are shown in section 5.3 to illustrate the application of response models. These models are easy to use and are very simple to construct, but in most cases they require the assumption of simplifications, which limit their applications.

## 5.1. APPLICATION OF STATIC MODELS

Fig. 3.1 illustrates a simple model and the corresponding equations are (3.1) and (3.2). If A and B are static, inputs must balance outputs:

Process (1) + process (3) = process (2) + process (4)
Process (4) + process (6) = process (5).

These two equations are derived from equation (3.3). In a static situation differential equations will therefore be reduced to algebraic equations, which are a more simple mathematical representation to use as model. The advantages are obvious: an analytical solution might be provided, in most cases fewer data are needed, a parameterization is most often easier and the computations are carried out more easily.

A static model can of course not be used in a transient state, see fig. 2.4, but it can be used to describe state A or an average situation of state C in fig. 2.4.

Fig. 4.6 - 4.8 illustrate static models, which give information on the mass flows. Fig. 2.9 gives an illustration of a static energy flow model in an ecosystem. As these examples show, input/output models are useful representations of static models and the matrix representation is a convenient computational tool for this type of models.

It is desirable to compare the results of different external inputs, process (1), process (3) and process (6), see fig. 3.1, the corresponding values of the state variables A and B and of the outputs process (5) and process (2) can be found. We compare with other words different static situations involved from different inputs. Such comparisons could be named response models, but we will in this context reserve the word response model to the relation between external factors and one state variable for static situations.

## 5.2. INPUT/OUTPUT ENVIRON ANALYSIS

The concepts and results developed are illustrated by a static five compartment model of nitrogen flow in a Puerto Rican Tropical Rainforest. The model is based on research of

Edminsten (1970) and is described in detail in Patten et al. (1976, p. 574 ff.). The dynamics of a compartment model such as fig. 5.1 can be described by systems of difference or differential equations, for example:

$$\frac{dx_i}{dt} = \dot{x}_i(t_i) = \sum_{\substack{j=0 \\ j \neq i}}^{n} f_{ij}(t) - \sum_{\substack{j=0 \\ j \neq i}}^{n} f_{ji}(t) \quad i = 1,\ldots,n, \quad t \in \tau \tag{5.1}$$

where $x_i$ represents the storage of a substance in compartment i, where it is conserved. $\dot{x}_i = dx_i/dt$ is the first derivate with respect to time of $x_i$. Time is defined on an interval $\tau$ with $t^o$ being the initial time and $t^+$ the final time. $f_{ij}$ and $f_{ji}$ are non-negative flows from compartment j to i and i to j, respectively. The environment of the system defined by the n compartments is denoted by subscript O. If $z_i = f_{io}$ and $y_i = f_{oi}$, environmental input and output respectively, of compartment i, equation (5.1) can be written:

$$\dot{x}_i(t) = z_i(t) + \sum_{\substack{j=1 \\ j \neq i}}^{n} f_{ij}(t) - \sum_{\substack{j=1 \\ j \neq i}}^{n} f_{ji}(t) - y_i(t) \quad i = 1,\ldots,n \tag{5.2}$$

In a static situation $\dot{x}_i = x_i(t) = 0$. Then (5.2) becomes:

$$\dot{x}_i = \frac{dx_i}{dt} = 0 = z_i + \sum_{\substack{j=1 \\ i \neq i}}^{n} f_{ij} - \sum_{\substack{j=1 \\ j \neq i}}^{n} f_{ji} - y_i, \quad i = 1,\ldots,n \tag{5.3}$$

where the time argument is omitted, because every flow is constant in the considered time interval, $\tau$.

As seen in fig. 5.1 it is assumed that the system is open ($z_i > 0$ or $y_i > 0$) and that the model is connected, i.e. that there are no compartments or groups of compartments isolated from others. It is often convenient to parameterize the model in terms of fractional transfer coefficients. Let $x_i$, $i = 1,2,\ldots,n$, denote constant steady state storage for each compartment.

It is then possible to set up a linear model by formulating each flow as a fraction of its donor compartment:

$$\dot{x}_i \equiv 0 = z_i + \sum_{\substack{i=1 \\ j \neq i}}^{n} a''_{ij} x_j - \sum_{\substack{j=0 \\ j \neq i}}^{n} a''_{ji} x_i$$

$$= z_i + \sum_{\substack{i=1 \\ j \neq i}}^{n} a''_{ij} x_j + a''_{ii} x_i \tag{5.4}$$

$$= z_i + \sum_{j=1}^{n} a''_{ij} x_j, \quad \ldots,n,$$

Another linear model can be set up by formulating an each flow as a fraction of its recipient compartment:

$$\dot{x}_i \equiv 0 = \sum_{\substack{j=0 \\ j \neq i}}^{n} a'_{ji} x_i - \sum_{\substack{j=1 \\ j \neq i}}^{n} a'_{ij} x_j - y_i \tag{5.5}$$

$$= -a'_{ii} x_i - \sum_{\substack{j=1 \\ j \neq i}}^{n} a'_{ij} x_j - y_i \qquad = - \sum_{j=1}^{n} a'_{ij} - y_i, \quad i = 1,\ldots,n.$$

144

The transfer coefficients are defined as:

$$a''_{ij} = f_{ij}/x_j, \quad i,j = 1,\ldots,n,; \; i \neq j$$

$$a''_{0i} = y_i/x_i, \quad i = 1,\ldots,n,$$

$$a''_{ii} = -\sum_{\substack{j=0 \\ j \neq i}}^{n} a''_{ji}, \quad i = 1,\ldots,n;$$

$$a'_{ji} = f_{ij}/x_i, \quad i,j = 1,\ldots,n,; \; i \neq j$$

$$a'_{i0} = z_i/x_i, \quad i = 1,\ldots,n,$$

$$a'_{ii} = -\sum_{\substack{j=0 \\ j \neq i}}^{n} a'_{ji}, \quad i = 1,\ldots,n.$$

With these definitions equations (5.4) and (5.5) can be written in matrix notation:

$$\dot{X} = 0 = A''.x^* + z \tag{5.6}$$

$$\dot{X} = 0 = -A'.x^* - y = A'.x^* + y \tag{5.7}$$

where

$$x^* = \begin{pmatrix} x_1 \\ \vdots \\ x_n \end{pmatrix}, \quad \dot{x} = \begin{pmatrix} \dot{x}_1 \\ \vdots \\ \dot{x}_n \end{pmatrix}, \quad z = \begin{pmatrix} z_1 \\ \vdots \\ z_n \end{pmatrix}, \quad y = \begin{pmatrix} y_1 \\ \vdots \\ y_n \end{pmatrix},$$

$$A'' = \begin{pmatrix} a''_{11} & \cdots & a''_{1n} \\ \vdots & & \vdots \\ a''_{11} & \cdots & a''_{nn} \end{pmatrix}, \quad A' = \begin{pmatrix} a'_{11} & \cdots & a'_{1n} \\ \vdots & & \vdots \\ a'_{11} & \cdots & a'_{nn} \end{pmatrix},$$

From equations (5.6) and (5.7) two expressions follow for the steady state compartment contents:

$$x^* = -\left(A''\right)^{-1} \cdot z \tag{5.8}$$

$$x^* = -\left(A'\right)^{-1} \cdot y \tag{5.9}$$

Patten and Auble (1980) have used system analysis as the one presented below as basis for introduction of the concept environ. Stimuli (inputs) are converted in systems to responses (outputs) according to the condition (state) of the system, when the stimulus is received. Patten (1982) considers as environ the entire system of interrelations associated with the entity. It is called environ, because it surrounds a set of influences localized within a defined system (Patten (1978)). He considers environ as an elementary particle of ecosystems.

Behind these more philosophical considerations lies, however, a practical application. Input environs express similarly the overall system response to a given input and output, environs express similarly the overall system response to a given output. Further computations on input/output models give as demonstrated by Matis and Patten (1981) a quantitative evaluation of input and output environs.

The concept unit environ, $E_o$, has also been introduced to visualize further the practical applicability of these considerations. A unit input environ is the input environ, that can produce a unit of output $y_i$ and correspondingly is a unit output environ, $E_j$, the output environ, that can produce a unit input $z_j$. For those, who are interested in the computational detail, can be referred to Matis and Patten (1981), but here we will only illustrate the meaning behind these concepts by presentation of some illustrative results.

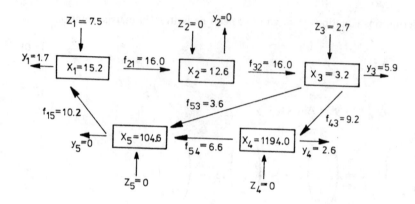

Fig.5.1: Static model of nitrogen flow in a tropical rainforest. Storages $x_i$, i = 1,..., 5, are in units of gN m$^{-2}$, and flows $z_i$, $y_i$ and $f_{ij}$, i, j = 1,..., 5, are in gN m$^{-2}$y$^{-1}$. The compartments are: $x_1$, leaves and epiphyllae; $x_2$, loose litter; $x_3$, fibrous roots; $x_4$, soil; $x_5$, wood. Input $z_1$ represents nitrogen dissolved in rainwater and fixation at atmospheric nitrogen by the epiphyllous complex; $z_3$ is nitrogen fixation by algae and bacteria associated with fibrous roots. Outputs are: $y_1$, free nitrogen release by denitrifying bacteria; denitrification and runoff associated with fibrous roots ($y_3$) and soil ($y_4$).

## ILLUSTRATION 5.1

The transfer and turnover coefficients of the fig. 5.1 model in units, 1/y, are:

$a''_{01}=1.7/15.2=0.11184$

$a''_{21}=16.0/15.2=1.05263$

$a''_{02}=0/12.6=0$

$a''_{32}=16.0/12.6=1.26984$

$a''_{03}=5.9/3.2=1.84375$

$a''_{43}=9.2/3.2=2.87500$

$a''_{53}=3.6/3.2=1.12500$

$a''_{04}=2.6/1194.0=0.00218$

$a''_{54}=6.6/1194.0=0.00553$

$a''_{05}=0/104.6=0$

$a''_{15}=10.2/104.6=0.09751$

$|a''_{11}|=(16.0+1.7)/15.2=1.16447$

$|a''_{22}|=16.0/12.6=1.26984$

$|a''_{33}|=(5.9+9.2+3.6)/3.2=5.84375$

$|a''_{44}|=(2.6+6.6)/1194.0=0.00771$

$|a''_{55}|=10.2/104.6=0.09751$

$a'_{01}=7.5/15.2=0.49342$

$a'_{02}=0/12.6=0$

$a'_{12}=16.0/12.6=1.26984$

$a'_{03}=2.7/3.2=0.84375$

$a'_{23}=16.0/3.2=5.00000$

$a'_{04}=0/1194.0=0$

$a'_{34}=9.2/1194.0=0.00771$

$a'_{05}=0/1194.0=0$

$a'_{35}=3.6/104.6=0.03441$

$a'_{45}=6.6/104.6=0.06310$

$a'_{51}=10.2/15.2=0.67105$

$|a'_{11}|=(7.5+10.2)/15.2=1.16447$

$|a'_{22}|=16.0/17.6=1.26984$

$|a'_{33}|=(2.7+16.0/3.2=5.84375$

$|a'_{44}|=9.2/1194.0=0.00771$

$|a'_{55}|=(3.6+6.6)/104.6=0.09751$

## ILLUSTRATION 5.2

For the model presented above we have:

$$\mathbf{A''} = \begin{pmatrix} -1.16447 & 0 & 0 & 0 & .09751 \\ 1.05263 & -1.26984 & 0 & 0 & 0 \\ 0 & 1.26984 & -5.84375 & 0 & 0 \\ 0 & 0 & 2.87500 & -.00771 & 0 \\ 0 & 0 & 1.12500 & .00553 & -.09751 \end{pmatrix}$$

$$\mathbf{A'} = \begin{pmatrix} -1.16447 & 1.26984 & 0 & 0 & 0 \\ 0 & -1.26984 & 5.00000 & 0 & 0 \\ 0 & 0 & -5.84375 & .00771 & .03441 \\ 0 & 0 & 0 & -.00771 & .06310 \\ 0.67105 & 0 & 0 & 0 & -.09751 \end{pmatrix}$$

$$-(\mathbf{A''})^{-1} = \begin{pmatrix} 1.69 & 0.92 & 0.92 & 1.21 & 1.69 \\ 1.40 & 1.55 & 0.77 & 1.01 & 1.40 \\ 0.31 & 0.34 & 0.34 & 0.22 & 0.31 \\ 113.86 & 125.95 & 125.95 & 241.46 & 113.86 \\ 9.97 & 11.03 & 11.03 & 14.51 & 20.23 \end{pmatrix}$$

$$-(\mathbf{A'})^{-1} = \begin{pmatrix} 1.69 & 1.69 & 1.45 & 1.45 & 1.45 \\ 0.77 & 1.55 & 1.33 & 1.33 & 1.33 \\ 0.19 & 0.19 & 0.34 & 0.34 & 0.34 \\ 95.46 & 95.46 & 81.68 & 211.46 & 165.66 \\ 11.66 & 11.66 & 9.97 & 9.97 & 20.23 \end{pmatrix}$$

147

**ILLUSTRATION 5.3**

Fig. 5.2 illustrates a model of water balance within the watershed of Okefenokee Swamp (Patten and Matis 1981 and 1982). The four compartments represent water storages in the swamp and adjacent uplands. The data in figs. 5.3 and 5.4 illustrate quantitative characteristics of environs. The bold arrow in each diagram identifies the unit output or input considered. As an example fig. 5.3 a depict the input environ E1 associated with each unit of loss from the upland surface water compartment (compare with fig. 5.3). It is shown that this unit output requires as support 0.023 units of storage in compartment 1, 0.0382 of storage in compartment 2, and internal flows of 0.1210 units from 1 to 2 and from 2 to 1, all originating as 1.0 unit of input to compartment 1. The output water has resided in compartment 1 a mean of 8 days (coefficient of variation 1.0) and in compartment 2 14 days (coefficient of variation 4.2) since entering the system. The other diagrams can be interpreted similarly.

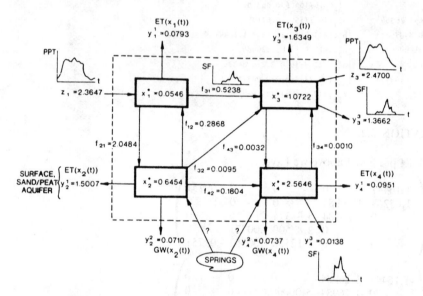

Fig. 5.2: Static water budget model of the watershed of Okefenokee Swamp. The compartments are: $x*1$ = upland surface storage, $x*2$ = upland groundwater storage, $x*3$ = swamp surface storage, $x*4$ = swamp subsurface storage. Inputs are: z1 = upland precipitation, z2 = swamp precipitation. Outputs are: y1i = evapotranspiration, i = 1,...,4, y22 = deep seepage, y33 = sheet and stream flow, y24 = percolation, deep seepage and lateral leakage, and y34 = baseflow. Intrasystem flows are: f21 = infiltration and percolation, f31 = channel and overland flow, f12 = baseflow and interflow, f32 = baseflow, f42 = lateral seepage, f43 = infiltration and percolation, and f34 = upwelling and water level rise. Storage units are $10^9$ m3, and input, output and internal flow units are $10^9$ m3 y-1. Areal basis is the entire watershed.

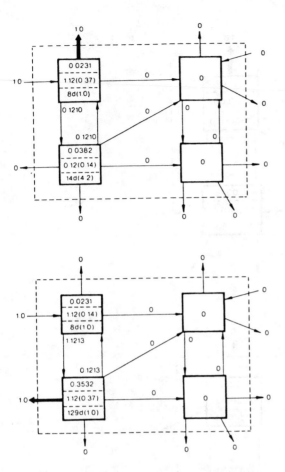

Fig.5.3: Unit input environs of Okefenokee water budget model: Each diagram shows water flows (billion of cubic meters per year)and storages (billions of cubic meters) required to generate one unit of output at the bold arrow.Flows are associated with arrows and storages appear as the upper numbers in each box. The middle numbers of each box represent,without parentheses,the mean number of past visits to that box of water presently in or leaving the compartment with the bold arrow, and within parentheses,associated standard deviations. Both these numbers are unitless. The bottom numbers in each box represent, without parentheses, the number expected past residence time in that box and within parentheses,associated coefficient of variation. The means are in h,d or y and the coefficients of variation are unitless.

149

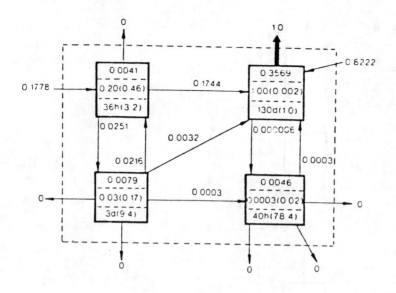

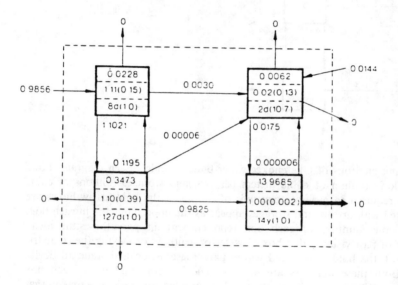

Fig.5.4: Explanation see fig.5.3.

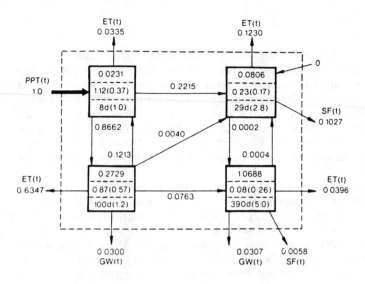

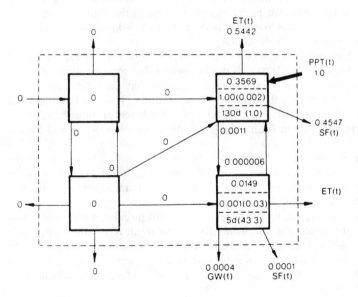

Fig.5.5: Unit output environs . Fig.5.5: Explanation see fig. 5.3.

## 5.3. RESPONSE MODELS

Response models attempt to relate one or a few external factors, which often are control functions to the state variable in focus. The response is expressed as a change in the considered state variable. If the relationship is simple it might be possible to cover it by use of an equation for a unit process, see chapter 3.

In most cases, however, more processes are involved. Then the best approach to the

151

problem might be an empirical or semiempirical expression. In the latter case it is possible on a theoretical basis to indicate the processes, that are involved in the problem and to give at least a semiquantitative description of these processes. It is with other words impossible to observe whether the empirical relation is theoreticallly reasonably.

Such types of models are usually very simple and they can therefore not take account of the high complexity of effects, that are usually observed in nature. The idea is however to find the most sensitive state variable to the external factors and use this state variable as a kind of indicator, which serves as an alarm for other undesired effects on the ecosystem level.

Fig. 5.6 illustrates an example. Adult laying hens were dosed orally each day for up to 60 days with leptophos, an organophosphorus insecticide that is relatively persistent in the environment at dose rates of 0, 0.5, 1.0, 5.0, 10.0 or 20.0 mg/kg body weight/day. Ataxia (paralysis of the legs) developed with all but the lowest dose rates. A rather complex biochemical model might account for the processes, which determines the effect as function of dose rate, but fig. 5.6 is the result of a toxicological study. The relationship seems reasonable, when decomposition and excretion processes are considered.

Sediments and soils accumulate heavy metals and other toxic substances, and are therefore often used as indicators for pollution studies. Due to the high concentrations found in sediments and soils it is possible to determine the concentrations of toxic substances in them with relatively high accuracy.

Fig. 5.7 and 5.8 show relationships between concentration of heavy metal in animal tissues and in sediments. Such simple models can be used to find concentrations of heavy metals of benthic animals in new sites. Notice in fig. 5.8 that the ratio lead/iron in sediment gives a better correlation than the lead concentration. The iron concentration expresses in this case indirectly the binding capacity of the sediment. It must be expected that pH plays an important role in freshwater (sea water has a pH close to 8.1) for the shown relations.

There are a great number of such examples, in which a sensitive indicator for the impact of pollution is used to construct a simple model. One more example shall be mentioned because it has been widely used in eutrophication management: the application of a Vollenweider plot or model.

Fig. 5.9 illustrates the original Vollenweider model (Vollenweider, 1969), which was based solely on phosphorus loading and mean depth. The result is a semiquantitatively trophic status of the lake, as seen from the figure. The model is semiempirical, as can be considered a specific derivative of general mass - balance equations.

A modified Vollenweider model has taken hydraulic residence into account.The equation used is as follows:

$$C_p = \frac{L_p}{z\left(\frac{1}{t_w} + s\right)}, \tag{5.10}$$

where $C_p$ is the phosphorus concentration of the lake, $L_p$ is phosphorus loading of the lake, z is the mean lake depth, s is the sedimentation coefficient and $t_w$ is the residence time for water (Vollenweider, 1980). The hydraulic residence time can be defined in a number of ways, which are essentially equivalent at steady state. As definition of tw is most often used the ratio of lake volume to total inflow.

By assuming a sedimentation rate of 10-20 m per year it is possible to simplify the equation:

$$C_p = \frac{L_p t_w}{z\left(1 + t_w^{0.5}\right)},$$

(5.11)

$C_p$ can be translated to chlorophyll concentration by use of the following emperical equation (Dillon and Rigler, 1974):

$$\log B = 1.449 \log(C_p) - 0.398$$

(5.12)

where B is chlorophyll a (July-October $\mu$g $1^{-1}$, 0-5m). Cp is here the total phosphorus in the same period (0-15m).

In chapter 7 more complex eutrophication models will be presented, which have the advantages compared with the above mentioned simple models, that they can describe the transient state and can give predictions as to the dynamics of the lake (f.inst. changes in the phytoplankton concentration on a week to week basis). The advantage of the Vollenweider models is, however, that they require a far smaller data base. The problem must determine which approach to use in each case. If f.inst. it is of great importance to be able to give more accurate predictions on the duration of the transient state, a more complex model is needed, which again implies that a more comprehensive data base is required.

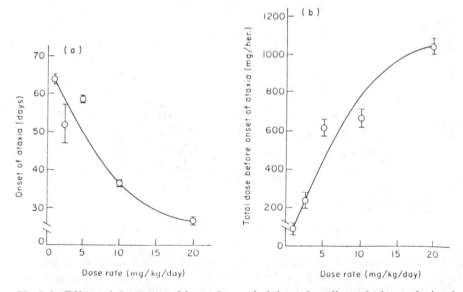

Fig.5.6: Effect of dose rate of leptophos, administered orally each day to laying hens (Gallus domesticus) on (a) time for symptoms of ataxia to appear and(b) total dose ingested before symptoms of ataxia appear. (From Abou-Donia and Preissig, 1976)

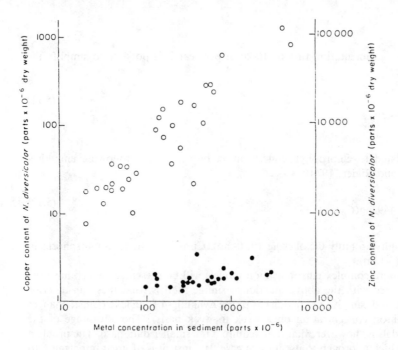

Fig.5.7: Concentration of zinc (o) and copper (o) in the tissues of the polychaete worm Nereis diversicolor (expressed as dry weights of tissue) taken from sites in more than 20 estuaries in Devon and Cornwall, and in the sediments at those sites. (From Bryan, 1976)

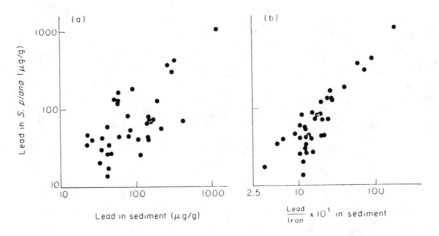

Fig.5.8: Concentrations of total lead ($\mu$g/g dry weight) in the soft tissues of 37 samples of the bivalve Scrobicularia plana collected from 17 estuaries in south and west England plotted against (a) the lead content of sediment particles and (b) the ratio of the concentration of lead to that of iron, multiplied by $10^3$ in sediment particles. (From Luoma and Bryan, 1978).

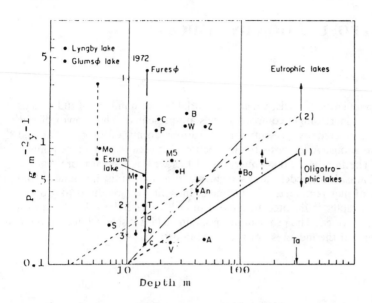

Fig.5.9: a,b and c correspond to removal of 90%, 95% and 99% (Furesoe) of P input respectively. Glumsoe, 1972 (Jørgensen, Jacobsen and Høi, 1973):, Lyngby lake, 1972 (F.L.Smidth/MT, 1973); Esrom lake, 1972.

G - Greifensee    V - Vanern
P - Pfaffikersee   F - Furesoe (1954)
B - Baldeggersee   M - Lake Mendota
W - Lake Washington T - Turlersee
Z - Zurichsee      Mo - Lake Moses
H - Halwillersee   An - Lac Annecy
Bo - Bodensee      L - Lac Leman
A - Aegersee       Ta - Lake Tahoe

## PROBLEMS CHAPTER 5

1. The following information exist for a lake Y: P-loading 0.5 g P $m^{-2}y^{-1}$, depth 3.5 m (average), retention time 2 years. N-loading 1.2 g N $m^{-2}y^{-1}$. What would you say is the trophic state of the lake? Would you prefer a 85% reduction of the P- or N-loading? And what would be the trophic state at the new steady state?

2. A model consists of 4 compartments A, B, C and D. There are constant inputs to A and B: 1 respectively 3 units per time unit. There is an output following a first order equation from D with a rate constant of 0.1 time unit-1. The following flows take place: from A to B and C, from B to C and D, and from C to A and D, and from D to A and B. All flows follow first reaction kinetics. The rate constants are respectively 0.1, 0.2, 0.1, 0.23, 0.15, 0.25, 0.18 and 0.12 all per time unit.Set up matrices models and find the steady state values of A, B, C and D.

# 6. MODELLING POPULATION DYNAMICS

This chapter covers population models, where state variables are numbers of individuals or species. Step by step increasingly complex models are presented. The growth of one population is first considered by presentation of various forms of exponential and logistic growth formulations. Afterwards the interactions between two or more populations are presented. The famous Lotka-Volterra model as well as several more realistic predator-prey and parasitism models are shown.Age distribution is introduced and computations by use of matrix models are illustrated, including the relations to growth. The last part of the chapter is devoted to harvest models, where the role of man on renewable resources is covered. The problem of optimum yield is discussed in relation to a multi species model, but the model is not presented in detail.

## 6.1. BASIC CONCEPTS

This chapter deals with biodemographic models, characterized by the use of numbers of individuals or species as units for state variables.

Already in the nineteen-twenties Lotka and Volterra developed the first population model, which is still widely used to-day, (see Lotka 1956 and Volterra 1926). A number of population models have been developed, tested and analyzed since and it will not be possible in this context to give a comprehensive review of all these models. The chapter will mainly focus on models of age distributions, growth, and species interactions. Only deterministic models will be mentioned. Those, who are interested in stochastic models can refer to Pielou (1969), which gives a very comprehensive treatment of this type of models.

A population is defined as a collective group of organisms of the same species. Each population has several characteristic properties, such as **polulation density (population size relative to available space), natality (birth rate), mortality (death rate), age distribution, dispersion growth forms and others.**

A population is a changing entity, and we are therefore interested in its size and growth. If N represents the number of organisms and t the time, then dN/dt = the rate of change in the number of organisms per unit time at a particular instant (t) and dN/(Ndt) = the rate of change in the number of organisms per unit time per individual at a particular instant (t). If the population is plotted against time a straight line tangential to the curve at any point represents the growth rate.

**Natality** is the number of new individuals appearing per unit of time and per unit of population.

We have to distinguish between absolute natality and relative natality, denoted $B_a$ and $B_s$ respectively:

$$B_a = \frac{\Delta N_n}{\Delta t} \tag{6.1}$$

$$B_s = \frac{\Delta N_n}{N \Delta t} \tag{6.2}$$

where $\Delta N_n$ = production of new individuals in the population.

**Mortality** refers to the death of individuals in the population. The absolute mortality rate, $M_a$, is defined as:

$$M_a = \frac{\Delta N_m}{\Delta t} \qquad (6.3)$$

where $\Delta N_m$ = number of organisms in the population, that died during the time interval $\nabla t$, and the relative mortality, $M_s$, is defined as:

$$(6.4)$$

$$M_s = \frac{\Delta N_m}{\Delta t \cdot N}$$

## 6.2. GROWTH MODELS

Growth models consider only one population. Its interactions with other populations are taking into consideration by the specific growth rate and the mortality, which might be dependent on the magnitude of the considered population but independent of other populations. In other words we consider only one population as state variable.

The simplest growth model assume *unlimited resources and exponential population growth*. A simple differential equation can be applied:

$$(6.5)$$

$$dN/dt = B_s \times N - M_s \times N = r \times N$$

where $B_s$ is the instantaneous birth rate per individual, $M_s$ the instantaneous death rate, $r = B_s - M_s$, N the population density and t the time. As seen the equation is equal to first order kinetics, see section 2.9. After integration we get:

$$(6.6)$$

$$N_t = N_o \times e^{rt}$$

where $N_t$ is the population density at time t and $N_o$ the population density at time 0. Figs. 6.1 and 6.2 illustrate the exponential growth in graphs.

By taking the logarithms, equation (6.6) can be transformed to:

$$(6.6a)$$

$$\ln(N_t/N_o) = r.t$$

The net reproductive rate, $R_o$, is defined as the average number of age class zero offspring produced by an average newborn organism during its entire lifetime.

Survivorship $l_x$ is the fraction surviving at age x. It is the probability that an average newborn will survive to that age, designed x. The number of offspring produced by an average organism of age x during the age period is designated $m_x$. This is termed **fecundity**, while the product of $l_x$ and $m_x$ is called the realized fecundity. In accordance with its definition $R_o$ can be found as:

$$R_0 = \int_0^\infty l_x m_x dx$$

(6.7)

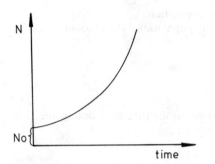

Fig.6.1: Growth in accordance with $dN/dt = r \times N$ $(r > o)$ corresponding to exponential growth.

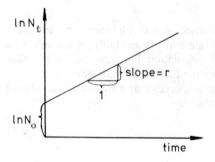

Fig.6.2: In $N_t$ versus the time t. In $N_t = \ln N_o + r \times t$.

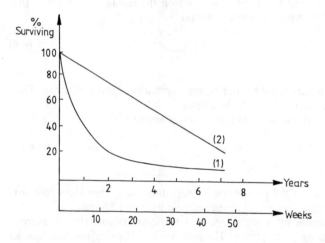

Fig.6.3: Survivorships of (1) the lizard Uta (the lower x-axis) and (2) the lizard Xantusia (the upper x-axis). After Reevey (1947), Tinkle (1967), Zweifeland Lowe (1966)

**TABLE 6.1**

**Estimated Maximal Instantaneous Rate of Increase (rmax, Per Capita Per Day) and Mean Generation Times (in Days) for a Variety of Organisms.**

| Taxon | Species | $r_{max}$ | Generation Time T |
|---|---|---|---|
| Bacterium | Escherichia coli | ca.60.0 | 0.014 |
| Algae | Scenedesmus | 1.5 | 0.3 |
| Protozoa | Paramecium aurelia | 1.24 | 0.33 - 0.50 |
| Protoza | Paramecium caudatum | 0.94 | 0.10 - 0.50 |
| Zooplankton | Daphia puxex | 0.25 | 0.8 - 2.5 |
| Insect | Tribolium confusum | 0.120 | ca. 80 |
| Insect | Calandra oryzae | 0.110(.09-.11) | 58 |
| Insect | Rhizopertha dominica | 0.085(.07-.10) | ca. 100 |
| Insect | Ptinus tectus | 0.057 | 102 |
| Insect | Gibbium psylloides | 0.034 | 129 |
| Insect | Trigonogenius globulus | 0.032 | 119 |
| Insect | Stethomezium squamosum | 0.025 | 147 |
| Insect | Mezium affine | 0.022 | 183 |
| Insect | Ptinus fur | 0.014 | 179 |
| Insect | Eurostus hilleri | 0.010 | 110 |
| Insect | Ptinus sexpunctatus | 0.006 | 215 |
| Insect | Niptus hololeucus | 0.006 | 154 |
| Octopus | - | 0.01 | 150 |
| Mammal | Rattus norwegicus | 0.015 | 150 |
| Mammal | Microtus aggrestis | 0.013 | 171 |
| Mammal | Canis domesticus 0.009 | ca. 1000 | |
| Insect | Magicicada septendecim | 0.001 | 6050 |
| Mammal | Homo sapiens | 0.0003 | ca. 7000 |

If we in equation (6.7) set $R_o$ to 1 and call the generation time T, then $R_o$ can be found as:

$$(6.8)$$

$$\ln R_o = \ln 1 + r \times T = r \times T$$

A curve which indicates $l_x$ as function of age is called a survivorship curve. Such curves differ significantly for various species as illustrated in fig. 6.3.

The so-called **intrinsic rate of natural increase, r,** is alike $l_x$ and $m_x$ dependent on the age distribution, and only time independent when the age distribution is stable. When $R_o$ is as high as possible - it means under optimal conditions and with a stable age distribution, the maximal rate of natural increase is realized and designated $r_{max}$. Among various animals it ranges over several orders of magnitude, see table 6.1.

The value of r is cumbersome to calculate, but can be determined by use of the equation:

$$\sum_x e^{-rx} \cdot l_x \cdot m_x = 1 \qquad (6.9)$$

Derivation of this equation can be found in Mertz (1970) or Eulen (1973).

If $R_o$ is close to one, r can be estimated using the following approximate formula

$$r \simeq \frac{\ln R_0}{T} \qquad (6.10)$$

**The Reproductive value,** $V_x$, is defined as the age-specific expectation of future off-spring. In a stable population at equilibrium, it is implied that:

$$V_x = \int_x^\infty \frac{l_t}{l_x} \cdot m_t \cdot dt \qquad (6.11)$$

The exponential growth is a simplification, which is only valid in a certain time interval. Sooner or later every population must encounter the limitation of food, water, air or space, as the world is finite. To account for this we introduce the concept of **carrying capacity, K,** defined as the density of organisms at which $R_o$ equals unity and r is zero. At zero density Ro is maximal and r becomes $r_{max}$. The simplest assumption we can make is that r decreases linearly with N and becomes zero, when N is equal to K. This leads to the classical Verhulst-Pearl **logistic growth equation:**

$$\qquad (6.12)$$

$$dN/dt = rN(K-N)/K$$

In fig. 6.4 the population growth is illustrated by the s- shape logistic expression. When N is little compared with K the growth will be approximately exponential.

The application of the logistic growth equation requires three assumptions:

1) That all individuals are equivalent. The constant K and r are independent on time and age distribution.

2) That K and r are immutable constants independent of time, age distribution etc.

3) That there is no time lag in the response of the actual rate of increase per individual to changes in N.

All three assumptions are unrealistic and can be strongly critized. Nevertheless a number of population phenomena can be nicely illustrated by use of the logistic growth equation.

Equation (6.12) can be solved analytically:

$$N = K/(1 + e^{(a-rt)}) \qquad (6.13)$$

where $a = \ln \frac{K-N}{N}$ when $t = 0$, it means $a = \ln \frac{K-N_0}{N_0}$

## Example 6.1

An algal culture shows a carrying capacity due to the self-shading effect. In spite of "unlimited" nutrients the maximum concentration of algae in a chemostate experiment was measured to be 120 g/m³. At time 0 was introduced 0.1 g/m³ of algae and 2 days after was observed a concentration of 1 g/m³. Set up a logistic growth equation for these observations.

**Solution:**

The first 5 days we are far from the carrying capacity and we have with good approximations:

$$\ln 10 = r_{max} \cdot 2$$

$$r_{max} = 1.2 \text{day}^{-1}$$

and since the carrying capacity is 120 g/m³, we have: (C = algae concentration)

$$\frac{dC}{dt} = 1.2 \cdot C \frac{120 - C}{C}$$

or

$$C = \frac{C}{1 + e^{a-1.2 \cdot t}}$$

where    $a = \ln((120 - 0.1)/0.1) = 7.09$

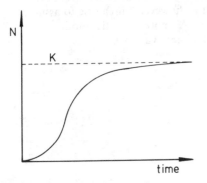

Fig.6.4: Logistic growth.

This simple situation in which there is a linear increase in the environmental resistance with density seems to hold good only for organisms that have a very simple life history.

**In populations of higher plants and animals, that have more complicated life histories, there is likely to be a delayed response.** Wangersky and Cunningham (1956 and 1957) have suggested a modification of the logistic equation to include two kinds of time lags: 1) the time needed for an organism to start increasing, when conditions are favourable, and 2) the time required for organisms to react to unfavourable crowding by altering birth- and death rates. If these time lags are t - $t_1$ and t - $t_2$ respectively, we get:

$$\frac{dN(t)}{dt} = rN_{(t-t_1)} \cdot \frac{K - N_{(t-t_2)}}{K} \qquad (6.14)$$

Population density tends to fluctuate as a result of seasonal changes in environmental factors or due to factors within the populations themselves (so-called intrinsic factors). We shall not go into details here but just **mention that the growth coefficient is often temperature dependent and since temperature shows seasonal fluctuations, it is possible to explain some of the seasonal population fluctuation in density in that way.**
Another modified logistic equation was introduced by Smith (1963):

$$\qquad (6.15)$$

$$\frac{dN}{dt} = r \cdot N \frac{K - N}{K + \frac{r}{C} \cdot N}$$

where

C = the rate of replacement of biomass in the population at saturation density.

## 6.3. INTERACTIONS BETWEEN POPULATION

The growth models presented in section 6.2 might have a constant influence from other populations reflected in the selection of parameters. It is, however, unrealistic to assume that interactions between populations are constant. A more realistic model must therefore contain the interacting populations (species) as state variables:

$$\frac{dN_1}{dt} = r_1 \cdot N_1 \left( \frac{K_1 - N_1 - \alpha_{12}N_2}{K_1} \right) \qquad (6.16)$$

$$\frac{dN_2}{dt} = r_2 \cdot N_2 \left( \frac{K_2 - N_2 - \alpha_{21} \cdot N_1}{K_2} \right) \qquad (6.17)$$

where $\alpha_{12}$ and $\alpha_{21}$ are competition coefficient. $K_1$ and $K_2$ are carrying capacities for species 1 resp. 2. $N_1$ and $N_2$ are numbers of species 1 and 2, while $r_1$ and $r_2$ are the corresponding maximum intrinsic rate of natural increase.
The steady-state situation is found by setting (6.16) and (6.17) equal to zero. We get:

$$N_1 = K_1 - \alpha_{12} \cdot N_2 \tag{6.18}$$

$$N_2 = K_2 - \alpha_{21} \cdot N_1 \tag{6.19}$$

These two linear equations are plotted in fig. 6.5, giving dN/dt isoclines for each species. Below the isoclines populations will increase, above them they decrease. Consequently four result, as illustrated in fig. 6.5. The four cases are also summarized in table 6.2.

The equation can also be written in a more general form for a community composed of n different species:

$$\frac{dN_i}{dt} = r_i N_i \left( \frac{K_i - N_i - \left( \Sigma_{j \neq i}^n \alpha_{ij} N_j \right)}{K_i} \right) \tag{6.20}$$

where iṡ and jṡ are subscript species and range from 1 to n. At steady state $dN_1/dt$ is equal to zero for all i and

$$N_i \equiv N_{ie} = K_i - \sum_{j \neq i}^n \alpha_{ij} N_j \tag{6.21}$$

Lotka-Volterra wrote a simple pair of **predation equations:**

$$\frac{dN_1}{dt} = r_1 \cdot N_1 - p_1 N_1 \cdot N_2 \tag{6.22}$$

$$\frac{dN_2}{dt} = p_2 \cdot N_i \cdot N_2 - d_2 \cdot N_2 \tag{6.23}$$

where $N_1$ is prey population density, $N_2$ predator population density, $r_1$ is the instantaneous rate of increase of the prey population (per head), $M_s$ is the mortality of the predator (per head) and $p_1$ and $p_2$ are predation coefficients.

**TABLE 6.2**

**Summary of the Four Possible Cases of Lotka-Volterra Competition Equations.**

| | Species 1 Can Contain Species 2 $(K_2/\alpha_{21} < K_1)$ | Species 2 Cannot Contain Species 2 $(K_2/\alpha_{21} > K_1)$ |
|---|---|---|
| Species 2 can contain Species 1 $(K_1/\alpha_{12} < K2)$ | Either species can win (Case 3) | Species 2 always wins (Case 2) |
| Species 2 cannot contain Species 1 $(K_1/\alpha_{12} > K_2)$ | Species 1 always wins (Case 1) | Neither species can contain the other: Stable coexistence (Case 4) |

Each population is limited by the other and in absence of the predator the prey population increases exponentially. By setting the two differential equations to zero, we find:

$$N_2 = \frac{r_1}{p_1} \tag{6.24}$$

$$N_1 = \frac{d_2}{p_2} \tag{6.25}$$

Thus each species isocline corresponds to a particular density of the other species. Below some threshold prey density, predator always decrease, whereas above that threshold they increase. Similarly prey increase below a particular predator density but decrease above it, see fig. 6.6. A joint equilibrium exists where the two isoclines cross, but prey and predator densities do not converge on this point.

Any given initial pair of densities results in oscillations of a certain magnitude. The amplitude of fluctuations depends on the initial conditions.

These equations are unrealistic since most populations encounter either *self-regulations or density -dependent feedbacks or both*.

By addition of a simple self-damping term to the prey equation results either in a rapid approach to equilibrium or in damped oscillations. Perhaps a more realistic pair of simple equations for modelling the **prey-predator relationship** is:

$$\frac{dN_1}{dt} = r_1 \cdot N_1 - z_1 \cdot N_1^2 - \beta_{12} \cdot N_1 \cdot N_2 \tag{6.26}$$

$$\frac{dN_2}{dt} = \gamma_{21} \cdot N_1 \cdot N_2 - \beta_2 \cdot \frac{N_2^2}{N_1} \tag{6.27}$$

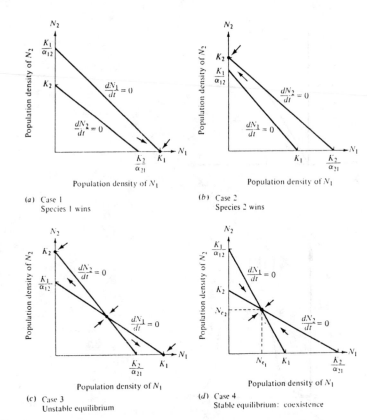

(a) Case 1
Species 1 wins

(b) Case 2
Species 2 wins

(c) Case 3
Unstable equilibrium

(d) Case 4
Stable equilibrium: coexistence

Fig.6.5 a,b,c,d

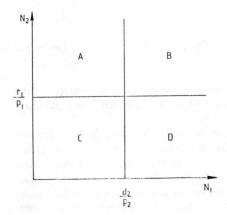

Fig.6.6: Prey-predator isoclines for Lotka-Volterra prey-predator equation. A: both species decrease B: predator increase, prey decrease, C: prey increase, predator decrease, p: both species increase.

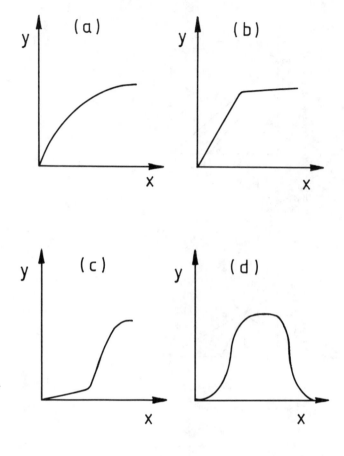

Fig.6.7: Four functional responses (Holling, 1959). y is number of prey taken per predator per day and x is prey density.

As seen the prey equation is a simple logistic expression combined with the effect of the predator. The predator expression considers a carrying capacity, which is dependent on the prey concentration,

However, these equations can also easily be critized. The growth term for the predator is as seen just a linear function of the prey concentration of density. Other possible relations are shown in fig. 6.7. The first relation (fig. 6.7.a) corresponds to a *Michaelis-Mentens expression*, while the second relation only approximates a Michaelis-Menten expression by use of *a first order expression in one interval and a zero order expression in another*. The third relation shown in fig. 6.7 corresponds to a *logistic expression:* with increasing prey density the predator density first grows exponentially and afterwards a damping takes place. This relation is observed in nature and might be explained as follows: the energy and time used by the predator to capture a prey is decreasing with increasing density of the prey. This implies that not only can the predator capture more prey due to increasing density, but also less of the energy consumed is used to capture the next prey. Thus, the density of the predator increases more than proportionally to the prey density in this phase. However, there is a limit to the food (energy) that the predator can consume and at a certain density of the prey, further decrease in the energy used to capture the prey cannot be obtained. Consequently the increase in predator density slows down as it reaches a saturation point at a certain prey density. The fourth relationship corresponds to the often found relation between growth and pH or temperature. It is here characteristic that the *predator density decreases above a certain prey density.* This response might be explained by effect on the predator of the waste produced by the prey. At a certain prey density the concentration of waste is sufficiently high to have a pronounced negative effect on the predator growth.

Holling (1959 and 1966) has developed more elaborate models of prey-predator relationships. He incorporated time lags and hunger levels to attempt to describe the situation in nature. These models are more realistic, but they are also more complex and require knowledge of more parameters.

In addition to these complications we have coevolution of predators and preys. The prey will develop better and better techniques to escape the predator and the predator will develop better and better techniques to capture the preys. To account for the coevolution it is necessary to have a current change of the parameters in accordance with the current selection, that takes place.

**Parasitism** is similar to that of predation, but differs from the latter in that members of the prey species affected are seldom killed, but may live for some time after becoming parasitized. This is accounted for by relating the growth and the mortality of the prey, $N_1$, to the density of the parasites, $N_2$. The carrying capacity for the parasites is furthermore dependent on the prey density.

The following equations account for these relations and include a carrying capacity of the prey:

$$\frac{dN_1}{dt} = \frac{r_1}{N_2} \cdot N_1 \left( \frac{K_1 - N_1}{K_1} \right) \tag{6.28}$$

$$\frac{dN_2}{dt} = r_2 \cdot N_2 \left( \frac{K_2 \cdot N_1 - N_2}{K_2 \cdot N_1} \right) \tag{6.29}$$

Symbiotic relationships are easily modelled with expressions similar to the Lotka-Volterra competition equations simply by changing the signs for the interaction terms:

$$\frac{dN_1}{dt} = r_1 \cdot N_1 \left( \frac{K_1 - N_1 + \alpha_{12}N_2}{K_1} \right) \tag{6.30}$$

$$\frac{dN_2}{dt} = r_2 \cdot N_2 \left( \frac{K_2 - N_2 - \alpha_{21}N_1}{K_2} \right) \tag{6.31}$$

In nature interactions among populations become often intricate. The expressions presented above might be of great help in understanding population reactions in nature, but when it comes to the problem of modelling entire ecosystems, they are in most cases insufficient. **Investigations of stability criteria for Lotka-Volterra equations are an interesting mathematical exercise, but can hardly be used to understand the stability properties of real ecosystems or even of populations in nature.**

The experience from investigations of population stability in nature indicates that it is needed to account for many interactions with the environment to be able to explain observations in real systems, see f.inst. Colwell (1973).

## ILLUSTRATION 6.1

This illustration concerns a anaerobic cultivation of two species of yeast, first described by Gause (1934). The two species are Saccharomyces cerevisiae (Sc) and Schizosacch aromyces (Kephir) (K). Gause cultivated both species in monocultures and also in mixture and the results suggest that the two species had a mutual effect upon each other. His hypothesis was that a production of harmful waste products (alcohols) was the only cause of interactions.

A conceptual diagram for the model to use is shown fig. 6.8. The model has three state variables: the two yeast species and the waste products. The amount of waste products depends on the growth of yeast. The growth of the yeast species depends on the amount of yeast and the growth rate of the yeast, which is again dependent on the species and a reduction factor, which account for the influence of the waste products on the growth. A CSMP-program partly after De Wit and Goudriaan (1974) is presented in table 6.3. The observed and computed values for growth of the two yeast species are shown table 6.4. As seen the fit between observed and calculated values is completely acceptable for the monoculture experiments, but is completely unacceptable for the mixed culture experiments. It can be concluded that the two species do not interfere solely through the production of alcohol. Additional biological knowledge about the interference between the two species must be introduced to the model to be able to explain the observations.

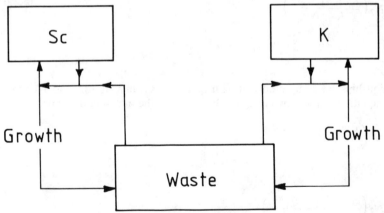

Fig.6.8: Conceptual diagram of the model presented in illustration 6.1. Waste is alcohol affecting the growth of two yeast species Sc and K.

**TABLE 6.3**

**CSMP-Program for Growth and Interference of two Yeast Species.**

```
TITLE MIXED CULTURE OF YEAST
     Y1 = INTGRL (IY1, RY1)
     Y2 = INTGRL (IY2, RY2)

INCON IY1 = 0.45, IY2 = 0.45
     RY1 = RGR1 * Y1 * (1. - RED1)
     RY2 = RGR2 * Y2 * (1. - RED2)

PARAMETER  RGR1 = 0.236, RGR2 = 0.049
     RED1 = AFGEN (RED1T, ALC/,MALC)
     RED2 = AFGEN (RED2T, ALC/MALC)

FUNCTION  RED1T = (0., 0.), (1., 1.)

FUNCTION  RED2T = (0., 0.), (1., 1.)

PARAMETER  MALC = 1.5
     ALC = INTGRL (ALC, ALCP1 + ALCP2)
     ALCP1 = ALPF1 * 1
     ALCP2 = ALPF2 * RY2

PARAMETER  ALPF1 = 0.122, ALPF2 = 0.270

INCON     IALC = 0.

FINISH     ALC = LALC
     LALC = 0.99 * MALC

TIMBER   FINTIM = 150., OUTDEL 2.

PRTPLT   Y1, Y2, ALC

END

STOP
```

## TABLE 6.4

**Observed and calculated values for growth of two species of yeasts in monocultures and mixtures.**

Schizosaccharomyces 'Kephir'.

| | Volume of yeast (arbitrary units) | | | |
| --- | --- | --- | --- | --- |
| | Monoculture | | Mixed | |
| Hours | Observed | Calculated | Observed | Calculated |
| 0 | 0.45 | 0.45 | 0.45 | 0.45 |
| 6 | - | 0.60 | 0.291 | 0.59 |
| 16 | 1.00 | 0.95 | 0.98 | 0.81 |
| 24 | - | 1.34 | 1.47 | 0.88 |
| 29 | 1.70 | 1.64 | 1.46 | 0.89 |
| 48 | 2.73 | 3.04 | 1.71 | 0.89 |
| 53 | - | 3.44 | 1.84 | 0.89 |
| 72 | 4.87 | 4.72 | - | - |
| 93 | 5.67 | 5.51 | - | - |
| 117 | 5.80 | 5.86 | - | - |
| 141 | 5.83 | 5.96 | - | - |

Saccharomyces cerevisiae.

| Hours | Observed | Calculated | Observed | Calculated |
| --- | --- | --- | --- | --- |
| 0 | 0.45 | 0.45 | 0.45 | 0.45 |
| 6 | 0.37 | 1.72 | 0.375 | 1.70 |
| 16 | 8.87 | 8.18 | 3.99 | 7.56 |
| 24 | 10.66 | 11.83 | 4.69 | 10.86 |
| 29 | 12.50 | 12.46 | 6.15 | 11.47 |
| 40 | 13.27 | 12.73 | - | 11.75 |
| 48 | 12.87 | 12.74 | 7.27 | 11.77 |
| 53 | 12.70 | 12.74 | 8.30 | 11.77 |

### 6.4. MATRIX MODELS

Another important aspect of modelling population dynamics is the influence of the age distribution, which indicates the proportion of the population belonging to each age class.

Fig. 6.9 shows typical age distribution curves.

Two populations with identical $l_x$- and $m_x$-values but with different age distributions, will grow differently. If a population has unchanged $l_x$ and $m_x$ schedules, it will eventually reaches a stable age distribution, it means that the the percentage of organisms in each age class remains the same. Recruitment into every age class is exactly balanced by its loss due to mortality and aging.

Equations (6.6), (6.12), (6.14) and (6.15) assume that the population has a stable age distribution. The intrinsic rate of increase, r, the generation time, T, and the reproductive value, $v_x$, are conceptually independent of the age distribution, but might of course be different for population of the same species with different age distributions. Therefore the models presented in the two previous paragraphs did not need to consider age distribution, although the parameters in actual cases reflect the actual age distribution.

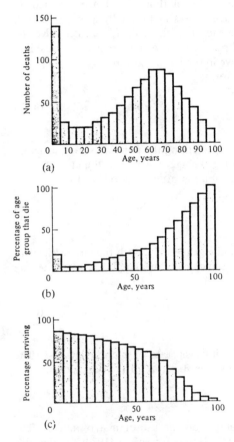

(a)

(b)

(c)

Fig.6.9: Typical age distribution curves (a), frequency distribution at death (b), force of mortality in various age growth (c), survivorship of various age groups.

A model predicting the future age distribution, was developed by Lewis and Leslie (1942). The population is divided into n + 1 equal age groups - group 0, 1, 2, 3,......n. The model is then presented by the following matrix equation:

$$
\begin{vmatrix}
f_o\ f_1\ f_2 & \cdots & f_{n-1}\ f_n \\
p_o\ o\ o & \cdots & o\ o \\
o\ p_1\ o & \cdots & o\ o \\
\cdots & \cdots & \cdots \\
\cdots & \cdots & \cdots \\
\cdots & \cdots & \cdots \\
o\ o\ o & \cdots & p_{n-1}\ o
\end{vmatrix}
\begin{vmatrix}
n_{t,o} \\
n_{t,1} \\
n_{t,2} \\
. \\
. \\
. \\
n_{t,n}
\end{vmatrix}
=
\begin{vmatrix}
n_{t+1,0} \\
n_{t+1,1} \\
n_{t+2,2} \\
. \\
. \\
. \\
n_{t+1,n}
\end{vmatrix}
\qquad (6.32)
$$

The number of organisms in the various age classes at time $t + 1$ are obtained by multiplying the numbers of animals in these age classes at time t by a matrix, which expresses the fecundity and survival rates for each age class. $f_0$, $f_1$, $f_2$......$f_n$ give the reproduction in the ith age group and $f_0$, $p_1$, $p_2$, $p_3$, $p_4$.....$p_n$ represent the probability that an organism in the $i^{th}$ age group will be alive in the $i + 1^{th}$ group.

The model can be written in the following form:

$$(6.33)$$

$$\mathbf{A} \cdot \mathbf{a}_t = \mathbf{a}_{t+i}$$

where A is the matrix, $a_t$ is the column vector representing the population age structure at time t and $a_{t+1}$ is a column vector representing the age structure at time $t + 1$. This equation can be extended to predict the age distribution after k periods of time:

$$(6.34)$$

$$\mathbf{a}_{t+k} = \mathbf{A}^k \cdot \mathbf{a}_t$$

The matrix A has $n + 1$ possible eigenvalues and eigenvectors. Both the largest eigenvalues,$\lambda$, and the corresponding eigenvectors are ecologically meaningful. $\lambda$ gives the rate at which the population size is increased:

$$(6.35)$$

$$\mathbf{A} \cdot \mathbf{v} = \lambda \cdot \mathbf{v}$$

where v is the stable age structure. $\ln\lambda$ is the intrinsic rate of natural increase. The corresponding eigenvector indicates as seen the stable structure of the population.

## Example 6.2

Usher (1972) has given a very illustrative example on the use of matrix models. The model is based upon data provided by Laws (1962) and Ehrenfeld (1970) for the blue whale before its extinction and sharp changes in survival rates.

The eigenvalue can be used to find the number of individuals that can be removed from a population to maintain the same number in each age class. It can be shown that the following equation is valid:

$$H = 100\left(\frac{\lambda - 1}{\lambda}\right),$$

where H is the percentage of the population, which can be removed.

The blue whales reach maturity at between four and seven years of age. They have a gestation period of about one year. A single calf is born and is nursed for about seven months. Averagely not more than one calf is born to a female every two years. The numbers of the two sexes are approximately equal. Survival rates are about 0.7 each two years for the first ten years and 0.78 for whales above 12 years. We divide the population into 7 groups with a two- year period for the first six groups and the age of 12 years and above as the seventh group. The fecundity for the first two groups is in accordance with the information about zero. The third group has a fecundity of 0.19 and the fourth group of 0.44. The maximum fecundity of 0.50 is reached at the age of 8-11 years. The fecundity of the last group is 0.45.

Find the intrinsic rate of natural increase, the stable structure of the whale population and the harvest, which can be taken to maintain a stable population size.

## Solution:

The eigenvalue can be found either by an iterative method or by plotting the number of whales (totally or for each age class separately) versus the period of time. The slope of this plot will after a stabilization period correspond to r, the intrinsic rate of increase, or $\ln\lambda$. We find by these methods that $r = 0.0036$ year$^{-1}$ or $\lambda$ = antilog $0.0036 = 1.0036$ (for one year) or $1.0036^2 = 1.0072$ for two years. The corresponding eigenvector is found by use of equation (6.35) to be:

$$a = [1000, 764, 584, 447, 341, 261, 885]$$

as the Leslie matrix is:

$$\begin{vmatrix} 0 & 0 & 0.19 & 0.44 & 0.50 & 0.50 & 0.45 \\ 0.77 & 0 & 0 & 0 & 0 & 0 & 0 \\ 0 & 0.77 & 0 & 0 & 0 & 0 & 0 \\ 0 & 0 & 0.77 & 0 & 0 & 0 & 0 \\ 0 & 0 & 0 & 0.77 & 0 & 0 & 0 \\ 0 & 0 & 0 & 0 & 0.77 & 0 & 0 \\ 0 & 0 & 0 & 0 & 0 & 0.77 & 0 \\ 0 & 0 & 0 & 0 & 0 & 0.77 & 0.78 \end{vmatrix}$$

The harvest that can be taken from the population is estimated to be:

$$H = 100 \frac{\lambda - 1}{\lambda}\%$$
$$= 0.71\% \text{ every two years or about } 0.355\% \text{ every year.}$$

If the harvest exceeds this value the population will decline. Population models of r-strategies might generally cause some more difficulties to develop than models of K-strategies due to the high sensitivity of the fecundity. The number of off-springs might be known quite well, but the number of survivors to be included in the first age class, the number of recruits, is difficult to predict. This is the central problem of fish population dynamics, since it represents natures regulation of population size (Beyer, 1981).

Beverton and Holt (1957) have suggested the following equation for recruitment:

$$R = \left(\frac{E}{E + yR_{max}}\right) R_{max} \qquad (6.36)$$

where the number of recruits (R) increases towards an asymtotic level $R_{max}$, when the egg production, E, increases.

Ricker (1954) suggests another equation, in which the number of recruits decreases from a maximum level towards zero as the production of eggs decreases:

$$R = R_1 \cdot E e^{-R_2 \cdot E} \tag{6.37}$$

Here $R_1$ and $R_2$ are constants. The decline of recruitment is explained by cannibalism by adults.

The graphic representation of the two different approaches is shown in fig. 6.10.

The need for models of recruitment is obvious and much effort is devoted to achieve a better quantitative description of this process in fishery management models, see Beyer and Sparre (1983).

Often are population models coupled with growth models, which attempt to predict the growth of the individuals. As the population model gives the number and the growth model the weight, the total biomass is easily found for the different age classes.

The von Bertalanffy equation is quite widely used to describe the growth of fish species and it constitutes the growth element of the classical Beverton and Holt (1957) one species model:

$$\frac{dw}{dt} = Hw^{2/3} - kw \tag{6.38}$$

w is the weight, H and K are constants.

The value 2/3 is based upon the assumption that the area of the intestine is proportional to the surface area and thereby to weight in 2/3. Similar relations have already been mentioned in section 2.7 under parameter estimation, as it has been found that many parameters are closely related to the surface of organisms, which again is approximately related to the weight 2/3.

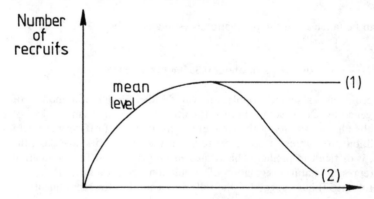

Fig.6.10: Recruitment curves. (1) is Beverton and Holt's equation, (2) is Ricker's.

Equation (6.38) can be solved analytically:

$$w(t) = w_\infty(1 - \exp(-K(t - t_0)))^3; w_\infty = \left(\frac{H}{k}\right)^3 \quad \text{and} \quad K = \frac{k}{3} \qquad (6.39)$$

Growth in weight can be derived into growth in length, l, in accordance to the similar-body assumption:

$$(6.40)$$

$$l(t) = L_\infty(1 - \exp(-K(t - t_0))); \quad L_\infty = \frac{H}{kq^{1/3}} \quad \text{and} \quad K = \frac{k}{3}$$

q is the condition factor. For fish of ordinary shape we find q = 0.01 g/cm$^3$. The von Bertalanffy equation is a special version of the more general growth model developed by Ursin 1967, 1979 and Andersen and Ursin 1977):

$$(6.41)$$

$$\frac{dw}{dt} = Hw^m - kw^n - \text{out}_{\text{span}}(t)$$

H, m, k and n are constants. OUT$_{\text{spawn}}$(t) gives the rate at which eggs flow out of the mature female at time t.

Fig. 6.11 illustrates the growth of North Sea herring with seasonal variations due to spawning and to variations in growth rate. The smooth curve is calculated from w(t) = 262(1 - exp(-0.46))$^3$. As seen these simple growth expressions describe the growth satisfactorily and if the spawning term is included, it is even possible to describe the observed seasonal variations quite accurately.

Observations show that the average spawning rate of the fish gradually increases to a maximum, occurring at some age ts, after which it begins to level off until it again reaches zero, see fig. 6.12. The result is an s-shaped curve for accumulated weight loss due to spawning, when plotted versus time.

Fig. 6.13 demonstrates the differences of five different growth curves among increasing complexity. As always the choice of model is dependent on the data available and the scope of the model. Curve A, B and C have already been mentioned above. Curve D incorporate continuous spawning by use of a Gauss expression with 68% of the spawning accomplished in 22 days. Finally curve E consider a temperature dependence of H and k, which is in accordance with observations. A temperature response similar to what is shown fig. 6.14 is introduced.

A fishery management model must in most cases include several species due to the interest in fishing for several species and due to the interactions between the species. This makes a fishery management model rather complex compared with most other models. In addition the model should also consider the optimum yield problem, which will be covered later in this chapter. However, the problem will be touched on in the illustration 6.2 and a simple approach to including the effect of fishing effort will be shown.

The mortality by fishing is accounted for by use of the following expression for the survival coefficient:

$$S = 1 - M - F, \tag{6.42}$$

where S is the survival coefficient, M the natural mortality and F the mortality by fishery. Both M and F are expressed as the fraction removed per unit of time steps.

This simplification is possible, because the year class and the time step cover the same period of time.

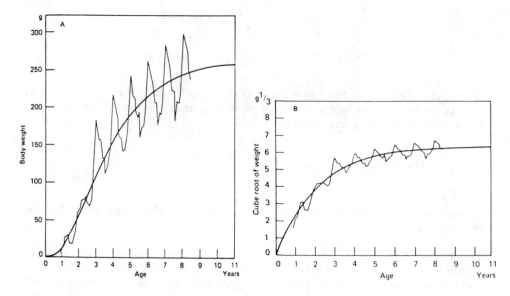

Fig.6.11: Growth of North Sea herring (Clupea harengus) with seasonal variation due to spawning and to variations in growth rate. A: smooth curve calculated from w(t) = 262 (1 - exp(-0.46t))³. B: smooth curve resembles a curve of growth in length.

The example includes also the effect of water quality on the fish mortality. The oxygen concentration, the ammonium concentration and the temperature influence the mortality rate. The influence is incorporated in the model bu use of either a table or an equation. This approach to cover the mortality process is deterministic. Stochastic mortality consideration is also possible, see Beyer and Sparre, 1983.

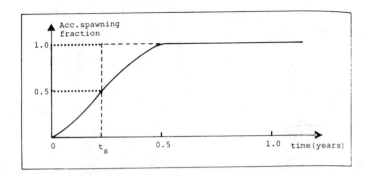

Fig.6.12: Fraction of total nomber of eggs spawned (during the spawning season) against time. Spawning considered as a continuous process is represented by the heavy outlined S-curve, whereas the approximation of spawning as a discontinuous event is the dashed line.

Fishery models are in principle not different from other models of renewable resources such as these used in forestry and agriculture. Some of the concepts have been illustrated by use of fishery models, because models have been widely used in this field and it is more illustrative to use concrete examples. The same basic ideas can, however, be used in all models of renewable resources and are also valid for the ideas and concepts, that will be presented in the next section by use of a more detailed harvest model than presented in illustration 6.2

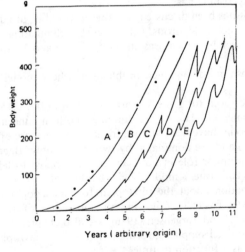

Fig.6.13: North Sea plaice, Pleuronectes platessa. Five growth curves of increasing complexity from left to right. A: von Bartalanffy curve from Beverton & Holt (1957). In curves B-E the exponents are adjusted to coincide with respiration experiments. B: spawning implicit in catabolism. C: instant spawning once a year. D: continuous spawning over a couple of months. E: as curve D, but taking into account seasonal temperature variation.

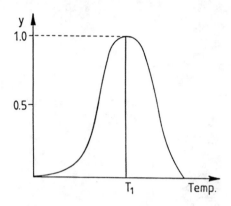

Fig.6.14: $y = \frac{H}{H_{max}}$ or $\frac{k}{k_{max}}$ as function of temperature. $T_1$ is the optimum temperature corresponding to $H = H_{max}$ and $k = k_{max}$. $T_1$ might be different for $H$ and $K$. Notice that the move is more steep for $T > T_1$ than for $T < T_1$.

## ILLUSTRATION 6.2

A Fishery model for Lake Victoria

Ref. Jørgensen et al., 1983.

For many years the fishery in Lake Victoria has been approximately in balance with a steady fish population. (Fishery Statistics, see Bergstrand et al., 1971.Chilver et al., 1974, Garrod (1960) and EAFRO annual reports). However, with the increasing population in the area it will be of great importance to obtain an optimum yield in the years to come and avoid overfishing.

In this context it has been discussed, whether it would be of advantage for the fishery to consentrate on the herbivorous fish types by eliminating the carnivorous types including the Nile Perch, which was introduced in Lake Victoria some decades ago (Hamblyn, 1966).

With the purpose of solving such problems a fishery model for Lake Victoria has been set up.

A simplified food web for Lake Victoria is shown in fig. 6.15. The two herbivorous species, Haplochromis and Tilapia are dominant both in concentration and in fishery. The carnivorous species have been lumped together, as they are of less importance to the fishery (about 12%) and are dominated by two species only: Bagrus and Nile Perch. Zooplankton play a minor role in Lake Victoria compared to lakes in temperature- or subtropic regions. The fishery model is a submodel of a more comprehensive model, which gives information about the phytoplankton and zooplankton concentrations (a eutrophication model basically following the principles of the model published by Jørgensen et al. (1978) and Jørgensen (1976), see also chapter 7.

The eutrophication model supplies information about the phytoplankton- and zooplankton concentrations as a function of time.

The state variables in the fish submodels are number per m3 and weight for each age class (1 age class = 1 ring year = 6 months) of Haplochromis, Tilapia and carnivorous fish averaged over one ring year.

Intensive studies during the last decades have provided information about the characteristics of the fish species in Lake Victoria, which are included in the model as table functions. A list of these tables is shown in table 6.5 with reference to the source of information. Table 6.6 contains a list of the symbols used.

A table is not used to show the relation between length and weight for Haplochromis, but the following equation is available.

(6.43)

$$WH(k) = 0.02 . LH(k)^{3.18}$$

The fish concentration as kg per m³ for each age class is of course easily found from the number of fish per m³ multiplied with the weight of one fish.

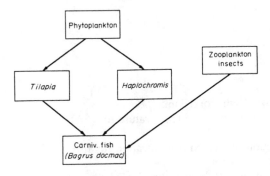

Fig.6.15: Relation between the classes of fish considered in the model and their feed.

The average number of Tilapia in age class, k, at time $t+1$ is found from:

(6.44)

$$NT(t+1,k) = NT(t,k-1) ST(t,k-1) \quad k > 2$$

where (compare with (6.42)

(6.45)

$$\underset{\text{(survival coefficient)}}{ST(t,k-1)} = 1 - \underset{\text{(mortality)}}{MT(t,k-1)} - \underset{\substack{\text{(predation} \\ \text{rate)}}}{PRT(t,k-1)} - \underset{\text{(catch rate)}}{CT(t,k-1)}$$

and for the first age class:

(6.46)

$$NT(t+1,1) = NT(t,2) . FT(t,2) + NT(t,3) . FT(r,3)....$$
$$= \sum_{l=2}^{i=k} NT(t,k)FT(t,k)$$

Corresponding equations are valid for Haplochromis and carnivorous fish, except that a predation term (PRT in eq. (6.45)) is missing in the carnivorous case.

The growth equations are:

**Tilapia**

$$LT(t,k) = 39(1 - exp)-FGPH.FGTT(k + 0,25))) \tag{6.47}$$

**Haplochromis**

$$LH(t,k) = 20(1-exp(-FGPH.FGTH(k + 0,18))) \tag{6.48}$$

**Carnivorous fish**

$$LC(t,k) = 82.5(1 - exp(-FGTC.FGPC(k + 0.415))) \tag{6.49}$$

## TABLE 6.5

**List of Program Tables.**

1)  Relation length/weight for Tilapia.
2)  Ratio of fertile females as function of weight for Tilapia/
3)  Mortality temperature coefficient as function of the temperature for Tilapia and carnivorous fish.
4)  Mortality oxygen coefficient as function of oxygen concentration for fish.
5)  Mortality pH-coefficient as function of pH for fish.
6)  Growth temperature coefficient as function of temperature for Tilapia.
7)  Growth feed coefficient as function of the ratio phytoplankton/fish for herbivorous species.
8)  Relation between net size and fish length for each of the 3 fish classes.
9)  Growth feed coefficient as function of ratio prey/predators for carnivorous fish.
10) Growth temperature coefficient as function of temperature for carnivorous fish.
11) Relation length/weight for carnivorous fish.
12) Ratio fertile females as function of length for carnivorous fish.
13) Ratio fertile females as function of length for Haplochromis.
14) Mortality temperature coefficient versus temperature for Haplochromis
15) Growth temperature coefficient as function of temperature for Haplochromis.

References:Cridland (1960) , Cridland (1962), Garrod (1959), Greenwood et al. (1965) , Jackson (1970), Roberts (1974), Ssebtongo (1972), Welcomme (1970) , Yanni (1970-1971).

## TABLE 6.6

## List of Symbols.

| Symbol | Unit | Definition |
|---|---|---|
| CATCH(t) | ring year$^{-1}$ | Actual Catch versus time |
| CC(K) | - | Catch rates, Carnivores |
| CH(K) | - | Catch rates, Haplochromis |
| CT(K) | - | Catch rates, Tilapia |
| FC(t,k) | - | Fecundity, Carnivores |
| FGPC | - | Growth factor versus prey/predator ratio, Carnivores |
| FGPH | - | Growth factor versus phyt./herb. ratio, Herbivores |
| FGTC | - | Growth factor versus temperature, Carnivores |
| FGTH | - | Growth factor versus temp., Haplochromis |
| FGTT | ring year$^{-1}$ | Growth factor versus temp., Tilapia |
| FH(t,k) | - | Fecundity, Haplochromis |
| FMN | - | Mortality factor, Ammonia |
| FMO | - | Mortality factor, Oxygen |
| FMPH | - | Mortality factor, pH |
| FMTH | - | Mortality factor versus temp., Haplochromis |
| FMTT | - | Mortality factor versus temp., Tilapia |
| FT(K) | - | Fecundity, Tilapia |
| HERB | gww m$^3$ | Total Herbivorus concentration |
| K | - | Class Indices |
| LC(K) | cm | Length, Carnivorus |
| LH(I) | - | Length, Haplochromis |
| LT(I) | - | Length, Tilapia |
| MC | - | Mortality, Carnivorus |
| MCC | - | Mortality coefficient, Carnivores |
| MCH | - | Mortality coefficient, Haplochromis |
| MCT | - | Mortality coefficient, Tilapia |
| MH | - | Mortality, Haplochromis |
| MT | - | Mortality, Tilapia |
| NC(t,k) | 1/m$^3$ | Number of Carnivores per volume |
| NH(t,k) | - | Number of Haplochromis per volume |
| NT(t,k) | - | Number of Tilapia per volume |
| NLC | Inch | Net size versus length, Carnivorus |
| NHL | - | Net size versus length, Haplochromis |
| NLT | - | Net size versus length, Tilapia |
| PCATCH(t) | gww/m$^3$ | Potential fish catch for a given strategy |
| PH | - | PH |
| PHYT | gDM/m$^3$ | Phytoplankton concentration |
| PREY | gww/m$^3$ | Prey eaten by K'th class of Carnivores |
| PRH(t,k) | - | Predation rate on Haplochromis |
| PRT(t,k) | - | Predation rate on Tilapia |
| RFLC | - | Ratio of fertiled females versus length, Carnivores |
| RFLH | - | Ratio of fertiled females versus length, Haplochromis |
| RFWT | - | Ratio of fertiled females versus weight, Tilapia |
| SURVC | - | Survival coefficient, Carnivores |
| SURVH | - | Survival coefficient, Haplochromis |
| SURVT | - | Survival coefficient, Tilapia |
| TEMP | DEG.C | Temperatures |
| t | y$_3$ | Chronological time |
| VOL | m | Lake volume |
| WC(t,k) | gww/F | Weight of one fish, Carnivores |
| WH(t,k) | gww/F | Weight of one fish, Haplochromis |
| WLC | - | Weight versus length, Carnivores |
| WLT | - | Weight versus length, Tilapia |
| WT(t,k) | - | Weight of one fish, Tilapia |

The fecundity for Tilapia of age class k can be found from:

$$FT(t,k) = 32.7.WT(t,k).NT(t,k).RFWT(t,k).SURVT \tag{6.50}$$

The equations for the two other fish classes are parallel, except that 33.0 is used for Haplochromis instead of 32.7. The mortality is computed from the following expression:

$$MT(t,k) = MCT.FMN.FMO.FMPH.FMTT \tag{6.51}$$

and a corresponding expression for Haplochromis and carnivorous fish PREY(k) is defined as:

$$PREY(t,k) = \sum_{i \in I} NH(t,k) \cdot WH(t,i) + \sum_{j \in J} NT(t,j) \cdot WH(t,j) \tag{6.52}$$

where the index sets I and J satisfy:

$$I = i \quad 0.11.WC(t,k) < WH(t,i) < 0.33.WC(t,k) \tag{6.53}$$

and

$$J = j \quad 0.11.WC(t,k) < WT(t,j) < 0.33.WC(t,k) \tag{6.54}$$

respectively, which quantifies the idea that only herbivores with a certain interval of weight are consumed by the kth class of carnivores. The predation rates are:

$$PRT(t,k) = \frac{4.8}{PREY(t,k)} \left( \sum_{m \in M} NC(t,m) \cdot WC(t,m) \right.$$

$$\left. - \sum_{n \in N} NC(t-1,n) \cdot WC(t-1,n) \right) \tag{6.55}$$

where the index sets M and N are

$$M = m \quad 0.11.WC(t,m) < WT(t,k) < 0.33.WC(t,m) \text{ and}$$
$$N = n \quad 0.11.WC(t-1,n) < WT(t,k) < 0.33.WC(t-1,n)$$

Expressing that the kth class of herbivores may only be consumed by carnivores within a certain weight interval. The bracket in equation (6.55) represents the rate of change of carnivorous biomass due to consumption. The number 4.8 is a yield factor. Analogue expressions are used for Haplochromis.

The fishery policy is the forcing functions of the model. It is expressed by the use of two functions:

1) The total catch during 6 months as a function of time (CATCH(t)),

2) the net size used for fishery. The net size is translated into length and weight by use of tables (see table 6.5).

Maximum potential fish catch for a given type of net is calculated as

$$\text{PCATCH}(t) = \sum_{k \in K} g \cdot \text{NT}(t, k) \cdot \text{WT}(t, k) + \text{corresp.} \tag{6.56}$$

where the index set K consists of two subsets:

$K = K_1 \cup K$ with
$K_1 = (k \; k_{min1} < k < k_{min2})$ and
$K_2 = (k \; k_{min2} < k)$

If $k \epsilon K_1$, g is set to 0.5, otherwise to 1.0 $k_{min1}$ and $k_{min2}$ are the lowest values of k that satisfy:

$\text{LT}(t,k) > f \text{ (mesh, size) - 3 cm}$
and $\text{LT}(t,k) > f \text{ (mesh, size)}$

respectively, i.e. fish with dimensions greater than or equal to the mesh size function are all caught, fish with demensions slightly smaller than the mesh size function are caught with 50% efficiency and even smaller fish are not caught at all. The actual catch rate in tons per ring-year is CATCH(t), so the relative fish catch may be expressed as

$$\frac{10^6 \cdot \text{CATCH}(t)}{\text{VOL} \cdot \text{PCATCH}(t)} \tag{6.57}$$

which equals CT(t.k) when $k \epsilon K_2$ and 2.CT(t,k) when $k \epsilon K_1$

Corresponding considerations can be set up for the other two fish types. Since it is a population model, mass conservation principles are not applied, but a mass balance can easily be computed for use in the total model.

The model was calibrated using published fishery and population data The following parameters were included in the calibration: MCC, MCH, MCT, SURVC, SURVH and SURVT.

The model is based on a comprehensive knowledge about the fishery and the species actually living in the lakes. Growth rates, mortalities, fecundity, sensitivity to pH, temperature and other external factors and other relevant information are available. Calibration is based in fishery statistics and results of bottom trawlings, which have been currently published by EAFRO.

## 6.5. HARVEST MODELS

Some decades ago when man did not threat the amount of renewable ressources, the optimum yield problem was easy to solve. With a good approximation the harvest, H, was proportional to the effort, e, and the abundance (density of fish), A:

$$H = k.e.A \qquad (6.58)$$

or the harvest per unit of effort was proportional with the abundance:

$$H/e = k\,A. \qquad (6.59)$$

The coefficient k is named the catchability coefficient in fishery models.

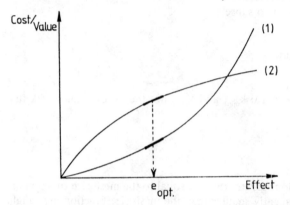

Fig.6.16: Cost of effort versus effort (1) and value of harvest versus effort (2). A + e = eopt the slopes of the two curves are equal, corresponding to optimum effort.

As long as H has no influence on the size of A, the problem is simple: will increased effort cost less than the value of the increased harvest? Often the cost of effort will increase more rapidly than the effort, see fig. 6.16, and the optimum harvest is found from the effort, that gives a slope on the cost/effort curve equal to the slope on the value of harvest/effort curve. If the value of one unit of harvest is V this latter curve corresponds to the equation:

$$H\,V = k\,e\,A\,V \qquad (6.60)$$

and the slope of the curve is k A V. See also fig. 6.16.

During the last decades mankind has increased his effort to exploit renewable resources with the result that latter have declined. The fishery in the North Sea is a very illustrative example.

Increased effort has had two effects:

1) the effort has reached a level at which it significantly influences the abundance of resources. This means, that H/e declines since A is reduced, see equation (6.58)
2) the effort has even reached a level, at which the rate of renewal is reduced. As the number of offspring is dependent on the number of fertile organism, this relation is obvious.

Unfortunately this effect has a certain time lag. If the feedback was immediate, the need for better management would be more convincing.

Fig. 6.17 illustrates clearly this effect on the North Sea herring and mackerel stocks. The first effect is relatively easy to consider and let us illustrate it by turning to the fishery management problem.

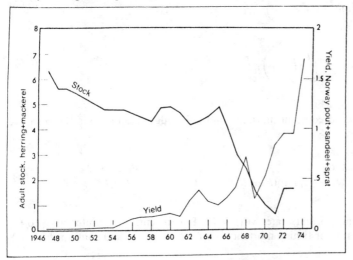

Fig.6.17: Stock depletion in North Sea herring and mackerel, and the development of fisheries for Norway pout, sand-eels and sprat (millions of tons). After Andersen and Ursin (1977).

Let R be the number of recruits at the time of recruitment, tr. The number of survivors at age t, N(t), can be found from the following differential equation:

$$dN(t)/dt = - Z(t) \times N(t),$$ (6.61)

where $Z(t)$ is the mortality rate of $N(t)$ at time t. We distinguish between rate of fishing mortality, $F(t)$, (or instantaneous coefficient of fishing mortality) and rate of natural mortality, $M(t)$, (or instanteneous coefficient of natural mortality). This means that:

$$Z(t) = F(t) + M(t) \text{ and}$$ (6.62)
$$dN(t)/dt = - (F(t) + M(t) N(t))$$ (6.63)

If (6.61) is integrated from tr to t, we get

$$N(t) = R \cdot \exp \left( - \int_{t_r}^{t} Z(t) dt \right)$$ (6.64)

or if $Z(t)$ is constant, Z:

$$N(t) = R.\exp(-Z(t-tr)) \tag{6.65}$$

Let $C(t)$ be the catch from $t_r$ to $t$, then

$$dC(t)/dt = F(t)N \cdot (t) \tag{6.66}$$

$$C(t) = R \int_{t_r}^{t} \exp\left(-\int_{t_r}^{t} z(t)dt\right)dt \tag{6.67}$$

With the simple model of constant mortality and knife-edge selection, see fig. 6.18, we have

$$\int_{t_r}^{t} z(t)dt = \begin{cases} M(t - t_r) & \text{if} & t_r \le t < t_c \\ M(t_c - t_r) + (F + M)(t - t_c) & & t_c \le t \le t_\lambda \end{cases}$$

M is as seen a constant mortality rate, $M(t) = M$. For the entire time interval, $t_r$ to $t_c$, considered, while the rate of fishing mortality is 0 for $t_r < t < t_c$ and constant $F(t) = F$ for $t_c < t < t\lambda$.

Thus the catch in number by knife-edge selection during the time $\lambda$
= t - $t_c$, is

$$C(t_\lambda) = R \exp(-M(t_c-t_r)).F(1-\exp)(-(F+M)\lambda)/(F+M) \tag{6.68}$$

The catch expressed in weight units is called the harvest or yield: $H(t)$. We get:

$$dH(t) = F(t).N(t).w(t) \, dt \text{ or} \tag{6.69}$$

$$H(t) = R \int_{t_r}^{t} F(t)\exp\left(-\int_{t_r}^{t} z(t)dt\right)w(t)dt \tag{6.70}$$

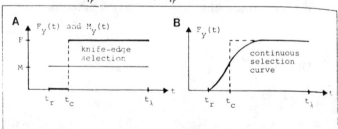

Fig.6.18: Models of mortality caused by fishing gear (fishing gear selection curves).

If we assume Bertalanffy's growth equation, constant natural mortality and knife-edge selection and rewrote the growth equation for mathematical convenience as

$$W(t) = W_\infty \sum_{n=0}^{3} \Omega_n \exp(-nK(x - t_0)) \qquad (6.71)$$

where $\Omega_n = 1$, $\Omega_1 = -3$ and $\Omega_3 = -1$,

we get the Beverton and Holt yield equation:

$$H(t) = W \cdot F \cdot R^+ \sum_{n=0}^{3} \Omega_n \exp(-nK(t_c - t_0))(1 - \exp(-(F + M + nK)\lambda)) \qquad (6.72)$$

The following symbols are used:

$R^+$ = number of recruits at age $t_c$ = $R.\exp(-M(t_c-t_r))$
$t_r$ = age of recruitment
$t_c$ = age at first capture
$t\lambda$ = maximum age $\lambda = t - t_c$ = fishable lifespan $W\infty$, K and to are Bertalanffy growth parameters $F$ = fishing mortality $t > tc$ (knife-edge selection) $M$ = natural mortality rate (constant)

The number of survivors at age t, $N(t)$ is

$$(6.73)$$
$$N(t) = R^+\exp(-Z(t-t_c))$$

These equations express the yield from a year class during its lifespan. However, in management you are faced with the problem of what harvest should be planned for all age groups in the following year. It is in most cases not possible to have a different harvest on different year classes. Only if the system is in a steady state situation equation (6.72) can be used for the management. Equation (6.72) has been used for several years in fishery biology and it has been assumed that the stock was close enough to a steady state to make the equation a reasonable approximation.

The Beverton and Holt yield per recruit equation is:

$$\text{yield/recruit} = W_\infty F \exp(-M(t - t_c)) \sum_{n-0}^{3} \frac{\Omega_n \exp(-nK(t_c - t_0))}{F + M + nK} \qquad (6.74)$$
$$\times (1 - \exp(-(F + M + nK)\lambda))$$

Usually (6.74) is applied to different models of mortality caused by harvest (fishery).

Fig. 6.19 shows H/Y versus F in accordance with eq.(6.74). A shows that fast growing species (with a large K value) should be fished more intensively than the slower growing species. B shows that it might be more profitable to catch only larger fish e.g. by choosing larger meshes for the fishing gears. Of course this statement is only true for a certain range of $t_c$ values.

187

The value of F at which H/R attains its maximum is called $F_{max}$ and the yield is named the maximum sustainable yield(MSY).

For a number of stocks of North Atlantic fish F is greatly in excess of $F_{max}$. Biologists have recommended to reduce the fishery and the international fishery management body has to a certain extent followed this advice. Fig. 6.20 shows the Y/R curve for the North Sea Whiting and in accordance to these results the fishery effort should be reduced to 40% of its size in 1978. Similar results have been obtained for other fish species such as the North Sea cod and haddock.

The Beverton-Holt model, that has been presented above, has mainly been criticized on the fact, that it does not take the interactions of fish species into considerations. The application of the model on such fish predators as cod and whiting is problematic due to the consumption of other important commercial species by these predators (Daan 1975, Jones 1978, Ursin 1979 and Sparre 1979).

It is therefore necessary to set up and use a multispecies model. Illustration 6.2 has already presented a multi species model but in a more simple version than the multi species model used for management of the North Sea fishery. This latter model will not be presented here as it would be too time consuming to present all the details of such a complex model. Those, who are interested in becoming acquainted with the details, we can refer to Beyer and Sparre (1983).

Much research effort has been devoted to the general problems of renewable resources and optimum control. A special issue of Ecological Modelling vol. 14 3-4 1982 reviews some of the latest contributions to this field. The problems are interesting and require complex mathematics to be solved properly, but it is of importance to solve them from a management point of view.

It would be to go too far in this context to present here the state of the art on the solution of these problems. However, it is appropriate to illustrate the importance of the stability concept in this relation, by use of a relatively simple example.

We consider a prey $X_1$ and a predator $X_2$ both harvested at constant yield. The following equations might be valid for this case:

$$\dot{X}_1 = r_1 \cdot X_1(1 - X_1 - rX_2) - r_1 \cdot Y_1 \tag{6.75}$$

$$\dot{X}_2 = r_2 \cdot X_2(1 - X_2/X_1) - r_2 \cdot Y_2 \tag{6.76}$$

where $r_1$, $r_2$ and v are parameters. $r_1Y_1$ and $r_2Y_2$ are the constant yield of prey and predator, respectively. The curve CD in fig. 6.20 gives for each value of $Y_2$ the absolute maximum sustainable value of $Y_1$ and vice versa.

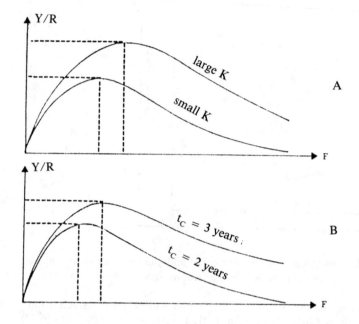

Fig. 6.19: Hypothetical examples on Beverton and Holt yield per recruit curves. See text for a detailed explanation.

The curve is adapted from Beddington and May (1980). It is obtained by maximizing $Y_1$ subject to the constraint $Y_2$ = constant or alternatively by maximizing $Y_2$ subject to the constraint $Y_1$ = constant. All pairs of values of $Y_1$ and $Y_2$ lying inside the curve CD correspond to jointly sustainable yields for which there exists at least one pair of equilibrium prey and predator populations but the equilibria will not necessarily be stable. If the equilibrium populations are $X_1'$ and $X_2'$ then the condition for the equilibrium being stable is that the eigenvalues of the stability matrix have negative real values and this is the case if

$$r_1(1-2X_1-vX_2) + r_2(1-2X_2/X_1) < 0 \text{ and} \tag{6.77}$$

$$vX_2^2 + (1-2X_1-rX_2)(X_1-2X_2) > 0 \tag{6.78}$$

189

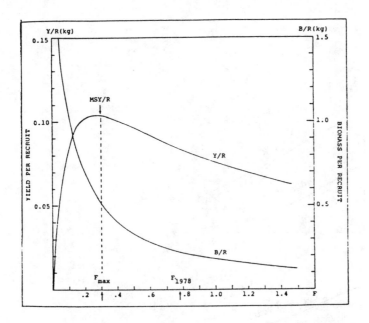

Fig.6.20:   Beverton and Holts biomass per recruit curve and yield per recruit curve. The present example deals with the North Sea whiting stock (Anon. 1980)

Let the curve AB represent the maximum prey yields that will produce a stable equilibrium if the predator yield is f.inst. $Y_2^+$, see fig. 6.21. If $r_2/r_1$ is small condition (6.77) becomes important. The smaller this ratio is, the more the curve AB lies to the left. The curve AB represents of course also the minimum predator yield that can produce a stable equilibria if the prey yield is between $Y_1^+$ and $Y^{++}$ see fig. 6.21.

Thus, for pairs of yields within the region ABEC shaded on fig. 6.21 there exists an equilibrium, but it is not stable.

From this we can learn that we need to distinguish between stable and unstable equilibrium and that it is not sufficient to find the possible pairs of yields by use of the equations (6.75) and (6.76). Furthermore, we can not necessarily shift from one harvest strategy with a stable equilibrium to another harvest, if the shift requires that we pass through an area where the equilibria are unstable.

It is indeed a matter for discussion, whether such stability investigations hold when practical ecosystem management is involved.

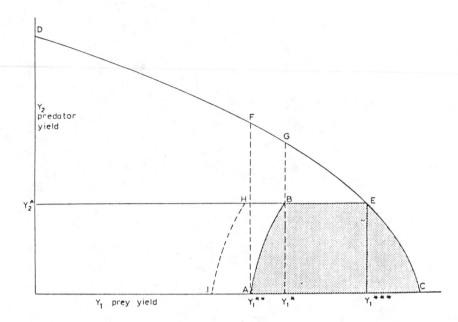

Fig.6.21: Prey yield.

The argument against the application of such stability studies is that in real ecosystem situations a great number of interactions and feedback regulations occur and that the stability in nature is quite different from the stability found by the use of two relatively simple differential equations. This argument is of course very strong, but on the other side will such a simple model as presented above account for the strongest influences on the prey and predator populations and it is also a fact that unstable equilibria exist in nature. More comparisons between model results and observations in nature are needed before a final conclusion can be drawn. Until then the results of stability studies should be carried out, but the results used preliminarily only and the conclusions should be subject to sound scepticism.

## PROBLEMS CHAPTER 6.

1. A population has a growth rate of 0.15 per day and a carrying capacity of 200 kg biomass per ha. The initial density is 10 kg biomass per ha. It is assumed that the population follows a logistic growth. How long a time would it require to reach 99% of the carrying capacity?

2. A fish population can be divided into 4 age classes. Age classes 2 and 3 produce 2000 recruits per year. The mortality is 0.3 for these age classes and 0.5 for age classes 1 and 4. Set up a matrix model for the fish population, find the eigenvalue and eigenvector and estimate the harvest, that can be taken from this population to maintain the initial population size. The initial number of fish in the four age classes are $10^5$, $5 \cdot 10^4$, $3 \cdot 10^3$ and $2.5 \cdot 10^3$.

# 7. DYNAMIC BIOGEOCHEMICAL MODELS

This chapter presents a great variety of dynamic biogeochemical models. A wide application and a pronounced development of this type of models has taken place during the last decade. The models are often formulated as a set of differential equations combined with some algebraic equations and a parameter list. Obviously the differential equations require the definition of an initial state.

The following biogeochemical models are included in the chapter: BOD/DO models, hydrodynamic models, eutrophication models, wetland models, toxic substance models, air pollution models, soil pollution models and plant growth models. The ideal presentation for all these would be to give details of a simple and a complex model and a general overview of the spectrum of models available in the area. This would, however, require several hundred pages and is therefore not feasible. However, an attempt has been made to present some simple models and to give the readers some key equations and other characteristic features of the most common models. In addition, several tables give a first impression of the spectrum of available models. Throughout the book eutrophication models have often been used as illustration for the modelling procedure and the modelling considerations. Therefore, one medium complex eutrophication model is discussed in more detail, although all equations are not presented. It is the hope that the reader thereby will get a good impression of how to develop and use a biogeochemical model and to get an idea of the advantages and disadvantages of this type of models. Hopefully, the reader will furthermore learn to be critical and will understand the considerations involved in modelling.

Hydrodynamic models are not presented in great detail, but the most common cases are included and it should be rather easy to use the presented equations as proper components in ecological models. For air pollution models and models of soil pollution, plant growth and crop production models, the most important submodels or model components are presented. Total models in the area can be developed by a suitable combination of the submodels presented, probably sometimes by use of submodels, which are not presented here, as they are more rarely applied.

Many of the models and model components presented are used in a management context. These aspects will be further treated in the following chapter.

## 7.1. APPLICATION OF DYNAMIC MODELS

Ecosystems are dynamic systems and it might therefore be the ultimate goal for a modeller to construct *dynamic models of ecosystems*. Models of population dynamics focusing on changes in the size of population caused by production of off-spring and various forms of mortality were presented in chapter 6. The growth of individuals or age classes was considered by use of growth dependence of various factors. Ecosystem management at the population level seems feasible by use of this type of models including the important management of renewable resources.

This chapter is devoted to another type of models, which has gained a wide application both in science and in management context. Biogeochemical models attempt to capture the dynamics of biochemical and geochemical compounds elements in ecosystems. When models are used as an instrument in pollution control, they must account for the fate and distribution of both pollutants and of natures own compounds. That will require the application of biogeochemical models, since they focus on the processes and transformation of various compounds in the ecosystem.

**Total ecosystem models** which couple population models with biogeochemical models have also been developed. This has been touched on in illustration 6.2. The food available for growth is dependent on the biogeochemical cycling in ecosystems and the growth rate is dependent on the general life condition in the ecosystem, which again is dependent on the biogeochemical cycling. The coupling between the two types of models takes place through such relations.

As pointed out in section 2.7 the construction of dynamic models requires data, which can elucidate the dynamics of the processes included in the model. Generally, a more comprehensive database is required to build a dynamic model than a static model. Therefore in a data poor situation it might be better to draw an average situation under different circumstances by use of a static model, than to construct a unreliable dynamic model, which contains uncertainty on the most crucial parameters.

The first biogeochemical model, that was constructed, was the Streeter-Phelps BOD-DO model in 1925 (see Streeter and Phelps 1925). It will be presented in detail in the next section, as it illustrates quite clearly the concepts of biogeochemical models. The Streeter-Phelps model consists of one differential equation, which can be solved analytically.

Hydrodynamic models can be considered as biogeochemical models, since these models describe the fate and distribution of the important compound water in ecosystems. Output from hydrodynamic models might often be used as forcing functions in ecological models. They are, however, not ecological models, as they do not account for any biological processes; they are, however, often used together with ecological models, as the distribution of chemical compounds and living organisms is dependent on the hydrodynamics. Some hydrodynamic submodels, which are used as components in ecological models will be presented in the following paragraphs, but hydrodynamic models are otherwise beyond the scope of this book.

Quite comprehensive hydrodynamic models were developed in the nineteen-fifties and sixties, but complex ecological management models were not developed before the late sixties. A new era then started with more complex BOD-DO models and the first eutrophication models. To a certain extent Chen and Orlobs paper on "a proposed ecological model for a eutrophic environment" from 1968 can be considered as a key paper in the initiation of this new ecological modelling era.

These models, although they are almost 20 years old, were ecologically quite sound as they had a reasonable balance between data and complexity. The further development in computer technology in the early nineteen-seventies rendered it quite easy to construct very complex models, but the quality and quantity of data did not develop to the same extent. The result was that some models developed in these years were not adequately calibrated and validated and did not contain a good ecological knowledge which would allow a reasonable simplification. It is quite easy to construct complex models with many state variables, parameters and equations. It is far more time consuming to provide the data required to calibrate the models. Some models developed in these years claim to be very general due to the great amount of state variables and process equations. However, the experience up through the seventies has shown that even very complex models cannot account for all the processes, that it is needed to include in models of a given ecosystem type f.inst. a lake, a river, a grassland etc. It was rather a case of very simple models, that could be applied more generally, as simple models eventually include the few processes, that are almost always the most significant.

The experience gained after 10 years of intensive application of ecological modelling, it means during the seventies, can be summarized in the following points:(see also the discussion in chapter 2)

1. **A good knowledge to the ecosystem is required to capture the essential features, which should be reflected in the model.**
2. **The scope of the model determines the complexity, which again determines the quality and quantity of the data needed for calibration and validation.**
3. **If good data are not available it is better to go for a somewhat too simple model than one, which is too complex.**

During the nineteen-seventies and the early eighties much experience was gained in modelling many different types of ecosystems and many different aspects including a number of pollution problems. The modellers learned also what *modifications* it was necessary to make, when a model was applied for the *same* problem, but on another ecosystem than that for which it was originally developed for. It was seen that the same model could hardly be applied to another ecosystem *without* some changes, unless as mentioned above the model was very simple. More and more models became well calibrated and validated. They could often be used as practical management tool, but in most cases it was necessary to combine the use of the model with a good knowledge to general environmental issues. Also in cases, when the model could not be applied to set up accurate predictions, the model was useful to enable the manager to see the reaction of the ecosystem to various management strategies qualitatively.Scientists, who applied models, found that they were very useful in indicating research priorities and also in capturing the system features of ecosystems.

Table 7.1 reviews types of ecosystems, which have been modelled by use of biogeo-chemical models up to the mid- eighties. An attempt has been made to indicate the modelling effort by use of a scale from 0-5. 5 means very intense modelling effort, 4 intense modelling effort, 3 some modelling effort, 2 a few models which have been fairly good studies, 1 one good study or a few not sufficiently well calibrated and validated models and 0 almost no modelling effort at all.

Table 7.2 similarly reviews the environmental problems, which have been modelled up to to-day. The same scale is applied to indicate the modelling effort on these pro-blems.

**TABLE 7.1**

**Review of Ecosystems modelled by Use of Biogeochemical Models.**

| Ecosystem | Modelling effort |
|---|---|
| Rivers | 5 |
| Lakes, reservoirs, ponds | 5 |
| Estuaries | 5 |
| Coastal zone | 3 |
| Open sea | 3 |
| Wetlands | 3 |
| Grassland | 4 |
| Desert | 1 |
| Forests | 4 |
| Agriculture land | 5 |
| Savanna | 2 |
| Mountain lands (above timberline) | 0 |
| Archic ecosystems | 0 |

**TABLE 7.2**

**Review of Environmental Problems modelled by Use of Biogeochemical Models.**

| Problem | Modelling effort |
|---|---|
| Oxygen balance | 5 |
| Eutrophiciation | 5 |
| Heavy metal pollution | 4 |
| Pesticide pollution | 4 |
| Protection of national parks | 3 |
| Ground water pollution | 4 |
| Carbon dioxide | 5 |
| Acidic rain | 4 |
| Total or regional distribution of air pollutants | 5 |
| Change in microclimate | 3 |

**7.2. BOD/DO MODELS**

**7.2.1. Simple BOD/DO Models**

This type of models is concerned with the oxygen concentration in rivers and streams. With minor modification it is, however, possible to use the same models for BOD/DO relations in other aquatic ecosystems.

BOD represents biodegradable matter. If oxygen is present, the biodegradation will require an amount of oxygen equal to the reduction in BOD.

The first water quality model considering the BOD/DO relationship in a river system was developed by in 1925 by Streeter and Phelps. It is based on the following assumptions:

1) **only one source of pollution exists**
2) **a constant load of pollution is discharged at a given point**
3) **there is no tributary inflow**
4) **flow rate is constant**
5) **the cross section of the river is uniform**
6) **the turbulence is sufficient to allow the concentration of BOD and dissolved oxygen (DO) to be uniform throughout the cross-section**
7) **the biodegradation and reaeration are first order reactions and they are the only processes to be considered.**

The following differential equations can be set up:

$$\frac{dD}{dt} = K_l \cdot L_t - K_a \cdot D \tag{7.1}$$

$D$ $= C_s - C_t$
$C_s$ = oxygen concentration at saturation
$C_t$ = oxygen concentration at time = t
$L_t$ = concentration of organic matter, measured as BOD, at time = t
$K_l$ = rate constant (day$^{-1}$)
$K_a$ = reaeration constant (day$^{-1}$)

Although the Streeter-Phelps' model is very simple, its use arises several questions:

1) How to estimate $K_l$ and $L_t$? 2) How to account for the influence of the temperature? 3) How to estimate the reaeration?

These three questions are discussed below:
Values for K1, Kn, Lo and No are given for some charateristic cases in table 7.3. Kn is the rate constant for nitrification:

$$NH_4^+ + 2O_2 \rightarrow NO_3^- + H_2O + 2H^+ \tag{7.2}$$

which often must be added to equation (7.1) to account for this oxygen consumption, see below. $N_o$ is the ammonium concentration and $L_o$ the concentration of organic matter, measured as BOD.

$K_l$ and $K_n$ are dependent on the temperature:

$$K_1 \text{ or } K_n \text{ at } T = (K_1 \text{ or } K_n \text{ at } 20^{\circ}C) \cdot K_T \tag{7.3}$$

where $K_T$ is a constant, see table 7.4

$K_a$ is dependent on the temperature, the flow rate of water and the water depth as formulated in the following equation:

$$K_a(20^\circ C) = \frac{2.26 \cdot v}{R^{2/3}} (day^{-1}) \tag{7.4}$$

$$K_a(T) = K_a(20) \cdot e^{\theta(T-20)} \tag{7.5}$$

$$R_a = K_a(T)(C_s - C_t) \tag{7.6}$$

where

| | |
|---|---|
| $K_a(20^\circ C)$ | = reaeration coefficient at 20°C (day$^{-1}$) |
| $K_a(T)$ | = reaeration coefficient at T°C (day$^{-1}$) |
| $v$ | = average flow (m.s$^{-1}$) |
| $R$ | = depth (m) |
| $R_a$ | = rate of reaeration (mg l$^{-1}$ day$^{-1}$) |
| $\theta$ | = constant = 0.024° C$^{-1}$, 15°$<$T$<$25°C |
| $C_s$ | = oxygen concentration at saturation (mg l$^{-1}$) |
| $C_t$ | = actual concentration at time = t (mg l$^{-1}$) |

TABLE 7.3

Characteristic Values, of $K_1$, $K_N$, $N_o$ and $L_o$ (20°C)

| | $K_1$(day$^{-1}$) | $K_N$(day$^{-1}$) | $N_o$ | $L_o$ |
|---|---|---|---|---|
| Municipal waste water | 0.35-0.40 | 0.15-0.20 | 80-130 | 150-250 |
| Mechanical treated municipal waste water | 0.35 | 0.10-0.25 | 70-120 | 75-150 |
| Biological treated municipal waste water | 0.10-0.25 | 0.05-0.20 | 60-120 | 10-80 |
| Potable water | 0.05-0.10 | 0.05 | 0-1 | 0-1 |
| River water | 0.05-0.15 | 0.05-0.10 | 0-2 | 0-5 |

TABLE 7.4

Temperature Dependence of $K_1$ and $K_N$

| | $K_1$ | $K_N$ |
|---|---|---|
| $K_T$ | 1.05 | 1.06-1.08 |

The concentration of organic matter in waste water is often indicated as BOD$_5$ or BOD$_7$

the oxygen consumption during 5 or 7 days respectively. $L_o$ is therefore not known, but as $BOD_5$ is equal to $L_o - L_5$, $L_o$ can easily be found. As mentioned above the decomposition is described as a first order reaction. We have:

$$\frac{dL_t}{dt} = -K_1 \cdot L_t \tag{7.7}$$

$$L_t = L_0 \cdot e^{-K_1 \cdot t} \tag{7.8}$$

$$L_5 = L_0 \cdot e^{-K_1 \cdot 5} \tag{7.9}$$

$$BOD_5 = L_0 - L_5 = L_0\left(1 - e^{-K_1 \cdot 5}\right) \tag{7.10}$$

The differential equation (7.1) can be solved analytically. The equation is reformulated:

$$\frac{dD}{dt} + K_a \cdot D = K_1 \cdot L_0 \cdot e^{-K_1 \cdot t} \tag{7.11}$$

When $D = D_o$ at $t = 0$, we get:

$$\tag{7.12}$$
$$D = \frac{K_1 \cdot L_0}{K_a - K_1}\left(e^{-K_1 \cdot t} - e^{-K_a \cdot t}\right) + D_0 \cdot e^{-K_a \cdot t}$$

For $K_1 = K_a$ the solution is: (7.12) is not valid in this case

$$D = (K_1 \cdot t \cdot L_0 + D_0) \cdot e^{-K_1 \cdot t} \tag{7.13}$$

From equation 7.12 it is possible to plot D, $C_t$ and the reaeration against time.

The point where the oxygen concentration is at a minimum is termed the critical point (see fig. 7.1). This point can be found from:

$$\tag{7.14}$$
$$\frac{dD}{dt} = 0 \qquad \frac{d^2D}{dt^2} < 0 \qquad \text{(minimum)}$$

$$\tag{7.15}$$
$$t_c = \frac{1}{K_a - K_1} \ln \left(\frac{K_a}{K_1} - \left(\frac{K_a}{K_1} - 1\right)\frac{D_0}{L_0}\right)$$

199

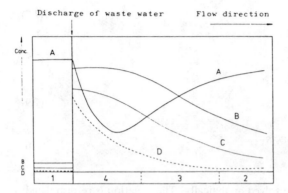

Fig.7.1:  A. Oxygen concentration, B. $BOD_5$, C. Nutrients, D. Suspended matter, versus distance in running water. The numbers correspond to the classification in the saprobic system (see table 7.5).

If nitrification is included in the oxygen balance equation (7.1) we get:

$$\frac{dD}{dt} = K_1 \cdot L_t + a \cdot K_n \cdot N_t - K_a \cdot D \qquad (7.16)$$

where a is the relation between ammonium concentration and oxygen consumption in accordance with (7.2). a is calculated to be $2.32/14 = 4.4$ mg $O_2$ per mg ammonium-N, but due to bacterial assimilation of ammonia, this ratio is corrected to 4.3 mg $O_2$ per mg ammonium-N in practice.

If nitrification is added to the Streeter-Phelps equation we get the following solution:

$$D = \frac{K_1 \cdot L_0}{K_a - K_1} \left( e^{-K_1 \cdot t} - e^{-K_a \cdot t} \right) + \frac{K_N \cdot N_0}{K_a - K_N} \left( e^{-K_N \cdot t} - e^{-K_a \cdot t} \right) + D_0 e^{-K_a \cdot t} \qquad (7.17)$$

In equation (7.1) and (7.16) time is used as independent variable, but if a river is considered, it might be useful to translate time to distance in flow direction, x, by use of the flow velocity, v,

$$ \qquad (7.18)$$

$$x = v \times t$$

**TABLE 7.5**

**One of the most commonly used ecological examinations is the application of the Saprobic System, which classifies running waters in 4 classes (Hynes, 1971):**

1) Oligo-saprobic water, in which the water is unpolluted or almost unpolluted.
2) Beta-mesosaprobic water, in which the water is slightly polluted.
3) Alfa-mesosaprobic water, in which the water is polluted.
4) Poly-saprobic water, in which the water is very polluted.

The classification is based on an examination of plant and animal species present in the water at a number of locations. The species can be divided into four groups:

1) Organisms characteristic of unpolluted water.
2) Species dominating in polluted water.
3) Pollution indicators.
4) Indifferent species.

The equations can be used to describe the oxygen profile in the flow direction, but if so, some hydraulic assumptions must be introduced: 1) that by discharge of polluted waste water into the river complete mixing with the flow will take place, 2) that the flowrate is the same throughout the entire cross section of the river.

This means that the concentration, C, just after discharge of waste water can be calculated from the following simple expression:

$$C = \frac{Q_w \cdot C_w + Q_r \cdot C_r}{Q_w + Q_r} \tag{7.19}$$

$Q_w$ = waterflow, waste water (1 s$^{-)}$)
$C_w$ = concentration in waste water (mg l$^{-1}$)
$Q_r$ = waterflow, river (1 s$^{-1}$)
$C_r$ = concentration in river water (mg l$^{-1}$)

### 7.2.2. Complex BOD/DO Models

Oxygen concentrations in river water can be determined by means of more complex ecological models, which take into account hydraulic components, growth of phytoplankton and zooplankton, oxygen consumption of the sediment, the spectrum of biodegradability of the components present in discgarded waste water, and the presence of toxic matter affecting the biodegradability. For further information on these more complicated models, see Rinaldi et al. (1979); Orlob (1981); Jørgensen (1981) and Gromiec (1983).

However, to give the readers an overview of the processes, that might influence the oxygen balance, Table 7.6 gives a survey of some useful models and the processes, that they consider in addition to decomposition of organic matter, nitrification and reaeration. Furthermore, illustration 7.1 in section 7.3 gives some detail of the so-called QUAL 1 model. It applies a more detailed description of hydrodynamics and is therefore included in the section on "Application of Hydrodynamics in biogeochemical Models".

Table 7.7 shows different approaches to description of the reaeration process.

**TABLE 7.6**

**Survey of some useful river and stream models.**

Processes in addition to decomposition of organic matter, nitrification and reaeration are indicated.

| Model and reference | Processes |
|---|---|
| Eckenfelder and OConnor (1961) | Turbulence of water, biological growth in river beds, acclimatization of microorganisms, toxicity. |
| Thomas (1961) | Settling. |
| Eckenfelder (1970) | Autocatalytic nitrification. |
| O'Connell and Thomas | Photosynthesis and respiration. |
| Fair et al. (1941) | Benthic oxygen uptake. |
| Edwards and Rolley | Benthic oxygen uptake. |
| O'Connor (1962) | Longitudinal mixing (dispersion). |
| Dobbins (1964) | Longitudinal mixing (dispersion). |
| Hansen and Frankel (1965) | Diurnal dissolved oxygen profiles. |
| O'Connor (1967) O'Connor and Di Toro (1970) | Temporal and spatial distribution in due dimension, photosynthesis, respiration, benthic respiration. |
| DOSAG I: Texas Water Development Board (1970) | Spatial and temporal variation under various conditions of flow and temperature. |
| DOSAG M: Armstrong (1977) | Benthic oxygen demand, coliform load. Improved version of DOSAG I. |
| QUAL I: Texas Water Development Board (1971) | Detailed hydrodynamics in one dimension. |
| QUAL II: Water Resources Engineers (1973) | as QUAL I + benthic oxygen demand, photosynthesis, respiration, 2 steps nitrification, coliforms, radioactive material, phosphorus. |
| RECEIV II: Raytheon Company (1974) | State variables: BOD, DO, $NH_4 + -N_2$ $NO_2$ -N, total $< 8x < 9 <$ p, coliforms, chlorophyll. |

**TABLE 7.7**

**Selected predictive Relationships for the Reaeration Coefficient $K_a$ at $20^\circ C$ days$^{-1}$**

| Reference | Relationship |
|---|---|
| O'Connor and Dobbins (1958) | $K_a = 4.8\ S^{0.25}\ D^{-1.25}$ |
| Krenkel and Orlob (1963) | $K_a = 2.6\ E_D\ D^{-2.087}$ |
| Thackston and Krenkel (1966) | $K_a = 18.6\ U_*\ H^{-1}$ |
| Tsivogluo (1967) | $K_a = 1.63\ \nabla H\ t^{-1}$ |
| Lau (1972) | $K_a = 1088.64\ U_*^3\ v^{-2}\ H^{-1}$ |
| Foree (1976) | $K_a = 0.116 + 2147.8\ S^{1.2}$ |

The notations used in this table are:

$D$   = mean depth of flow, m
$\nabla H$ = the change in water surface elevation, m
$V$   = mean flow velocity, m sec$^{-1}$
$U_*$  = shear velocity, m sec$^{-1}$
$E_D$  = mean vertical eddy diffusivity, m$^2$ sec$^{-1}$
$t$   = time of flow, d
$S$   = channel slope, m m$^{-1}$

However, relatively simple models, which consider only 2-4 state variables are adequate for most water pollution control tasks. The more complex models should only be used in such cases, where the water is heavily polluted and extensively used for water supply or in other cases, where expensive and difficult pollution control decisions must be made.

It should be remembered that BOD is not a completely right measure of organic pollution, see f.inst. Jørgensen and Johnsen (1981).

## EXAMPLE 7.1

A municipal waste water plant discharges secondary effluent to a surface stream. The worst conditions occur in the summer months, when stream flow is low and water temperature high. The waste water has a max. flow of 12000 m$^3$/day, a BOD$_5$ of 40 mg d$^{-1}$ (at $20^\circ C$), a dissolved oxygen concentration of 2 mg e$^{-1}$ and a temperature of $25^\circ C$. The stream has a minimum flow of 900 m$^3$ h$^{-1}$, a BOD$_5$ of 3 mg e$^{-1}$ and dissolved oxygen concentration of 8 mg e$^{-1}$. Use $K_1 = 0.15$ (see table 2.18 for the mixed flow). The temperature in the stream reaches a maximum of $22^\circ C$. Complete mixing is almost instanteneous. The average flow (after mixing) of the stream is 0.2 m sec$^{-1}$ and the depth of the stream 2.5 m.

**Solution:**

Find critical oxygen concentration.

$$K_a(20^\circ C) = \frac{2.26 \cdot 0.7}{2.5^{2/3}} = 0.25 d^{-1}$$

Mixture: 500 m³/h(WW) + 1900 m³/h (stream) = totally 2400 m³/h

$$BOD_5 \text{ of mixture: } \frac{500 \cdot 40 + 1900 \cdot 3}{2400} = 10.7 \text{ mg l}^{-1}$$

$$L_o = 20.3 \text{ mg l}^{-1}$$

$$\text{Dissolved oxygen} = DO_{mixture} = \frac{500 \cdot 2 + 1900 \cdot 8}{2400} = 6.7 \text{ mg l}^{-1}$$

$$\text{Temperature}_{mixture} = \frac{500 \cdot 25 + 1900 \cdot 22}{2400} = 22.6^\circ$$

$$K_1 \text{ at } 22.6^\circ = 0.15 \cdot 1.05^{22.6-20} = 0.17$$

$$K_a \text{ at } 22.6^\circ = 0.25 \cdot e^{0.024(22.6-20)} = 0.27$$

Initial oxygen deficit = 8.7 - 6.7 = 2 mg l$^{-1}$

$$\text{Critical location(time)} = \frac{1}{0.27 - 0.17} \ln\left(\frac{0.27}{0.17} - \left(\frac{0.27}{0.17} - 1\right)\frac{2}{20.3}\right) = 4.25 d$$

$$D_c = \frac{0.17 \cdot 20.3}{0.27} e^{-0.17 \cdot 4.25} = 6.2 \text{ mgl}^{-1}$$

These conditions occur at 0.2 . 3600 . 24 . 4.25 m = 73440 m

Oxygen concentration at critical point 8.7 - 6.2 = 2.5 mg l$^{-1}$

### 7.3. APPLICATION OF HYDRODYNAMICS IN BIOGEOCHEMICAL MODELS

#### 7.3.1. Introduction.

For many biogeochemical models of aquatic ecosystems it is *absolutely necessary to include hydrodynamic processes.* It is in principle a combination of the transfer processes reviewed in chapter 3, but as is the case for ecological models, hydrodynamic models cannotot just be constructed by a combination of the unit processes, but it is required to know the system to combine the essential processes and exclude those, which are insignificant in this context.

The idea is not here to present a several pages textbook in hydrodynamics models, but to present only the elements of hydrodynamics, that are most important for the biogeochemical models.

The following issues have been selected and are presented below:

1) The hydrodynamics of completely mixed systems with dynamic waste input forms.
2) Hydrodynamic models of streams and rivers, considering only the most simple cases - it means the steady state conservative case, the steady state non conservative case and dynamic response of single stream systems.
3) Estuarine models, considering only the steady state case and dynamics of single system estuarines.
4) Few remarks to multidimensional models.
5) Thermocline models applicable to lakes, impoundment and reservoirs.

Biogeochemical models are based on the mass conservation principle as presented in section 3.1. Hydrodynamic models use the same principles, but combine transfer processes to obtain the mass balance in stead of chemical and biological and other physical processes. The possible couplings to biogeochemical models are easy to establish (in many of the hydrodynamic models presented below). A first order decay is included for the considered component. This expression might often be replaced by a more complex one deduced from the ecological model component.

   When two or more coupled state variables are modelled the combination of hydrodynamic models and ecological models is, of course much more cumbersome but in principle there is no difference between the simple hydrodynamic-ecological model and the more complex one.

### 7.3.2. The Hydrodynamic of complete mixed Systems.

A completely mixed aquatic system of volume V ($L^3$) is considered. Waste discharge W (M/Tn) results in a concentration C ($M/L^3$). The flow through the system is Q ($L^3/T$). C is decay in accordance with a first reaction. The decay coefficient is K ($T^{-1}$). It is assumed that all system parameters are constant in time. A mass balance gives:

$$V \frac{dC}{dt} = W(t) - Q \cdot C - K \cdot V \cdot C \qquad (7.20)$$

The boundary condition is:

$$(7.21)$$

$$C = C_o \text{ at } t = 0.$$

If $K' = Q + K \cdot V$, the equation can be rewritten:

$$(7.22)$$

$$V \frac{dC}{dt} + K' \cdot C(t) = W(t)$$

If W(t) = 0, equation (7.22) gives the solution:

$$(7.23)$$

$$C(t) = C_o \exp [-(K'/V)t]$$

This case is illustrated in fig. 7.2.

If $W(t) = W_r$, where $W_r$ is a constant waste input, we get the equation:

$$V \frac{dC}{dt} = W_r - Q \cdot C - K \cdot V \cdot C \tag{7.24}$$

which has the solution:

$$C(t) = \frac{W_r}{Q + K \cdot V}\left(1 - \exp\left(-\left(\frac{Q}{V} + K\right)t\right)\right) \tag{7.25}$$

which at steady state gives:

$$C = \frac{W_r}{Q + K \cdot V} \tag{7.26}$$

This case is illustrated in fig. 7.3.

If $W(t) = 0$ up to time $t = t_o$ and then a constant input occur W and is valid for t $\geqslant$ 't$_o$, we get what is called a step input and equation (7.23) can be used to describe C(t) for t $\leqslant$ to, while equation /7.25) is valid t$\geqslant$to. A graphic representation of a step input is shown in fig. 7.4.

W(t) is often a periodic forcing function given by:

$$W(t) = W_r + W_a \sin(\omega t - \alpha); \quad \omega = \frac{2\pi}{T_p} \tag{7.27}$$

where $W_a$ is the amplitude, $\alpha$ is the phase shift expressed in radians, $\omega$ is the frequency of the input (radians/T) and $T_o$ is the period of the input. The differential equation takes the form of:

$$V \cdot \frac{dC_p}{dt} + K' \cdot C_p(t) = W_0 \sin(\omega t - \alpha) \tag{7.28}$$

where the subscript p denotes the periodic output response. The solution is given by:

$$C_p = W_0 \cdot A_m(\omega) \sin\left[\omega t - \alpha - \theta(\omega)\right] \tag{7.29}$$

where

$$A_m(\omega) = \frac{1/V}{\left[(K'/V)^2 + \omega^2\right]^{1/2}} \tag{7.30}$$

$$\theta(\omega) = \arctan\left(\frac{\omega}{K'/V}\right) \quad \text{radians} \tag{7.31}$$

It is seen from equation (7.29) that $C_p$ also is a periodic function.

As the frequency increases, the phase lag between the input and output varies from zero to an asymptotic value of $\pi/2$ radians $= 90°$, see fig. 7.5.

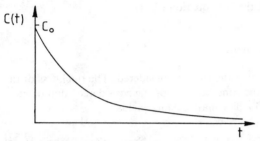

Fig.7.2: C(t) plotted versus t for W(t) = 0

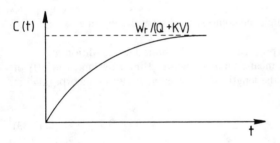

Fig.7.3: C(t) plotted versus t for w(t) = Wr (constant).

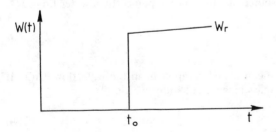

Fig.7.4: Step input. W(t) versus t.

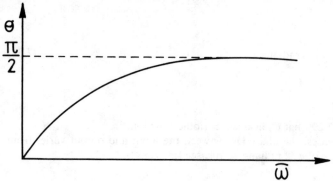

Fig.7.5: $\theta$ versus $\omega$. Graphic representation of equation (7.31)

### 7.3.3. Hydrodynamic Models of Streams and Rivers.

A mass balance for a cross sectional slice of the river is considered. The thickness of the slice is $\Delta$ x, the volume V.C(M/L$^3$) is the concentration of the considered component, t (T) is the time and Q the flow rate (L$^3$/T). The mass balance yields:

$$V\Delta C = QC \cdot \Delta - (Q + \Delta Q)\left(C + \frac{\partial C}{\partial x}\Delta x\right)\Delta t \pm KVC\Delta t \pm \text{ Sources and Sinks} \qquad (7.32)$$

where K is the decay coefficient (T$^{-1}$).
   The source and sink expression might be composed of several components:

1) a point source, which can be incorporated through the boundary condition at x = 0.
2) a distributed source or sink, that means input of waste, f.inst. by land run-off, or removal , f.inst. by settling along the length of the stream. It can be incorporated in the mass balance as:

$$(7.33)$$

$$C_d \cdot \Delta t \cdot V$$

where $C_d$ (M/L$^3$T) is the amount of material C added or withdrawn from the stream per unit of time. In case the distributed source or sink is accompanied by a water flow $\Delta Q$, the expression is changed to:

$$C_w \cdot \Delta Q \cdot \Delta t$$

where $C_w$ is the concentration of the considered component in the waterflow $\Delta Q$. If both expressions are included in the mass balance equation, we get:

$$(7.34)$$

$$V \cdot \Delta C - Q \cdot C\Delta t - (Q + \Delta Q)\left(C + \frac{dC}{x}\Delta x\right)\Delta \pm K \cdot V \cdot C \cdot \Delta t \pm C_w \cdot \Delta Q \cdot \Delta T \pm C_d \cdot \Delta t \cdot V$$

This equation is divided through by $\Delta t$ and V = A$\Delta$x:

208

$$\frac{\Delta C}{\Delta t} = -\frac{Q}{A}\frac{\partial C}{\partial x} - \frac{C}{A}\frac{\Delta Q}{\Delta x} - \frac{\Delta Q}{A}\partial C/\partial x \pm K \cdot C \pm \frac{C_w}{A}\frac{\Delta Q}{\Delta x} \pm C_d \tag{7.35}$$

If the infinitesimally small terms are allowed to approach to zero, we obtain:

$$\frac{\partial C}{\partial t} = -\frac{Q}{A}\frac{\partial C}{\partial x} - \frac{C}{A}\frac{\partial Q}{\partial X} \pm K \cdot C \pm C_w\left(\frac{\partial Q}{\partial X}/A\right) \pm C_d \tag{7.36}$$

The boundary condition is usually:

$C(t) = C_o(t)$  at x = 0

where $C_o(t)$ is the concentration of C at a waste discharge point x = 0.

This equation is used to consider the three cases mentioned in the introduction; section 7.3.1.

In the steady state conservative case K is zero, Q and A are constants in time, $\delta C/\delta t = 0$ and $C = C_o$. As Q is a constant $\delta C/\delta t = 0$. Equation (7.36) then becomes:

$$0 = -\frac{Q}{A} \cdot \frac{dC}{dx} \tag{7.37}$$

The stream velocity U(L/T) is introduced:

$$0 = -U\frac{dC}{dx} \tag{7.38}$$

The solution to this equation is a concentration, which simply can be computed from a mass balance at the discharge point so that:

$$C_0 = \frac{C_s \cdot Q_s + C_w \cdot Q_w}{Q_s + Q_w} = \frac{W}{Q_s + Q_w} \tag{7.39}$$

where the subscript s denotes concentration and flow in streams and the subscript w denotes the concentration and flow in waste water. $W = C_s.Q_s + C_w.Q_w$.

In the steady state non-conservative case, we obtain the following equation:

$$0 = -\frac{Q}{A}\frac{dC}{dx} - \frac{C}{A}\frac{dQ}{dx}$$
$$- K \cdot C + \left(\frac{dQ}{dx}/A\right) \cdot C_w \pm C_d \tag{7.40}$$

with the boundary condition:

C = C$_o$ at x = 0

Integration of the differential equation yields:

209

$$C = C_o \exp[-(K/U) \times x] \qquad (7.41)$$

Finally we consider the fixed parameter/dynamic input case. Q, A and K are constants in time, but may vary in space and the waste discharge is a function of time. We consider also that $C_o = 0$. The general equation is then:

$$\frac{\partial C}{\partial t} = -\frac{Q}{A}\frac{\partial C}{\partial x} - (K + \Upsilon)C \qquad (7.42)$$

where

$\Upsilon = (\partial Q/\partial x)/A$
$C = C_o$ at $x = 0$ and $C_o$ can be found from (7.39)

The solution is:

$$C(x,t) = C_0(t - t^+)\exp\left(-\int_0^x \frac{k+r}{U}\cdot dx\right) \qquad (7.43)$$

$$t^+ = \text{time of travel to } x = \int_0^x \frac{dx}{U}$$

## EXAMPLE 7.2

N is the number of coliform bacteria/l remaining after some time and we assume a first order die off equation. At 20°C K is falling in the range 0.5-3.5 day⁻¹. Let us assume a K = 1.O day⁻¹. How many km after the discharge of waste water in a river is the concentration of coliform bacteria 0.1% of the concentration at the discharge point? The flow rate of the river is 0.1 m/sec.

**Solution:**

The following equation is valid:

$$U\frac{dN}{dx} + K \cdot N = 0$$

At $x = 0$  $N = 100\%$ and $x/U$ is equal to the travel time.

We get:

$$N = N_0 \exp(-(K/U)x)$$

or

$$\ln \frac{N}{N_0} = -K \cdot t^+$$

or

$$\ln(10^{+3}) = t^+$$
$$t^+ = 6.9$$

$$x = 6.9 \times 0.1 \times 3600 \times 24 \text{ m} = 59.6 \text{ km}.$$

### 7.3.4. Estuarine Models

We consider a one-dimensional estuary, which is subject to tidal dispersion. The major difference between this case and a river/stream model is inclusion of this process by use of the following expression, see also chapter 3):

$$\left(-E \cdot A \frac{\partial C}{\partial x}\right) \Delta t \tag{7.44}$$

where $E(L^2/T)$ is the tidal dispersion coefficient. The minus sign indicates that the dispersion is assumed to occur in the direction of decreasing concentration. The full mass balance equation for the estuary is then:

$$V \cdot \Delta C = \left[Q \cdot C \cdot \Delta t - (Q + \Delta Q)\left(C + \frac{\partial C}{\partial x}\Delta x\right)\right]\Delta t + \left(\left(-E \cdot A \frac{\partial C}{\partial x} \cdot \Delta t\right)\right. \tag{7.45}$$
$$\left. -(-EA + \Delta E \cdot A)\left(\frac{\partial C}{\partial x} + \frac{\partial}{\partial x}\left(\frac{\partial C}{\partial x}\right)\Delta x\right)\right)\Delta t \pm K \cdot V \cdot C \cdot \Delta t \pm C_w \cdot \Delta Q \cdot \Delta t \pm V \cdot C_d \cdot \Delta t$$

The same symbols are used in this equation as in the river equation, see section 7.3.3.

Dividing through by t and $V = A$ x we get the following equation, when infinitesimally small terms are used:

$$\frac{\partial C}{\partial t} = -\frac{Q}{A}\frac{\partial C}{\partial x} - \frac{C}{A}\frac{\partial Q}{\partial x} + \frac{1}{A}\left[EA\frac{\partial}{\partial x}\left(\frac{\partial C}{\partial x}\right) + \frac{\partial EA}{\partial x} \cdot \frac{\partial C}{\partial x}\right] \tag{7.46}$$
$$\pm K \cdot C \pm C_w\left(\frac{\partial Q}{\partial x}/A\right) \pm C_d$$

For the constant coefficient, point source case with first order decay the equation is:

$$O = -\frac{Q}{A}\frac{dC}{dx} + E\frac{d^2C}{dx^2} \pm K \cdot C \tag{7.47}$$

The solution in this case is:

$$C = C_0 \exp(j_i x) \quad \text{for} \quad x < 0 \tag{7.48}$$
$$= C_0 \exp(j_2 x) \quad \text{for} \quad x > 0$$

$$j_i = \frac{U}{2E}\left[1 + \sqrt{1 + \frac{4K \cdot E}{U^2}}\right] \tag{7.49}$$

$$j_2 = \left[\frac{U}{2E}\right]\left[1 - \sqrt{1 - \frac{4KE}{U^2}}\right] \tag{7.50}$$

$$C_0 = \frac{W}{Q\sqrt{1 + \frac{4KE}{U^2}}} \tag{7.51}$$

This solution, but without the advection term is used in illustration 7.4 in section 7.6.3.

If C is a conservative variable, f.inst. chloride, the differential equation is:

$$0 = U\frac{dC}{dx} + E\frac{d_2C}{dx^2} \tag{7.52}$$

with the boundary condition:

$C = 0$    at    $x = -\infty$    and    $C = C_0$    at    $x = 0$

The solution is simply:

$$\begin{aligned} C &= C_0 \exp(Ux/E) \quad \text{for} \quad x < 0 \\ &= C_0 \quad \text{for} \quad x > 0 \end{aligned} \tag{7.53}$$

For the distributed waste load case we obtain the following equation:

$$E\frac{d^2C}{dx^2} - U\frac{dC}{dx} \pm K \cdot C + \frac{W}{A} = 0 \tag{7.54}$$

where W represents the constant addition of material along the length of the estuary. This equation has an analytical solution, which is not be presented here.

## ILLUSTRATION 7.1

If the longitudinal chloride profile is available for an estuary, it is possible to find the eddy diffusion coefficient. Fig 7.6 illustrates a semilogarithmic plot of chloride versus distance. As we have

$$C = C_0 \exp\left(\frac{UX}{E}\right) \quad \text{or} \quad \ln\frac{C}{C_0} = \frac{UX}{E} \tag{7.54A}$$

in flow (Q = 15000000 m³/day) and A (= 20000 m²) is known (see fig. 7.6):

$$\begin{aligned} \text{Slope} &= \frac{\ln\left(\frac{5}{2.8}\right)}{10} = 0.058 \text{Km}^{-1} \\ U &= \frac{15 \cdot 10^6}{2 \cdot 10^4} = 750 \text{m/day} = 0.75 \text{Km/day} \\ E &= \frac{0.75}{0.058} \text{Km}^2/\text{day} = 12.9 \text{Km}^2/\text{day} \end{aligned} \tag{7.54B}$$

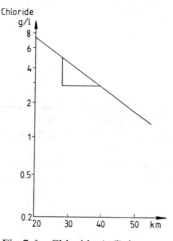

Fig.7.6: Chloride (g/l) is measured and plotted versus distance. From estuary mouth. Fresh water flow is 15 . 106 m³ day-1 and the area is determined to 20000 m².

### 7.3.5. Multidimensional Models and numerical Methods.

Almost all the laws of nature can be described as partial differential equations (PDEŚ) Examples are the Schrodinger equation in quantum mechanichś, Maxwellś equations in electromagnetic theory and Newtonś law in mechanicś. As most models of ecological systems are considered as biogeochemical models they are inherently partial differential equations in their mathematical embodiment. Most transport phenomena described in this book are special cases of the PDE:

$$\frac{\partial \bar{c}}{\partial t} = \frac{i}{A}\frac{\partial}{\partial \bar{r}}\left[A \cdot D\frac{\partial \bar{c}}{\partial \bar{r}} - \bar{v}{:}\bar{c}\right] + \bar{s} \tag{7.55}$$

where $\bar{c}$ denotes concentrations of the tranported components, A is a cross section area, D a generalized diffusivity, $\bar{v}$ is velocity, $\bar{s}$ a source or sink term that may include chemical reactions and t and $\bar{r}$ are the independent variables, such as time and the spatial coordinates (x, y, z).

Hydrodynamic equations describe the spatial and temporal distribution of the velocity of a fluid and are basically derived from Newton's law of momentum conservation:

$$\varrho\frac{\partial \bar{v}}{\partial t} = -\varrho\bar{v} \cdot \frac{\partial \bar{v}}{\partial \bar{v}} - \frac{\partial}{\partial \bar{v}} \cdot \bar{\bar{p}} + F \tag{7.56}$$

Here $\varrho$ is mass density of the fluid, $\bar{\bar{P}}$ the stress tensor and F the external forces acting on a volume element.

Except for very rare cases PED's do not have analytical solutions. For simple ordinary differential equations like

$$\frac{dc}{dt} = -c \qquad (7.57)$$

analytical solutions (c = k.e-t) contain arbitrary constants (K), which are fixed by the initial conditions. The analogue for these in -PEDś are arbitrary functions, which are fixed be boundary and initial conditions. Boundary conditions are often form the real challenge when PDEś are to be solved.

For real-life boundary conditions and PDEś some type of discretization is necessary to obtain a solution. Basically there are two techniques for this: finite difference and finite element. These methods will briefly be described.

For fairly general PDE's, like

$$\bar{f}\left(\bar{r}, t, \bar{c}, \frac{\partial \bar{c}}{\partial r}, \frac{\partial^2 \bar{c}}{\partial \bar{r}^2}, \frac{\partial \bar{c}}{\partial t}, \bar{s}\right) = \bar{0} \qquad (7.58)$$

(same notation as above), boundary conditions are usually formulated by specifying

(1) $\bar{c}(\bar{r}, t), \bar{r} \in \partial\Omega, t \in [t_1, t_2]$

(2) $\frac{\partial \bar{c}}{\partial \bar{r}}, \bar{r} \in \partial\Omega, t \in [t_1, t_2]$

(3) a combination of (1) and (2).

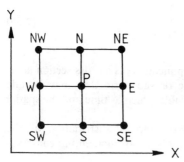

Fig.7.7: Discretization on a grid.

Here $\delta\Omega$ is the set of points at the boundary of the region $\Omega$ of interest. $t_1$ and $t_2$ are the initial and final time of interest. Initial conditions are specification of $c(r, t_1)$ for all $r \in \Omega$.

**Finite difference methods** are based on approximating derivatives by difference quotients evaluated on a grid. An example of a rectangular grid is shown in fig. 7.7 with a possible way of approximating spatial derivatives.

$$\frac{\partial C(P)}{\partial x} = \frac{C(E) - C(W)}{2\Delta x} + \cdots$$

$$\frac{\partial C(P)}{\partial y} = \frac{C(N) - C(X)}{2\Delta y} + \cdots$$

$$\frac{\partial^2 C(P)}{\partial x^2} = \frac{C(NE) - C(NW) - C(SE) + C(SW)}{4\Delta x \Delta y} + \cdots$$

$$\frac{\partial^2 C(P)}{\partial y^2} = \frac{C(N) - 2C(P) + C(S)}{(\Delta y)^2} + \cdots$$

The difference approximation shown in fig. 7.7 are centered difference formulae, which are accurate if c is a polynomial of 2nd degree or lower; other types of approximations, like one-sided difference formulae are only accurate to the 1st order.

If $\partial c / \partial t$ in eq. (7.58) is missing the problem is referred to as a boundary value problem. Initial value problems may be formulated as:

$$\frac{\partial \bar{c}}{\partial t} = \bar{f}\left(\bar{r}, t, \bar{c}, \frac{\partial \bar{c}}{\partial \bar{r}}, \frac{\partial^2 \bar{c}}{\partial \bar{r}^2}, \bar{s}\right) \tag{7.59}$$

With the same discretization technique as for boundary value problems, eq. (7.59) may be approximated by:

$$\tag{7.60}$$

$$\frac{\partial \bar{c}}{\partial t} = \bar{A} \, \bar{c} + \bar{B}$$

where $\bar{A}$ is a matrix and $\bar{B}$ a vector.

A common - but not always the best - practice is to solve (7.60) by integrating a set of ordinary differential equations:

$$\tag{7.61}$$

$$\frac{d\bar{c}(\bar{r}_i, t)}{dt} = \bar{A}_i \cdot \bar{c}(\bar{r}_i, t) + \bar{B}_i$$

This may be done with one of the standard algorithms, such as Runge-Kutta with fixed or variable step size, but fixed order, or a variable order, variable step size method, like Gears algorithm for stiff problems. Order in this context refers to the degree of the approximating polynomial. This technique to solve (7.60) is called the Method of Lines.

If c is varying almost as much or even more in space as in time a better way is to discretize the whole system; (7.60) may be approximated by:

$$\frac{\overline{c}(\overline{r}, t + \Delta t) - \overline{c}(\overline{r}, t)}{\Delta t} = \overline{A}\left[(1 - \alpha)\overline{c}(\overline{r}, t) + \alpha\overline{c}(\overline{r}, t + \Delta t)\right] \tag{7.62}$$

where $\alpha\epsilon(0,1)$.

If $\alpha = 0$, the method is explicit, and it is straight forward to calculate c at the next time step by repeated use of (7.62). Unfortunately the solution may easily blow up (be numerical unstable) for certain values of $\Delta t$.

If $\alpha > 0$, the method is implicit, and it is necessary to solve a set of algebraic equations given by (7.62) at each time step. For $\alpha = 1$ the method is unconditionally stable. A very popular method to solve the diffusion/transport equation in one dimension, is to set $= 1/2$. This is the Crank-Nicholson scheme. Since the matrix A in this case tri-diagonal, eq. (7.62) may be solved by looping two times only.

It may be noticed that the Method of lines actually is a finite difference method - explicit or implicit - with very narrow rectangles. (Compare fig. 7.7).

There are methods known as ADI (Alternating Direction Integration) which possess the good qualities of both explicit and implicit methods.

Finite Element Methods (FEM) were originally developed to solve complex problems in mechanical engineering. Mechanical systems were represented as series of discrete polygons called elements interconnected only at a finite number of locations called nodal points.

The success of any approximate method - whether finite element of finite difference - lies in the construction of a proper set of approximating basis functions, such as polynomia or spline functions. FEM offers a systematic way of constructing such sets. The basic idea is to selct the set such, that some measure of discrepancy between the exact and the approximate solution is minimized. To do this various types of variational principles and other functional analysis concepts are used in tranforming the original PDEs to relations among parameters of the basis functions.

Although FEM-solvers are usually more difficult to implement on a computer than finite difference techniques, they are more promising tools for most hydrodynamic and related problems, in particular if the boundary of the modelled system is complicated. For certain types of non-linearities and time dependencies finite difference methods may be a better choice, but many high quality FEM program packages that handle these problems are emerging to day.

### 7.3.6. Modelling a Stratified Lake.

The simulation of an annual temperature cycle in a lake can be accomplished with reasonably good agreement between model and prototype. This one-dimensional model is based upon mass and heat conservation:

$$\frac{\partial V_j}{\partial t} = Q_j - Q_{j+1} - Q_{oj}$$

$V_j$ = volume of j the control element $(L^3)$

$Q_j$ = vertical flow rate $(L^3 T^{-1})$

$Q_i$ = flow advected to the control volume in the horizontal plane ($L^3T^{-1}$)
$Q_o$ = flow advected from the control volume in the horizontal plane ($L^3T^{-1}$)

and all flows are functions of time. It is noted that except for the surface element

$$\frac{\partial V_j}{\partial t} = 0 \tag{7.63}$$

the general equation for heat energy stored in a control volume within the reservoir is, expressed in terms of temperature:

$$\frac{d(V_j T_j)}{dt} = \underbrace{(U_i T_i - U_0 T_0)_j}_{\text{local advection}} + + \underbrace{\left(\frac{1}{c\varrho}\int_Z^{Z+\Delta Z} q_{sz} a_{z} sz\right)_j}_{\text{solar radiation}}$$

$$\underbrace{- Q_j T_j + Q_{j+1} T_{j+1}}_{\text{vertical advection}} + \underbrace{\left(Ea\frac{\partial T}{\partial Z}\right)_j - \left(Ea\frac{\partial T}{\partial Z}\right)_{j+1}}_{\text{vertical diffusion}} \tag{7.64}$$

where

| | | |
|---|---|---|
| c | = | specific heat (J kg $-1C^{o-1}$ |
| $q_{sz}$ | = | solar radiation intensity at depth Z |
| | = | $(1- q_n)e^{-\eta z}$, in which |
| $q_n$ | = | net solar radiation penetrating the surface |
| $\eta$ | = | bulk extinction coefficient |
| $a_Z$ | = | horizontal area at depth Z |
| $E_Z$ | = | coefficient of diffusion |

Lam and Simons (1976) developed an advection-diffusion model of Lake Erie.
This model has been applied for the simulation of chloride distributions in Lake Erie for conditions corresponding to summer and fall 1970. The governing equation for the model is a statement of mass conservation for a layer bounded by two horizontal planes, 1 and 2. It was concluded that the models could also be readily applied to the simulation of non-conservative water quality parameters, including nutrients, and that they should serve as a basis for modelling ecosystem of large lakes. A general procedure to set up a multi-layer, multi-segment lake model could be:

1. To set up the mass conservation equations.
2. The temperature is included as a state variable, and the equation mentioned above is set up.
3. The temperature submodel is calibrated to fit the observations.
4. Adjustments are made in diffusion coefficients to calibrate the model to observed water quality conditions.

The MIT reservoir model (Ryan and Harleman 1971) uses a similar approach, applying the following basic heat-transport equation for an internal element

$$\frac{\partial T}{\partial t} + \underbrace{\frac{1}{A}\frac{\partial(Q_v T)}{\partial Z}}_{\substack{\text{vertical}\\\text{advection}}} = \underbrace{\frac{E}{A}\frac{\partial}{\partial Z}\left(A\frac{\partial T}{\partial Z}\right)}_{\substack{\text{vertical}\\\text{diffusion}}} + \underbrace{\frac{B U_i T_i}{A} - \frac{B U_o T}{A}}_{\substack{\text{local}\\\text{advection}}} - \underbrace{\frac{1}{\varrho c}\frac{\partial \Phi Z}{\partial Z}}_{\substack{\text{solar}\\\text{radiation}}} \qquad (7.65)$$

where

| | | |
|---|---|---|
| T | = | is the temperature at the depth Z |
| A | = | area of the "element" |
| B | = | width of the "element" |
| $U_i$ | = | horizontal inflow velocity |
| $T_i$ | = | temperature of the inflow |
| $U_o$ | = | horizontal outflow velocity |
| $Q_v$ | = | vertical flow rate |
| $\Phi$ | = | internal short wave solar radiation flux oer unit horizontal area |
| E | = | vertical turbulent diffusion coefficient |
| c | = | heat capacity of water |
| $\varrho$ | = | density of water |

The equation is solved using an explicit finite difference scheme.

However, there is no satisfactory representation of the vertical eddy diffusivity as a function of depth and time. To meet this problem the MIT reservoir model has been expanded to include wind mixing (Octavia et al., 1977). The mixing rule is based on the ratio of turbulent kinetic input by the wind to the potential energy of the isothermal wind-mixed layer relative to the element immediately below it. The potential energy, PE of the ith element above the jth element is defined as

$$PE = g \sum_1 A(i)\Delta Z(\varrho(j_1 t) - \varrho(i,t))D(i,t) \qquad (7.66)$$

| | | |
|---|---|---|
| A(i) | = | area of the ith element |
| $\varrho(i,t)$ | = | density of the ith element at time t |
| $\varrho(j,t)$ | = | density of the element immediately below the mixed layer at time t |
| $\Delta Z$ | = | thickness of the layer |
| D(i,t) | = | distance between the ith element and jth element at the time t |
| g | = | gravitation constant |

Since the mixed layer is assumed isothermal, it can be shown that

$$PE = g\Delta\varrho V_m \delta Z\frac{m}{Z} = g\Delta\varrho \overline{A}H\Delta Z\frac{m}{Z} \qquad (7.67)$$

where

m     = the number of elements in the mixed layer

m Z   = depth of the mixed layer = H
$V_m$   = volume of the mixed layer = $\bar{A}H$, where
$\bar{A}$   = the average cross sectional area of the mixed layer and
$\Delta \varrho$   = density difference between the mixed layer and the jth
      element immediately below.

A measure of the energy input by the wind, KE, is

$$KE = \varrho v A_{surface} \Delta t, \text{ where}$$
v is the friction velocity and $A_{surface}$ the surface area.

The critical ratio is

(7.68)

$$PE/KE = 1$$

and the mixing algorithm states that if PE/KE < 1 there will be no mixing, while if PE/KE > 1, layers will be mixed one by one and the ratio rechecked after mixing each layer until the ratio becomes < 1.

The schematic of the iterative heating-wind mixing procedure is shown on fig. 7.8.

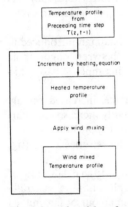

Fig.7.8: Algorithm for model of temperature profile.

The same procedure can be applied when the concentrations of components in ecological models are considered. The approach is attractive, as the calibration of E is omitted. The model has been successfully been applied to the calculation of temperature profiles in lakes (Octavia et al., 1977), and has also been applied to ecological constituents. Parker (1978) has introduced a model which takes the spatial heterogeneity of phytoplankton and dissolved nutrient by the nutrient-phytoplankton interaction into consideration. Dubois (1975) and Parker (1976) have coupled the dynamics of kredator-prey populations with the consequences of physical processes in a turbulent medium. Halfon and Lam (1978) have simulated the spatial movement of phosphorus by using the computed currents from a 3-dimensional hydrodynamic model. A biological submodel describes the phosphorus dynamics and primary production in each grid cell of 20 x 20 km² in Lake Superior.

As seen from these references the basic work for generating 3-dimensional ecological models has been carried out. However, complete 3-dimensional ecological models, consisting of several ecological state variables, have not yet been generated. The question is whether more is lost than gained with the application of such a complex model at this stage. However, multilayer models might be a reasonable compromise for ecological modelling of stratified lakes.

## 7.4. EUTROPHICATION MODELS

### 7.4.1. Eutrophication

From a thermodynamic point of view, a lake can be considered as an open system which exchanges material (waste water, evaporation, precipitation) and energy (evaporation, radiation) with the environment. However, in many great lakes the input of material per year is not able to change the concentration measurably. In such cases the system can be considered as almost closed, which means that it is exchanging energy, but not material with the environment.

The flow of energy through the lake system leads to at least one cycle of material in the system (provided that the system is in a steady state, see Morowitz, 1968). As illustrated in figs. 2.1, 2.8, 3.6 and 7.11 the important elements all participate in cycles that control eutrophication.

The word eutrophy is generally taken to mean "nutrient rich". Nauman introduced in 1919 the concepts of oligotrophy and eutrophy, distinguishing between oligotrophic lakes containing little plantonic algae and eutrophic lakes containing much phytoplankton.

The eutrophication of lakes in Europe and North America has increased rapidly during the last decade due to increased urbanization and consequently increased discharge of nutrient per capita. The production of fertilizers has grown exponentially in this century (fig. 7.9) and the concentration of phosphorus in many lakes reflects this (fig. 7.10) (from Ambuhl, 1969)

The word eutrophication is used increasingly in the sense of artificial addition of nutrients, mainly nitrogen and phosphorus, to waters. Eutrophication is generally considered to be undesirable, but this is not always true. The green colour of eutrophied lakes makes swimming and boating less safe due to the increased turbidity, and from an aesthetic point of view the chlorophyll concentration should not exceed 100 mg m-3. However, the most critical effect from an ecological point of view is the reduced oxygen content of the hypolimnion, caused by the decomposition of dead algae. Eutrophic lakes sometimes show a high oxygen concentration, during the summer time, at the surface, but a concentration of oxygen in the hypolimnion, that is lethal to fish.

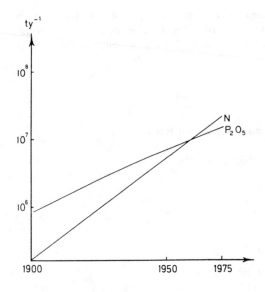

Fig.7.9: The production of fertilizers (t y$^{-1}$) has, as demonstrated for N and P2O5, grown exponentially (the y-axis is logarithmic).

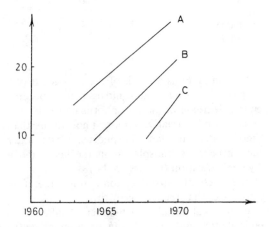

Fig.7.10: The total P-concentration (μg per 1) in 3 segments of Vierwaldstatter-see as a function of time. A Kreutzrichter, B Lake Gersauer, C Lake Urner.

About 16-20 elements are necessary for the growth of freshwater plants, and Table 7.8 lists the relative quantities of essential elements in plant tissue. The present concern about eutrophication relates to the rapidly increasing amount of phosphorus and nitrogen, which are normally present at relatively low concentrations. Of the two,

221

phosphorus is considered the major cause of eutrophication, as it was formerly the growth-limiting factor for algae in the majority of lakes, but, as mentioned previously and demonstrated in fig. 7.1, its usage has increased tremendously during the last decade.

**TABLE 7.8**

**Average Freshwater Plant Composition on a wet weight Basis.**

| Element | Plant content% |
|---|---|
| Oxygen | 80.5 |
| Hydrogen | 9.7 |
| Carbon | 6.5 |
| Silicon | 1.3 |
| Nitrogen | 0.7 |
| Calcium | 0.4 |
| Potassium | 0.3 |
| Phosphorus | 0.08 |
| Magnesium | 0.07 |
| Sulfur | 0.06 |
| Chlorine | 0.06 |
| Sodium | 0.04 |
| Iron | 0.02 |
| Boron | 0.001 |
| Manganese | 0.0007 |
| Zinc | 0.0003 |
| Copper | 0.0001 |
| Molybdenum | 0.00005 |
| Cobalt | 0.000002 |

Nitrogen is limiting in a number of East African lakes as a result of the nitrogen depletion of soils by intensive erosion in the past. However, today nitrogen may become limiting in lakes as a result of the tremendous increase in the phosphorus concentration caused by discharge of waste water, which contains relatively more phosphorus than nitrogen. While algae use 4-10 times more nitrogen than phosphorus, waste water generally contains only 3 times as much nitrogen as phosphorus in the lakes and a considerable amount of nitrogen is lost by denitrification (nitrate -> $N_2$).

The growth of phytoplankton is the key process of eutrophication and it is therefore important to understand the interacting processes that regulate growth.

Primary production has been measured in great detail in a number of lakes. This process represents the synthesis of organic matter and over-all can be summarized as follows (for details see section 3.4.1):

$$\text{Light} + 6CO_2 + 6H_2O \rightarrow C_6H_{12}O_6 + 6O_2$$

The composition of phytoplankton is not constant (note that Table 7.8 gives only an average concentration), but is to a certain extent reflecting the concentration of the water. If, e.g., the phosphorus concentration is high, the phytoplankton will take up relatively more phosphorus - this is called luxury uptake.

As seen from Table 7.8 phytoplankton consists mainly of carbon, oxygen, hydrogen, nitrogen and phosphorus: without these elements no algal growth will take place. This leads to the concept of limiting nutrient mentioned above and in section 3.4.1 and which has been developed by Liebig as the law of the minimum. This states that the yield of any organism is determined by the abundance of the substance that in relation to the needs of the organism is least abundant in the environment (Hutchinson, 1970). However, the concept has been considerably misused due to oversimplification. First of all, growth might be limited by more than one nutrient. The composition as mentioned above is not constant, but varies with the composition of the environment. Furthermore, growth is not at its maximum rate until the nutrients are used, and is then stopped, but the growth rate slows down when the nitrients become scattered.

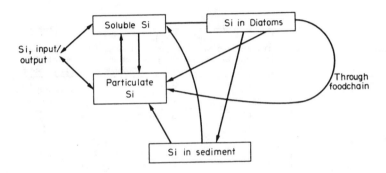

Fig.7.11: The silica cycle.

In 3.4.1 is discussed, how this can be considered in a relation between phytoplankton growth and nutrient concentrations. Here consideration is also given to how the interaction of several limiting nutrients simultaneously can be taken into account.

Another side of the problem is consideration of the nutrient sources. It is of importance to set up mass balances for the most essential nutrients.

The sequences of events leading to eutrophication has often been as follows: Oligotrophic waters will have a ratio of N:P greater than or equal to 10, which means that phosphorus is less abundant than nitrogen for the needs of phytoplankton. If sewage is discharged into the lake the ratio will decrease, since the N:P ratio for municipal waste water is 3:1, and consequently nitrogen will be less abundant than phosphorus relative to the needs of phytoplankton. In this situation, however, the best remedy for the excessive growth of algae is not the removal of nitrogen from the sewage, because the mass balance might then show that nitrogen-fixing algae will give an uncontrollable input of nitrogen into the lake. It is necessary to set up a mass balance for the nutrients and this will often reveal that the input of nitrogen from nitrogen-fixing bluegreen algae, precipitation and tribuaries is contributing too much to the mass balance for the removal of nitrogen from the sewage to have any effect. On the other hand, the mass balance may reveal that the phosphorus input (often more than 95%) comes mainly from sewage, which means that it is better management to remove phosphorus from the sewage than nitrogen. Thus, it is not important which nutrient is most limiting, but which nutrient can most easily be made to limit the algal growth.

### 7.4.2. Eutrophication Models an Overview.

Several eutrophication models with a wide spectrum of complexity have been developed. As for other models the right complexity of the model is dependent on the available data and the ecosystem. Table 7.9 reviews various eutrophication models.

## TABLE 7.9

### Various Euthophication Models

Table 7.9

| Model | Number of state variables per layer or segment | Nutrient considered | Seg- ments | Dimension (D) or Layers (L) | Constant Stoicho- metrics (CS) or independent Nutri- ent Cycle (NC) | Calibrated (C) and/or Validated (V) | Number of case studies in litterature |
|---|---|---|---|---|---|---|---|
| Vollenweider | 1 | P (N) | 1 | 1L | CS | C + V | many |
| Imboden | 2 | P | 1 | 2L, 1D | CS | C + C | 3 |
| O'Melia | 2 | P | 1 | 1D | CS | C | 1 |
| Larsen | 3 | P | 1 | 1L | CS | C | 1 |
| Lorenzen | 2 | P | 1 | 1L | CS | C + C | 1 |
| Thomann 1 | 8 | P,N,C | 1 | 3L | CS | C + V | 1 |
| Thomann 2 | 10 | P,N,C | 1 | 7L | CS | C | 1 |
| Thomann 3 | 15 | P,N,C | 67 | 7L | CS | – | 1 |
| Chen & Orlob | 15 | P,N,C | sev. | 7L | CS | C | min. 2 |
| Patten | 33 | P,N,C | 1 | 1L | CS | C | 1 |
| Di Toro | 7 | P,N | 7 | 1L | CS | C + V | 1 |
| Biermann | 14 | P,N,Si | 1 | 1L | NC | C | 1 |
| Canale | 25 | P,N,Si | 1 | 2L | CS | C | 1 |
| Jørgensen | 17 | P,N,C | 1 | 1–2L | NC | C + V | 17 |
| Cleaner | 40 | P,N,C,Si | sev. | sev. L | CS | C | many |
| Nyholm | 7 | P,N | 1–3 | 1–2L | NC | C + C | 13 |

The table indicates the characteristic features of the models, the number of case studies it has been applied (of course with some modification from case study to case study, as a general model is nonexisting and as site specific properties should be reflected in the selected modification) and whether the model has been calibrated and validated.

Some of the most simple models, that can be used in a data-poor situation are presented below. These models give the reader a good impression of problems involved in modelling the eutrophication process.

Simple eutrophication models are based upon three steps:

1) Determination or calculation of nutrient loading.
2) Prediction of the nutrient concentration (usually only one nutrient is considered)
3) Prediction of eutrophication.

These three steps are presented below.

It is of course not possible to treat all the more complex models in detail. Therefore one model among the more complex models has been selected and presented in more detail. Eutrophication models demonstrate quite clearly the ideas behind biogeochemical models and therefore it has been found fruitful to go into some illustrative details about validation of the model and its prognosis. The presented results obtained by use of a relatively complex eutrophication model, demonstrate what can be achieved to-day by use of ecological models, provided that sufficient effort is expended to obtain good data and good ecological background knowledge about the modelled ecosystem.

### 7.4.3. Some relatively simple Eutrophication Models.

Determination of nutrient balances is the basic of all eutrophication models. It is possible by measurement of the concentrations and flow rates of in- and outputs. Alternatively it is possible to calculate the nutrient loading as demonstrated below, although it is only recommendable to use the calculation method, when data are not available.

### I Calculation of the nutrient loading of lakes.
The first step is to set up a nutrient balance for the lake system. Even with a lack of data it is possible to give some general lines.

### a) Natural P and N loads from land
Table 7.10 shows a phosphorus-, $E_p$, and nitrogen-, $E_n$, export scheme based on a geological classification.

The figures are based on an interpretation of the following references: Dillon and Kirchner (1975), Loenholdt (1973) and (1976), Vollenweider (1968) and Loehr (1974).

To calculate the natural nutrient loading to a lake, one must know 1) the area A1 of the watershed of each tributary to the lake, and 2) classify each as to geology and land use.

The total amount of phosphorus, $I_{P\ell}$, and nitrogen, $I_{N\ell}$, supplied to the lake from the land is therefore calculated by the use of the following equations:

$$I_{P\ell}\left(mgy^{-1}\right) = \sum_{i=1} A_\ell\left(m^2\right) E_P\left(mgm^{-2}y^{-1}\right) \tag{7.69}$$

$$I_{N\ell}\left(mgy^{-1}\right) = \sum_{i=1} A_\ell\left(m^2\right) E_n\left(mgm^{-2}y^{-1}\right) \tag{7.70}$$

**TABLE 7.10**

**Export Scheme of Phosphorus, $E_P$, and Nitrogen, $E_N$. (mg m$^{-2}$ y$^{-1}$)**

| Land use | Geological classification Igneous ($E_P$) | Sedimentary | Geological classification Igneous ($E_N$) | Sedimentary |
|---|---|---|---|---|
| Forest runoff | | | | |
| Range | 0.7-9 | 7-18 | 130-300 | 150-500 |
| Mean | 4.7 | 11-7 | 200 | 340 |
| | | | | |
| Forest + Pasture | | | | |
| Range | 6-12 | 11-37 | 200-600 | 300-800 |
| Mean | 10.2 | 23.3 | 400 | 600 |
| | | | | |
| Agricultural Areas | | | | |
| Citrus | 18 | | | 2240 |
| Pasture | 15-75 | | | 100-850 |
| Cropland | 22-100 | | | 500-1200 |

## b) Natural P and N loads from precipitation

Table 7.11 is a compilation of the following references. Schindler and Nighswander (1970), Armstrong and Schindler (1971), Barica and Armstrong (1971), Dillon and Rigler (1974a). Lee and Kluesener (1971) and Jørgensen et al. (1973). Based upon the annual precipitation of P (mm y$^{-1}$) it is possible to find the supply of phosphorus Ipp and nitrogen INP from precipitation:

$$I_{PP} \text{ (mg Y}^{-1}) = P\, C_{PP} A_S$$

$$I_{NP} \text{ (mg y}^{-1}) = P\, C_{NP}\, A_S$$

(7.71)

where $A_S$ is the surface area of the lake, and $C_{pp}$ and $C_{NP}$ are the phosphorus and nitrogen concentrations in rain water (see table 7.11)

## c) Artificial P and N loads

The calculation of the artificial nutrient supply to a lake must necessarily be based on per capita and yearly figures, and great care must be taken in selecting the appropriate value. The following points must be taken into consideration:

1) **The discharge per capita and per year is approx. 800-1800 g P and 3000-3800 g N.**
2) **Mechanical treatment removes 10-15% of the nutrients.**
3) **Biological treatment removes 10-15% of the nutrients.**
4) **Chemical precipitation removes 80-90% of the phosphorus.**
5) **The retention coefficients, R, of total phosphorus for septic tile filter beds of different characteristics are shown in table 7.12 (after Brandes et al. (1974). The retention coefficients of total nitrogen for septic tile filter beds are of the order 0.01-0.1.**

## TABLE 7.11

### Nutrient Concentration in Rain Water (mg l⁻¹)


| | $C_{PP}$ | $C_{NP}$ |
|---|---|---|
| Range | 0.025 - 0.1 | 0.3 - 1.6 |
| Mean | 0.07 | 1.0 |

## TABLE 7.12

### Retention Coefficients. (Brandes et al., 1974)   D = grain size

| Filter bed | $R_S$ |
|---|---|
| 4% sed. mud 96% sand (70 cm) | 0.76 |
| 75 cm sand D = 0.3 mm | 0.34 |
| 75 cm sand D = 0.6 mm | 0.22 |
| 75 cm sand D = 0.24 mm | 0.48 |
| 75 cm sand D = 1.0 mm | 0.01 |
| 10% sed. mud 90% sand (37 cm) | 0.88 |
| 50% limestone 50% sand (37 cm) | 0.73 |
| Silty sand (70 cm) | 0.63 |
| 50% clay silt 50% sand (37 cm) | 0.74 |

Based upon the considerations indicated above, the P load ($I_{Pw}$) and N load ($I_{Nw}$) can be found.

### II Prediction of the nutrient concentration in a lake

Vollenweider (1969) assumed that the change in concentration of phosphorus in a lake with time is equal to the supply added per unit volume minus loss through sedimentation and the loss by outflow.

$$d\frac{[P]}{dt} = \frac{I_{P\ell} + I_{pp} + I_{Pw}}{V} = s[P] - r[P] \tag{7.72}$$

where [P] represents the total phosphorus concentration (mg l⁻¹), V is the lake volume (l), s is the sedimentation rate (y⁻¹) and r the flushing rate (y⁻¹); r is equal to $\frac{Q}{V}$, where Q is the total volume of water flowing out per year (l.y⁻¹).

This equation can be solved analytically:

$$[P] = \frac{I_P}{V(s + r)} - (1 - \frac{V(r + s)[P_o]}{I_p})e^{-(r+s)t} \tag{7.73}$$

$$\tag{7.74}$$

$$I_p = I_{P\ell} + I_{PP} + I_{PW}$$

The equations for [N], the nitrogen concentration, are parallel to those for [32P]. The steady state solution is:

$$[P] = I_p/r + s \qquad (7.75)$$

$$[N] = I_N/r + s \qquad (7.76)$$

where

$$I_N = I_{Nl} + I_{NP} + I_{Nw} \qquad (7.77)$$

As seen it is necessary to calculate or measure Q. In some cases the long-term average inflow, $Q_{in}$, can be calculated as

$$Q_{in} = A_\ell \cdot [P](1 - k') \qquad (7.78)$$

where k' is the ration of evapotranspiration to precipitation. k' is often known for a given geographical area, and Q can be found on the basis of a water balance:

$$Q = Q_{in} + A_s \times P - A_s \times E_V \qquad (7.79)$$

where
$E_V$ represents the evaporation in mm $y^{-1}$ $m^{-2}$.

The only alternative to these calculations is to measure Q or $Q_{in}$. It is rather difficult to determine s, the sedimentation rate. However, an alternative retention coefficient, R (equal to the fraction of the loading that is not lost via the outflow) may be used, Kirchner and Dillon (1975) determined by multiple regression analysis that R was highly correlated with $Q/A_s$, the areal water loading.

The equation for the prediction of R is:

$$R = 0.426 \exp\left(-0.271\frac{Q}{A_s}\right) + 0.574 \exp\left(-0.00949\frac{Q}{A_s}\right) \qquad (7.80)$$

If the lake in question has one or more lakes upstream that are sufficiently large to retain a significant amount of the total nutrient exported from their respective portion of the water shed, this can be taken into account by calculating the supply to the upstream lake, the lakes retention coefficient, $R_A$, and multiplying the supply by $(1-R_A)$ to give the fraction transfer to the downstream lake.

The above mentioned retention coefficient was generated for phosphorus. Calculations carried out in 18 lake studies in Scandinavia have shown that R is relatively 10-20% lower (average 16%) for nitrogen than for phosphorus.

The use of R instead of s leads to the following basic equation:

$$[P] = \frac{I_P(1 - R)}{Vr} - \left(1 - \frac{VrP_o}{1(1 - R)}\right)e^{-\frac{r}{1-R}t} \qquad (7.81)$$

and for the steady state situation:

$$[P] = \frac{I_P(1 - R)}{Vr} \qquad (7.82)$$

The equations for nitrogen are parallel; only a 16% lower R value is used.

Imboden (1974) suggested a two-compartment model for phosphorus content. The model considers a stratified lake and includes input, output, and exchange between hypolimnion and epilimnion, as well as sediment exchange. Four coupled differential equations for dissolved and particulate phosphorus are applied. The model has been improved (Imboden and Göhter, 1978, Imboden, 1979) by describing nutrient and biomass concentrations as continuous functions of time and depth and by replacing the first-order kinetic by Michaelis-Menten kinetics: O'Melia (1974) and Snodgrass and O'Melia (1975) developed a similar model, but did not include release of phosphorus from the sediment; however, depth-dependent rates of turbulent diffusion were considered.

Larsen et al. (1974) found that the Vollenweider and Snodgrass-O'Melia models underestimated the actual amount of epilimnetic phosphorus, when applied to Lake Shagawa in Minnesota. They then applied a slightly more complex model consisting of a three-compartment epilimnetic model, which includes algae as a sink for soluble reactive phosphorus and conversion of particulate phosphorus to the soluble form. The basic equations for this model are:

$$\frac{dPA}{dr} = \text{MYMAX}(T) \cdot \text{LIGHT} \cdot \frac{\text{PS}}{\text{KP} + \text{PS}} \cdot \text{PA} \qquad (7.83)$$
$$- (\text{CONR1} + \text{SETTL1} + \varrho_w) \cdot \text{PA}$$

$$\frac{dPS}{dr} = \frac{\text{PSIN}}{\text{VE}} - \text{MYMAX}(T) \cdot \text{LIGHT} \cdot \frac{\text{PS}}{\text{KP} + \text{PS}} \cdot \text{PA} \qquad (7.84)$$
$$+ \text{CONR2} \cdot \text{PP} + \varrho_w \text{PS}$$

$$\frac{dPP}{dr} = \frac{\text{PPIN}}{\text{VE}} + \text{CONR1} \cdot \text{PA} + (\text{CONR2} - \text{SETTL2} + \varrho_w \cdot \text{PP} \qquad (7.85)$$

PA          is the concentration of algal phosphorus [M L$^{-3}$],
Light        is the fractional reduction of MYMAX(T) in the epilimnion due to the availability of light,
MYMAX(T) is the maximum specific growth rate of phytoplankton as a function of temperature [T$^{-1}$],
KP          is the half-saturation constant for phosphorus [M L$^{-3}$],
CONR1      is the rate constant for conversion of algal phosphorus to particulate phosphorus [T$^{-1}$],
CONR2      is the rate constant for conversion of particulate phosphorus to soluble phosphorus [T$^{-1}$],

PP          is the concentration of particulate (non-algal) phosphorus [M L$^{-3}$],
PPIN        is the rate of supply of particulate phosphorus to the epilimnion [M T$^{-1}$],
PS          is the concentration of soluble phosphorus [M L$^{-3}$],
PSIN        is the rate of supply of soluble phosphorus to the epilimnion [M T$^{-1}$],
SETTL1      is the rate constant for settling of algal phosphorus (corresponding to a settling velocity of 0.02 m day$^{-1}$),
SETTL2      is the rate constant for settling of non-algal particulate phosphorus (corresponding to a settling velocity of 0.04 m day$^{-1}$),
T           is the temperature,
VE          is the volume of the epilimnion [L$^3$],

Lorenzen et al. (1976) developed a model consisting of two differential equations only, one for soluble and one for exchangeable phosphorus in the sediment:

$$\frac{d\text{PS}}{dt} = \frac{\text{PSIN}}{\text{VL}} + \frac{K_2 \cdot \text{AREA} \cdot \text{PSED}}{\text{VL}} - \frac{K_1 \cdot \text{AREA} \cdot \text{PS}}{\text{VL}} - \frac{Q}{\text{VL}} \cdot \text{PS} \qquad (7.86)$$

$$\frac{d\text{PSED}}{dt} = \frac{K_1 \cdot \text{AREA} \cdot \text{PS}}{\text{VS}} - \frac{K_2 \cdot \text{AREA} \cdot \text{PSED}}{\text{VS}} - \frac{K_1 K_3 \cdot \text{AREA} \cdot PS}{\text{VS}}, \qquad (7.87)$$

where

AREA is the lake surface area [L2],
$K_1$    is the rate of transfer of phosphorus to the sediment [L T$^{-1}$],
$K_2$    is the rate of transfer of phosphorus from the sediment [L T$^{-1}$],
$K_3$    is the fraction of total phosphorus input to the sediment that is not available for exchange,
PSED is the total concentration of exchangedable phosphorus in the sediment [M L$^3$],
VS    is the sediment volume [L$^3$],
Q     is the outflow [L$^3$ T$^{-1}$],
VL    is the lake volume [L$^3$],

The purpose of the model is to predict long-term changes in lakes that have undergone significant changes in loading rates. PSIN is therefore understood as the annual loading of PS, Q the annual outflow, and $K_1$ and $K_2$ are measured in m r$^{-1}$.

The equations can be solved analytically and the steady state solution of PS is

$$\text{PS}_\infty = \frac{\text{PSIN}}{Q + K_1 K_3 \cdot \text{AREA}}. \qquad (7.88)$$

A characteristic feature of this model is that, in spite of its simplicity, it considers the sediment-accumulated phosphorus and that only a fraction of the total phosphorus input to the sediment is available for the exchange process. More complex models do not include this important property of the phosphorus in the sediment, although it is of great importance for the long-term change in lakes because a substantial part of the phosphorus in a lake system is accumulated in the sediment.

The parameters of this model are estimated by the following procedure. When reasonably good data on loading rates and average aqueous and sediment concentrations are known:

(1) $K_3$ is estimated.
(2) Since

$$K_1 K_3 = \frac{PSIN - PS_\infty \cdot Q}{PS \cdot AREA},$$ (7.89)

$K_1$ can be calculated.
(3) $K_2$ is calculated from
    PS

$$K_2 = \frac{PS_\infty}{PSED_\infty} \cdot K_1(1 - K_3)$$ (7.90)

as the ratio of steady state aqueous to sediment phosphorus concentrations is given by (analytical solution)

$$\frac{PS_\infty}{PSED_\infty} = \frac{K_2}{K_1} \frac{1}{1 - K_3}$$ (7.91)

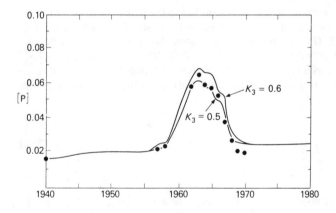

Fig. 7.12: Calculated and observed ($\bullet$) annual average total phosphorus concentrations (mg l$^{-1}$) in Lake Washington.

The model was used on Lake Washington by applying data from 1941-1950 to calculate a consistent set of model constants, based upon $K_3 = 0.6$. K3 can be found on the basis of sediment analysis (a more detailed examination of mud-water exchange of phosphorus was reported by Kamp-Nielsen, 1975). The observations during 1955-1970, which showed that the phosphorus loading increased up to 1964 and decreased thereafter, were well predicted by the model. However, $K_3 = 0.5$ gave a better result (fig. 7.12).

Lappalainen (1975) improved Volleweider's approach by considering the state of a lake as a function of lake volume, discharge and phosphorus input. In this model a regression equation that relates the net sedimentation of phosphorus and the oxygen concentration of the hypolimnion is determined. The model includes a relationship between the sedimentation of phosphorus and volume, discharge, and phosphorus input. This sedimentation submodel and the regression expression were used to construct a model for the prognosis of the oxygen concentration of the hypolimnion, which is used to determine the boundary phosphorus input, comparable with loads given earlier in the literature.

## III Predictions of eutrophication

Dillon and Rigler (1974) and Sakamoto (1966) have developed a relationship for estimating the average summer chlorophyll in a concentration (chl.a) with the N:P ratio of the water > 12:

$$(7.92)$$

$$\log_{10} (chl.a) = 1.45 \log_{10}(P) \cdot 1000 - 1.14$$

In case the N:P ratio is < 4 the following equation, based upon eight case studies was evolved:

$$(7.93)$$

$$\log_{10} (chl.a) = 1.4 \log_{10}(N) \cdot 1000 - 1.9$$

(N) and (P) are expressed as $mg^{-1}$ while (chl.a) is found in $ug l^{-1}$. If the $\frac{N}{P}$ ratio is between 4 and 12, the use of the smallest value of (chl.a) found on the basis of the two equations is recommended. Dillon et al. (1975) have set up a relationship between the Secchi disc transparency, SE and (chl.a), which is shown in fig. 7.13. The use of equation (7.81) involves that (P) or/and (N) as a function of time is known. It might, however, be more convenient since the expression is exponential, to use half life time, t1/2. As seen from (7.72) and (7.81) we have that

$$t_{1/2} = \frac{(ln2)(1 - R)}{r} \qquad (7.94)$$

The simple model presented above will never be as good a predictive tool as a model based on more accurate data and taking more processes into consideration.

However, the semiquantative estimations, which can be obtained by use of the simple model presented, are better than none at all.

Methods to estimate the input of P and N to a lake, from natural and artificial sources, are considered. From equation (7.81) it is possible to estimate the P concentrations in the lake water as a function of time.

The N concentration can be estimated by a parallel set of equations.

These considerations can be translated into chlorophyll, a concentration can be determined by means of equation (7.92) and/or (7.93) and in fig. 7.13 the transparency can be found when (chl.a) is known.

It is possible in this way to test different waste water treatment programmes and answer such questions as should N or P be removed? What would be the increase in efficiency, if the transparency should be improved by twice or more?

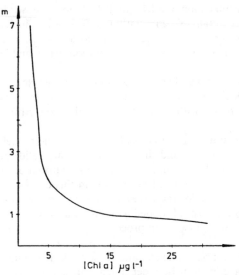

Fig.7.13: Transparency versus (m) (chl.a).

### 7.4.4. Complex Eutrophication Model.

The model presented figs. 2.1, 2.9 and with modifications in fig. 2.10 has been selected as an illustrative example of a medium complex eutrophication model. The results presented table 2.9 and fig. 2.14 are also related to this model.

The model was developed for Lake Glumsoe - a case study, which has the following advantages:

(1) The lake is shallow (mean depth 1.8 m) and no formation of a thermocline takes place. The case study is relatively simple;

(2) The lake is small (volume 420,000 m³) and well mixed, which implies that a model need not consider hydrodynamics but can concentrate on ecological processes;

(3) Retention time is short (< 6 months), which means that any change due to a management action must be observed rather rapidly;

(4) A radical change in nutrient input occurred in April 1981, and the water quality changes that followed have been observed (Jørgensen et al., 1984) ;

(5) Unique, in that a prognosis of change was published before any changes actually took place (Jørgensen et al., 1978). It has since been possible to validate this prognosis;

(6) The lake has been intensely studied during 1973-1984. The model is therefore based on comprehensive data.

However, the model has also been applied in 15 other case studies - of course with the necessary modifications,which will be presented below.

Due to the comprehensive investigation of the applicability of the model to Lake

Glumsoe, to the unique feature, that the model prediction under radically changed loadings was validated to the wide application of the same model with modifications, it is probably the most well examined eutrophication model, published up to now. It implies that the results represent, what is obtainable in relation to validation under almost unchanged loadings, to accuracy in predictions, (see chapter 8) and to general applicability. Therefore emphasis is laid on these results in the presentation below and in section 8.3.2.

The ecology of Lake Glumsoe was investigated before the model was developed, (Jørgensen et al., 1973). The phases in modelling presented in section 2.2 were followed rather carefully so as to be able to obtain a model with the needed predictive power for it to be used as a management instrument.

Figs. 2.1 and 2.9 are the conceptual diagrams of the N- and P-flows of the model. The last figure gives the equations of the P-processes and similar equations are valid for N- and C-processes. Many of the equations can be found in other eutrophication models and in the chapter on unit processes. It seems therefore of little value in this context to present all the equations of the model, and the following pages are devoted to the most characteristic features of the model to illustrate typical modelling considerations.

The growth of phytoplankton is described as a two-step process:

(1) Uptake of nutrients in accordance with Monod's kinetics, and (2) growth determined by the internal substrate concentration. In other words, independent nutrient cycles of phosphorus, nitrogen and carbon are considered. Phytoplankton biomass as well as carbon, phosphorus and nitrogen in algal cells must be included as state variables, all expressed in the units $g/m3$. This is complex compared with the constant stoichiometric approach, but as Jørgensen (1976) has shown, it was impossible to obtain an accurate time at which the maximum phytoplankton concentration and production occured using the simpler noncausal Monod's kinetic for growth of phytoplankton. The proportions of nitrogen and phosphorus in both zooplankton and fish are included in the model to secure element conservation.

The growth of phytoplankton is described by use of a growth rate coefficient $\mu$, which is limited by four factors:

A temperature factor,

$$FTI = \exp(A(T-T_{opt})) (T_{max}-T)/T_{MAX}-T_{OPT}) A(T_{max}-T_{opt})$$

where

$A$, $T_{opt}$ and $T_{max}$ are species dependent constants. T is temperature

A factor for intracellular nitrogen, NC:

$$FN3 = 1 - NC_{min}/NC;$$

(7.95)

A parallel factor for intracellular phosphorus:

$$FP3 = 1 - PC_{min}/PC; \text{ and similarly}$$

(7.96)

A factor for intracellular carbon:

$$FC3 = 1 - CC_{min}/CC.$$ (7.97)

NC, PC and CC are determined by nutrient uptake rates:

$$UC = UC_{max}.FC1.FC2.FRAD$$ (7.98)

$$UN = UN_{max}.FN1.FN2$$ (7.99)

$$UP = UP_{max}.FP1.FP2$$ (7.100)

where

$UC_{max}$, $UN_{max}$ and $UP_{max}$ are species dependent constants (maximum uptake rates); generally, UCmax will be greater the smaller the size of the considered phytoplankton. FC1, FN1 and FP1 are expressions that give the limitations in uptake:

$$FC1 = (FCA_{max} - FCA)/(FCA_{max} - FCA_{min})$$ (7.101)

$$FN1 = (FNA_{max} - FNA)/(FNA_{max} - FNA_{min})$$ (7.102)

$$FP1 = (FPA_{max} - FPA)/(FPA_{max} - FPA_{min})$$ (7.103)

where

$FCA_{max}$, $FCA_{min}$, $FNA_{max}$, $FNA_{min}$, $FPA_{max}$ and $FPA_{min}$ are constants indicating the maximum respectively minimum contents of nutrients in phytoplankton. FCA, FNA and FPA are determined as CC/PHYT, NC/PHYT and PC/PHYT. FC2, FN2 and FP2 give the limitations in uptake caused by the nutrient level in the lake water:

$$FC2 = C/KC + C$$ (7.104)

$$FN2 = NS/NS + KN$$ (7.105)

$$FP2 = PS/PS + KP$$ (7.106)

As seen, these expressions are in accordance with the Michaelis-Menten formulation. FRAD is a complex expression, covering the influence of solar radiation. This influence is integrated over depth and the self-shading effect is included. The intracellular nitrogen, phosphorus and carbon can now be determined by differential equations:

$$\frac{dNC}{dt} = UN \cdot PHYT - \left(SA + \frac{GZ}{F} + \frac{Q}{V}\right)NC$$ (7.107)

$$\frac{dPC}{dt} = UP \cdot PHYT - \left(SA + \frac{GZ}{F} + \frac{Q}{V}\right)PC \qquad (7.108)$$

$$\frac{dCC}{dt} = (UC - RC) \cdot PHYT - \left(SA + \frac{GZ}{F} + \frac{Q}{V}\right)CC \qquad (7.109)$$

where

PHYT is the phytoplankton concentration, GZ the grazing rate corresponding to gross growth of zooplankton, F a yield factor (approximately 2/3), Q the outflow rate and V the volume. RC is the respiration rate, found as

$$RC = RC_{max} \cdot \left(\frac{CC}{CC_{max}}\right)^{2/3} \qquad (7.110)$$

As the sediment accumulates nutrients it is important to describe quantitatively the processes determining the mass flows from sediment to water. To what extent will accumulated compounds in the sediment be redissolved in the lake water? The exchange processes between mud and water of phosphorus and nitrogen have been extensively studied, as these processes are important for the eutrophication of lakes. Chen and Orlob (1975) have igmored the exchange of nutrients between mud and water and, as pointed out by Jørgensen et al. (1975), this will inevitably give a false prognosis. Ahlgreen (1973) applied a constant flow of nutrients between sediment and water, and Dahl Madsen et al. (1974) are using a simple first order kinetic to describe the exchange rate.

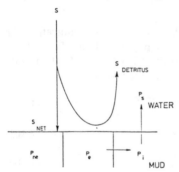

Fig.7.14: S sedimentation, s divided in Sdetritus and Snetto; Pnet non-exchangeable phosphorus in unstabilized sediment; PE exchangeable phosphorus in unstabilized sediment; PI phosphorus in interstitial water; PS dissolved phosphorus in water.

A more comprehensive submodel (fig. 7.14) for the exchange of phosphorus has been developed by Joergensen et al. (1975). The settled material, S, is divided into Sdetritus and $S_{net}$, the first being mineralized by microbiological activity in the water body, and the latter being the material actually transported to the sediment. $S_{net}$ can also be divided into two flows:

$$S_{net} = S_{net,s} + S_{net,e}$$

where
$S_{net,s}$ = flow to the stable nonexchangeable sediment, and $S_{net,e}$ = mass flow to the exchangeable unstable sediment.

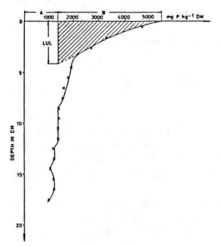

Fig.7.15: Analysis of core from Lake Esrom. mg P per g dry matter is plotted against the depth. The hatched area represents exchangeable phosphorus. $f =$ (B.A⁻¹), LUL is the unstabilized layer.

Correspondingly, $P_{ne}$ and PE, non-exchangeable and exchangeable phosphorus concentrations, both based on the total dry matter in the sediment, can also be distinguished. An analysis of the phosphorus profile in the sediment (fig.7.15) will give the ratio, (f), of the exchangeable and non-exchangeable part of the settled phosphorus,

$$f = \frac{S_{net,e}}{S_{net,s}} \quad \text{and} \quad \frac{d\text{PE}}{dt} = \alpha \cdot f \cdot \text{PS}_{net,s} \qquad (7.112)$$

where $\text{PS}_{net,s}$ = the phosphorus settling (mg m⁻³ 24h⁻¹) to the stabilized sediment, and a = factor converting water concentration units to concentration units in the sediment (mg P kg⁻¹ DM). $\text{PS}_{net,s}$ can be found from sediment profile studies. The increases of the stabilized sediment can be found by numerous methods. The application of lead isotopes is, for example, a fast and reliable method.

Exchangeable phosphorus is mineralized similarly to detritus in a water body, and a first order reaction gives a reasonably good description of the conversion of $P_e$ into interstitial phosphorus PI:

$$\frac{d\text{PE}}{dt} = \alpha \cdot F \cdot \text{PS}_{net,s} - K5 \cdot \text{PE} \cdot K6^{T-20} \qquad (7.113)$$

where K5 = a rate coefficient, K6 = a temperature coefficient, and T = temperature.

Finally, the interstitial phosphorus, PI, will be transported by diffusion from the pore water to the lake water. This process, which has been studied by Kamp-Nielsen (1974), can be described by means of the following empirical equation (valid at 7°C):

$$\text{Release of P} = 1.21(\text{PI} - \text{PS}) - 1.7 \ (\text{mg P m}^{-2} \ 24\text{h}^{-1}) \tag{7.114}$$

where PS is the dissolved phosphorus in the lake water. It means that:

$$\frac{d\text{PI}}{dt} = K5 \cdot \text{PE} \cdot K6^{\,T-20} - \beta \cdot (1.21(\text{PI} - \text{PS}) - 1.7) \cdot \frac{T}{280} \tag{7.115}$$

where $\beta$ converts concentration units in the sediment to those for lake water, and T is the absolute temperature as the release rate was found to be proportional to T. This sub-model was validated in three case studies (Jørgensen et al., 1975) examining sediment cores in the laboratory. Kamp-Nielsen (1975) has added an adsorption term to these equations.

A similar submodel for the nitrogen release has been set up by Jacobsen et al., (1975). The nitrogen release from sediment is expressed as a function of the nitrogen concentration in the sediment and the temperature, considering both aerobic and anaerobic conditions.

The predation of phytoplankton by zooplankton Z, and that of zooplankton by fish F are both expressed by modified Monod expressions:

$$\mu Z = \frac{\text{PHYT} - 0.5}{\text{PHYT} - KA} \quad \text{and} \quad \mu F = \frac{\text{ZOO} - \text{KS}}{\text{ZOO} + \text{KZ}} \tag{7.116}$$

where KA, KS and KZ are constants. These expressions are in accordance with Steele (1974).

The following pont in the model were changed during 1979-83 and this gave a better validation:

(1) FC3, FN3 and FP3 were changed to

$$FC3 = \frac{\text{FCA} - \text{FCA}_{min}}{\text{FCA}_{max} - \text{FCA}_{min}} \tag{7.117}$$

and parallel for FN3 and FP3;

(2) The $T_{opt}$ in the temperature factor was changed to the actual temperature in the lake water during the summer months to allow for temperature adaptation;

(3) The temperature dependance of phytoplankton respiration was changed to an

exponential expression in accordance with Wetzel (1975);

(4) FC4 was changed to

$$FC4 = \frac{CC}{CC_{max}}$$ (7.118)

The exponent 1/2 is valid for individual cells as the surface is approximately proportional to the weight or volume of the cells, but since phytoplankton concentration is used here, application of the exponent 2/3 is irrelevant;

(5) As mentioned above, only part of the settled phosphorus is exchangeable. In the case study referred to, it was found that 15% of the settled phosphorus was nonexchangeable to be able to account for the observed phosphorus profile in the sediment. In the new version exchangeable and nonexchangeable nitrogen were also distinguished. It is possible (based upon the nitrogen profile in the sediment) to estimate that nonexchangeable nitrogen is 4-5 times higher than nonexchangeable phosphorus. As algae contain on average seven times as much nitrogen as phosphorus, the exchangeable part of the settled nitrogen, called KNEX, can be estimated by the following equation:

$$KNEX = \left(\frac{5}{7}KEX + \frac{2}{7}\right)$$ (7.119)

where KEX is the exchangeable fraction on the settled phosphorus; in this study KEX = 0.85, which means that

$$KNEX = \left(\frac{5}{7} \cdot 0.85 + \frac{2}{7}\right) = 0.89$$ (7.120)

These changes gave a better correspondance between the modelled and the observed nitrogen balance; and finally

(6) A carrying capacity of zooplankton was introduced to give a better simulation of zooplankton and phytoplankton. Carrying capacities are often observed in ecosystems (see section 6.2), but their necessity in this case may be due to too simple a simulation of the grazing process. Phytoplankton might not be grazed by all zooplankton species present, and some species might use detritus as a food source. The zooplankton growth rate, z, is computed in accordance with these modifications as

$$\mu_z = \mu_{z,max} \cdot FPH \cdot FT2 \cdot F2CK \tag{7.121}$$

where FPH is defined above, FT2 is a temperature regulation expression, and F2CK accounts for carrying capacity,

$$F2CK = 1 - \frac{ZOO}{CK} \tag{7.122}$$

where CK = 26 mg/l was chosen in this case.

An intensive measuring period was applied to improve parameter estimation as described in section 2.7. The results of this effort can be summarized as follows:

(1) Different expressions of simultaneous limiting factors were tested and only two expressions gave an acceptably maximum growth rate for phytoplankton and an acceptable low standard deviation. These are (a) multiplication of the limiting factors, and (b) averaging the limiting factors;

(2) The previously applied expression for the influence of temperature on phytoplankton growth gave unacceptable parameters with too high standard deviations. A better expression is:

$$FT7 = \exp (T - TOPT) ((TMAX - TOPT))A(TMAX\text{-}TOPT) \tag{(7.123)}$$

where A is a constant found to be 0.14;

(3) It was possible to make another parameter estimation (table 2.13), which gives, for some of the parameters, more realistic values. Whether this would give an improved validation when observations from a period with drastic changes in the nutrients loading is available could not be stated

(4) The other expressions applied for process descriptions were confirmed.

It is important to validate models against independently measured values. Therefore, another set of measurements was made from October 15, 1974 to October 15, 1975. No general method of validation is available, but almost the same method suggested by WMO (1975) for validation of hydrological models was applied.

Table 7.13 gives results of the validation improved as described above. The following numerical validation criteria were applied:

(1) Y, coefficient of variation of the residuals of errors for the state variables for the validation period, defined as

$$Y = \frac{\left[\Sigma(y_c - Y_m)^2\right]^{1/2}}{\overline{Y}_m} \qquad\qquad (7.124)$$

where $y_c$ = calculated values of the state variables, $Y_m$ = measured values of the state variables, n = number of comparisons, and $\overline{Y}_m$ = average of measured values over the validation period;

(2) R, the relative error of mean values:

$$\qquad\qquad\qquad (7.125)$$

$$R = \frac{\overline{Y}_c - \overline{Y}_m}{\overline{Y}_m}$$

where $\overline{Y}_c$ is the average of measured values over the validation period;

(3) A, relative error of maximum values:

$$\qquad\qquad\qquad (7.126)$$

$$A = \frac{Y_{c,max} - Y_{m,max}}{Y_{m,max}}$$

where $Y_{c,max}$ = maximum value of the calculated state variable in the validation period, and $Y_{m,max}$ = maximum value of the measured state variable in the validation period; and

(4) TE, timing error:

$$\qquad\qquad\qquad (7.127)$$

$$TE = \text{Date of } Y_{c,max} - \text{date of } y_{m,max}$$

Y, R and A give the errors in relative terms. By multiplication by 100, the errors are obtained as %. The standard deviation, Y, for all measured state variables, is as seen 31%. It is the standard deviation for one comparison of model value and measured value. As the standard deviation for a comparison of n sets of model values and measured values is √n times smaller and n is in the order of 200, the overall average picture of the lake is given with a standard deviation of about 2%, which is very acceptable. Y is generally 5 times larger for hydrodynamic models (WMO 1975). The relative errors of mean values, R, are 3% for production, 10% for phytoplankton and 2% for nitrogen - very acceptable values, but the relative error for total phosphorus is 26% and for zooplankton 27%, which total must be considered a little too high. The relative errors of the maximum values, A, are from 0% to 18%, which is acceptable. The model ability to predict maximum production and maximum phytoplankton concentration has special interest for a eutrophication model; the relative errors are 8 respectively 15% - fully

241

acceptable. The ability to predict the time, when maximum values occur, is expressed by use of TE. Production and phytoplankton (use of susp. matter 1-60 ) give fully accordance between model values and measured values. TE for nitrogen total and soluble are also acceptable, while the zooplankton and phosphorus values are on the high side. All in all the validation has demonstrated, that the model has value as predictive tool, although the dynamics of phosphorus and zooplankton could be improved.

As mentioned previously in the introduction to this model, it has been applied with modifications to 15 other case studies. These were all based on ecological observations of the system under consideration. Table 7.14 reviews the modifications needed in the 15 case studies to get an workable model. By calibration carried out in accordance with section 2.7, it was found that the most crucial parameters were all approximately in the range of values found in the literature, see Table 7.15. Note that the parameters shown here all were found by (1) using literature values as initial guesses, (2) using of intensive measuring periods to get good first estimations of parameters, (3) a first coarse calibration of the model to improve parameter estimations, (4) use of an automatic calibration procedure which allows a finer calibration of 6-8 of the most important (most sensitive to phytoplankton concentration) parameters. This procedure was repeated at least twice and only when the same parameter values were found, was the calibration considered satisfactory.

**TABLE 7.13**
**Numerical Validation of the described Model.**

| Validation Criteria | State variable | Value |
|---|---|---|
| Y | all | 0.31 |
| R | Ptotal (P4) | 0.26 |
| R | Psoluble (PS) | 0.16 |
| R | Ntotal (N4) | 0.02 |
| R | Nsoluble (NS) | 0.14 |
| R | Phytoplankton (CA) | 0.10 |
| R | Zooplankton (Z) | 0.27 |
| R | Production | 0.03 |
| A | Ptotal (P4) | 0.12 |
| A | Psoluble (PS) | 0.18 |
| A | Ntotal (N4) | 0.07 |
| A | Nsoluble (NS) | 0.03 |
| A | Phytoplankton (CA) | 0.15 |
| A | Zooplankton (Z) | 0.00 |
| A | Production | 0.08 |
| TE | Ptotal (P4) | 105 days |
| TE | Psoluble (PS) | 60 days |
| TE | Ntotal (N4) | 15 days |
| TE | Nsoluble (NS) | 15 days |
| TE | Phytoplankton (CA) | 0 days * 120 days * * |
| TE | Zooplankton (Z) | 60 days |
| TE | Production | 0 days |

* ) based on measuring suspended matter 1-60 $\mu$. * * ) based on chlorophyll.

242

**TABLE 7.14**

Survey of eutrophication studies based upon the application of a modified Glumsø model.

| Ecosystem | Modification | Level |
|---|---|---|
| Glumsø, version A | basis version | 6 |
| Glumsø, version B | nonexchangeable nitrogen | 6 |
| Ringkøbing Firth | boxes, nitrogen fixation | 5 |
| Lake Victoria | boxes, thermocline, other foodchain | 4 |
| Lake Kyoga | other foodchain | 4 |
| Lake Kobuto Sese Seko | boxes, thermocline other foodchain | 4 |
| Lake Fure | boxes, nitrogen fixation thermocline | 3 |
| Lake Esrom | boxes, Si-cycle thermocline | 4 |
| Lake Gyrstinge | level fluctuations sediment exposed to air | 4-5 |
| Lake Lyngby | basis version | 6 |
| Lake Bergunda | nitrogen fixation | 2 |
| Broia Reservoir | macrophytes, 2 boxes | 1 |
| Lake Great Kattinge | resuspension | 5 |
| Lake Svogerslev | resuspension | 5 |
| Lake Bue | resuspension | 5 |
| Lake Kornerup | resuspension | 5 |
| Lake Balaton | adsorption to suspended matter | 2 |

Level 1: Conceptual diagram selected.
Level 2: Verification carried out.
Level 3: Calibration using intensive measurements.
Level 4: Calibration of entire model.
Level 5: Validation. Object function and regression coefficient are found.
Level 6: Validation of a prognosis for significant changed loading.

# TABLE 7.15
## Comparison of important parameters from 12 eutrophication studies

| Parameter | Unit | Glumsø version A | Glumsø version B | Rindkøbing Firth | Lake Vic. | Lake Fure |
|---|---|---|---|---|---|---|
| FNAMIN | Min kg N pr kg phytoplankton biomass | | 0.015 | 0.035 | 0.007 | 0.015 | 0.015 |
| FNAMAX | Max kg N pr. kg. phytoplankton biomass | - | 0.10 | 0.11 | 0.055 | 0.10 | 0.11 |
| FPAMIN | Min kg P pr. kg phytoplankton biomass | - | 0.001 | 0.0015 | 0.0052 (0.0011) | 0.001 | 0.001 (0.0013) |
| FPAMAX | Max kg P pr. kg phytoplankton | - | 0.013 | 0.01 | 0.08 | 0.013 | 0.02 (0.0015) |
| CDRMAX | Max growth rate phytoplankton | $(24h)^{-1}$ | 2.53 | 1.00 | 2.5 (1.25) | 1.47 | 3.4 (3.6) |
| KP | Michaelis constant P-uptake | $mgl^{-1}$ | 0.02 | 0.02 | 0.02 (0.005) | 0.02 | 0.02 (0.008) |
| KN | Michaelis constant N-uptake | $mgl^{-1}$ | 0.2 | 0.84 | 0.2 | 0.2 | 0.2 |
| RCMAX | Max respiration rate phytoplankton | $(24h)^{-1}$ | 0.13 | 0.16 | 0.13 (0.1) | 0.112 | 0.1 |
| SVS | Settling rate phytoplankton | m/24h | 0.19 | 0.15 | 0.19 (0.02) | 0.16 | 0.20 |
| UCMAX | Max rate C-uptake | $(24h)^{-1}$ | 0.55 | 1.75 | 0.96 (0.17) | 0.55 | 0.58 |
| UPMAX | Max rate P-uptake | $(24h)^{-1}$ | 0.02 | 0.0025 | 0.0014 | 0.0014 | 0.002 |
| UNMAX | Max rate N-uptake | $(24h)^{-1}$ | 0.015 | 0.023 | 0.011 (0.018) | 0.015 | 0.016 |
| ZOOMAX | Max growth rate zooplankton | $(24h)^{-1}$ | 0.188 | 0.244 | 0.188 | 0.158 | 0.2 |
| RZMAX | Max respiration rate zooplankton | $(24h)^{-1}$ | 0.028 | 0.06 | 0.01 | 0.039 | 0.04 |
| DENIT | Denitrification rate | $(24h)^{-1}$ | 0.03 | 0.1 | 0.08 | 0.0012 | 0.05 |
| BETA | Specific extintion coefficient phytoplankton | $m^2/g$ | 0.18 | 0.18 | 0.18 (0.6) | 0.18 | 0.18 |
| KDP | Decomposition rate of detritus(P) | $(24h)^{-1}$ | 0.4 | 0.8 | 0.4 | 0.4 | 0.5 |

| Lake Esrom | Lake Gyrstinge | Lake Lyngby | Lake Great Kat-tinge | Lake Svogerslev | Lake Bue | Lake Kornerup | Literature Range |
|---|---|---|---|---|---|---|---|
| 0.015 | 0.015 | 0.015 | 0.015 | 0.015 | 0.015 | 0.015 | 0.015-0.04 |
| 0.10 | 0.10 | 0.10 | 0.10 | 0.10 | 0.10 | 0.10 | 0.08-0.15 |
| 0.005 | 0.004 | 0.001 | 0.001 | 0.001 | 0.001 | 0.001 | 0.001-0.005 |
| 0.02 | 0.02 | 0.013 | 0.013 | 0.013 | 0.013 | 0.013 | 0.013-0.03 |
| 2.5 | 2.5 | 1.8 | 3.0 | 4.5 | 4.0 | 7.5 | 1-5 |
| 0.005 | 0.02 | 0.02 | 0.02 | 0.02 | 0.02 | 0.02 | 0.005-0.03 |
| 0.05 | 0.2 | 0.2 | 0.2 | 0.2 | 0.2 | 0.2 | 0.05-0.5 |
| 0.1 | 0.15 | 0.2 | 0.13 | 0.13 | 0.12 | 0.13 | 0.1-0.5 |
| 0.2 | 0.2 | 0.05 | 0.16 | 0.5 | 0.12 | 1.2 | 0.1-0.8 |
| 0.55 | 0.6 | 0.4 | 0.55 | 0.55 | 0.55 | 0.55 | 0.2-1.0 |
| 0.0014 | 0.02 | 0.008 | 0.003 | 0.003 | 0.003 | 0.003 | 0.002-0.01 |
| 0.015 | 0.02 | 0.012 | 0.035 | 0.16 | 0.13 | 0.16 | 0.01-0.05 |
| 0.036 | 0.19 | 0.2 | 0.15 | 0.65 | 0.21 | 0.29 | 0.1-0.8 |
| 0.05 | 0.03 | 0.04 | 0.028 | 0.028 | 0.028 | 0.028 | 0.02-0.1 |
| 0.2 | 0.2 | 0.03 | 0.03 | 0.03 | 0.03 | 0.03 | 0.001-0.1 |
| 0.4 | 0.18 | 0.18 | 0.18 | 0.18 | 0.18 | 0.18 | 0.1-0.3 |
| 0.5 | 0.4 | 0.25 | 0.4 | 0.4 | 0.4 | 0.4 | 0.1-1.0 |

## 7.5. WETLAND MODELS

### 7.5.1. Introduction.

Wetland is defined by Cowardin et al. (1979) as an ecosystem transitional between aquatic and terrestial ecosystems, where the water table is usually at or near the surface or the land is covered by shallow water. The modelling of wetlands is relatively new compared to the modelling of aquatic and terrestrial ecosystems. Recently, however, several models of wetlands have been developed. Models of forested swamps, bogs, marshes and tundra have appeared in the literature during the last 5-6 years.

Mitsch (1983) has given a more comprehensive review of wetland models, than it is possible to give here. He distinguish between energy/nutrient models, hydrological models, models of spatial ecosystem, models of tree growth, process models, causal models and regional energy models.

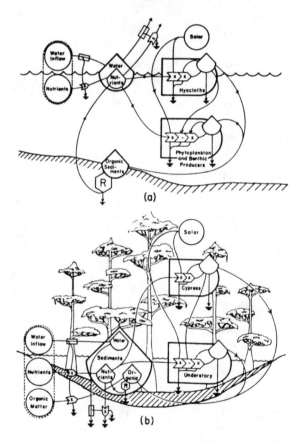

Fig.7.16: Conceptual wetland models for a) water gyacinth marsh and b) cypress swamp in Florida. Models are the energy/nutrient type.

246

Fig. 7.16 gives conceptual diagrams of two wetland models of the first type. Odum's energy language is used. The symbols used are explained in fig. 4.13.

A cypress dome simultation model has been selected to illustrate a typical wetland model. It is presented in the next paragraph and demonstrates several characteristic features of wetland models.

### 7.5.2. Cypress Dome Simultation Model

Cypress domes have been studied, mainly in the U.S, for their value in waste water renovation, timber production, storm water retention, wildlife protection and ground water recharge. These values make the cypress dome an ideal system to demonstrate management options with the use of simulation models.

The conceptual diagram of the model presented here is shown fig. 7.17. It deals with questions of optimum harvest, possible effects of fire and disposal of secondary sewage.

The model considers cypress trees, understory plants, 4 state variables in the sediment (nitrogen, phosphorus, organic peat and water) and dead trees can take place in wet or dry conditions. Fire is simulated as a pulse of varying frequency and only has an effect on the dome if the water level ($Q_6$) is low compared with the organic deposits ($Q_7$).

The burning of trees is given as:

$$(7.128)$$

$$I = K_8 \cdot Q_1 (Q_2 Q_7/Q_8)$$

where $K_8$ is a coefficient, $Q_1$ is the cypress biomass, $Q_2$ is the understory biomass, $Q_7$ is the organic peat, $Q_6$ is the water level and I the fire intensity. When the fire intensity exceeds a certain threshold cypress trees are killed and converted to dead standing trees according to a linear relationship.

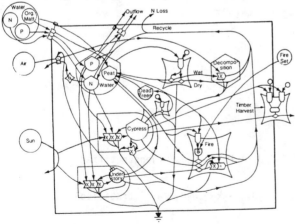

Fig.7.17: Simulation model of Florida cypress dome. Model diagram was modified by H.T. Odum.

Tables 7.16, 7.17 and 7.18 give the details of the model flows such as litterfall and gross primary production were obtained from yearly averages. The model was designed to run for 100 years.

Fig. 7.18 shows some simulation results.

**TABLE 7.16**

**Cypress Dome Simulation Models.**

Cypress

$$\dot{Q}_1 = k_3 Q_1 Q_3 Q_4 J_r - k_5 Q_1^2 - k_7 Q_1 - k_8 Q_1 \left[ \frac{Q_2 Q_7}{Q_6} \right] - k_9 Q_1 - k_{56} Q_1$$

Understory

$$\dot{Q}_2 = k_4 Q_2 Q_3 Q_4 J_r - k_{10} Q_2 - k_{11} Q_2 - k_{13} Q_2$$

Nitrogen

$$\dot{Q}_3 = J_5 - k_{29} Q_3 Q_6 - k_{30} k_3 Q_1 Q_3 Q_4 J_r - k_{34} k_4 Q_2 Q_3 Q_4 J_r + k_{31} k_5 Q_1^2 +$$

$$k_{35} k_{10} Q_2 + k_{39} k_{15} Q_7 J_7 + k_{40} k_{16} Q_7 - k_{32} Q_2 Q_3 + k_{36} k_8 Q_1 \left[ \frac{Q_2 Q_7}{Q_6} \right]$$

$$+ k_{37} k_{13} Q_2 + k_{38} k_{17} Q_7$$

Phosphorus

$$\dot{Q}_4 = J_4 - k_{21} Q_4 Q_6 - k_{22} k_3 Q_1 Q_3 Q_4 J_r - k_{24} k_4 Q_2 Q_3 Q_4 J_r + k_{23} k_5 Q_1^2 +$$

$$k_{25} k_{10} Q_2 + k_{41} k_{15} Q_7 J_7 + k_{42} k_{16} Q_7 + k_{26} k_8 Q_1 \left[ \frac{Q_2 Q_7}{Q_6} \right]$$

$$+ k_{27} k_{13} Q_2 + k_{28} k_{17} Q_7$$

Water

$$\dot{Q}_6 = J_3 - k_{18} Q_6 - k_{19} Q_6$$

Organic Peat

$$\dot{Q}_7 = J_6 + k_7 Q_1 + k_{11} Q_2 - k_{15} Q_7 J_7 - k_{16} Q_7 - k_{17} Q_7 + k_{57} Q_8$$

Dead Cypress

$$\dot{Q}_8 = k_{56} Q_1 - k_{57} Q_8$$

Sunlight

$$J_0 = J_r + k_1 Q_1 Q_3 Q_4 J_r + k_2 Q_2 Q_3 Q_4 J_r \quad \text{(nonstratified)}$$

$$J_0 = J_{r2} + k_1 Q_1 Q_3 Q_4 J_{r1} + k_2 Q_2 Q_3 Q_4 J_{r2} \quad \text{(stratified)}$$

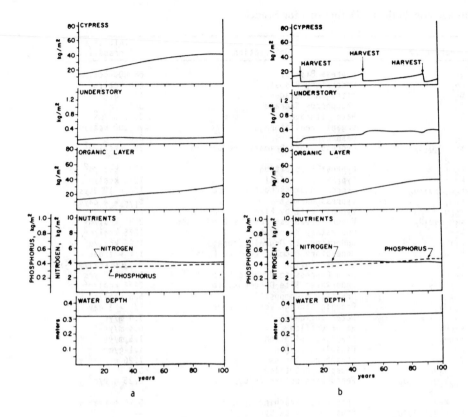

Fig.7.18: Simulation of cypress dome model (a) for undisturbed conditions for 100 years, beginning with a cypress biomass of 15 kg/m², and (b) for harvesting cypress trees when biomass reaches 15 kg/m² for yield of 10 kg/m².

**TABLE 7.17**

**Storage and Pathway Definitions for Models.**

| Parameter | Description | Initial or Average Value |
|---|---|---|
| $Q_1$ | Cypress Biomass | 68,000 kcal/m$^2$ |
| $Q_2$ | Understory Biomass | 400 kcal/m$^2$ |
| $Q_3$ | Nitrogen Storage | 4160 g/m$^2$ |
| $Q_4$ | Phosphorus Storage | 340 g/m$^2$ |
| $Q_6$ | Water Storage | 0.32 m$^3$/m$^2$ |
| $Q_7$ | Organic Peat Storage | 78,000 kcal/m$^2$ |
| $Q_8$ | Dead Cypress | 0.0 |
| $k_3 Q_1 Q_3 Q_4 J_r$ | Cypress Gross Primary Production | 3277 kcal/m$^2$ yr |
| $k_5 Q_1^2$ | Cypress Respiration | 1721 kcal/m$^2$ yr |
| $k_7 Q_1$ | Cypress Litterfall | 1214 kcal/m$^2$ yr |
| $k_8 Q_1 Q_2 Q_7 / Q_6$ | Cypress Biomass Lost to Fire | 1/k = 10 days |
| $k_9 Q_1$ | Cypress Harvest | $t_{1/2}$ = 4 days |
| $k_{56} Q$ | Cypress Kill by Fire | --- |
| $k_4 Q_2 Q_3 Q_4 J_r$ | Understory Gross Primary Production | 2591 kcal/m$^2$ yr |
| $k_{10} Q_2$ | Understory Respiration | 1295 kcal/m$^2$ yr |
| $k_{11} Q_2$ | Understory to Organic Storage | 1295 kcal/m$^2$ yr |
| $k_{13} Q_2$ | Understory Biomass Lost to Fire | 1/k = 1 day |
| $J_6$ | Organic Inflow | 0-1700 kcal/m$^2$ yr |
| $k_{57} Q_8$ | Dead Cypress to Organic Storage | 1/k = 20 yrs |
| $k_{15} Q_7 J_7$ | Underwater Site Decomposition | 2464 kcal/m$^2$ yr |
| $k_{16} Q_7$ | Dry Site Decomposition | 536 kcal/m$^2$ yr |
| $k_{17} Q_7$ | Organic Storage Lost to Fire | 1/k = 1 day |
| $J_3$ | Water Inflow | 7.9 m/yr |
| $k_{18} Q_6$ | Water Outflow | 6.4 m/yr |
| $k_{19} Q_6$ | Evapotranspiration | 1.5 m/yr |
| $J_7$ | Dissolved Oxygen | 1.1 g/m$^2$ |
| $J_4$ | Phosphorus Inflow | 1.26 g-P/m$^2$ yr |
| $k_{21} Q_4 Q_6$ | Phosphorus Outflow | 0.0 |
| $k_{22} k_3 Q_1 Q_3 Q_4 J_r$ | Phosphorus Uptake by Cypress | 0.36 g-P/m$^2$ yr |
| $k_{23} k_5 Q_1^2$ | Phosphorus Leaching by Cypress | 0.19 g-P/m$^2$ yr |
| $k_{26} k_8 Q_1 Q_2 Q_7 / Q_6$ | Phosphorus from Cypress Fire | --- |
| $k_{24} k_4 Q_2 Q_3 Q_4 J_r$ | Phosphorus Uptake by Understory | 0.46 g-P/m$^2$ yr |
| $k_{25} k_{10} Q_2$ | Phosphorus Leaching by Understory | 0.23 g-P/m$^2$ yr |
| $k_{27} k_{13} Q_2$ | Phosphorus from Understory Fire | --- |
| $k_{41} k_{15} Q_7 J_7$ | Phosphorus Recycle--Wet Decomposition | 0.34 g-P/m$^2$ yr |
| $k_{42} k_{16} Q_7$ | Phosphorus Recycle--Dry Decomposition | 0.07 g-P/m$^2$ yr |
| $k_{28} k_{17} Q_7$ | Phosphorus from Organic Storage Fire | --- |
| $J_5$ | Nitrogen Inflow | 10.3 g-N/m$^2$ yr |
| $k_{29} Q_3 Q_6$ | Nitrogen Outflow | 0.0 |
| $k_{30} k_3 Q_1 Q_3 Q_4 J_r$ | Nitrogen Uptake by Cypress | 5.6 g-N/m$^2$ yr |
| $k_{31} k_5 Q_1^2$ | Nitrogen Leaching by Cypress | 2.9 g-N/m$^2$ yr |
| $k_{36} k_8 Q_1 Q_2 Q_7 / Q_6$ | Nitrogen from Cypress Fire | --- |
| $k_{34} k_4 Q_2 Q_3 Q_4 J_r$ | Nitrogen Uptake by Understory | 6.9 g-N/m$^2$ yr |
| $k_{35} k_{10} Q_2$ | Nitrogen Leaching by Understory | 3.4 g-N/m$^2$ yr |
| $k_{37} k_{13} Q_2$ | Nitrogen from Understory Fire | --- |
| $k_{39} k_{15} Q_7 J_7$ | Nitrogen Recycle--Wet Decomposition | 12.3 g-N/m$^2$ yr |
| $k_{40} k_{16} Q_7$ | Nitrogen Recycle--Dry Decomposition | 2.6 g-N/m$^2$ yr |
| $k_{38} k_{17} Q_7$ | Nitrogen from Organic Storage Fire | --- |
| $k_{32} Q_2 Q_3$ | Denitrification | 8.1 g-N/m$^2$ yr @$Q_2$ = 4000 |

**TABLE 7.18**

**Parameter Values for Cypress Dome Models**

| Parameter | Value | Parameter | Value |
|---|---|---|---|
| $k_1$ | $0.577 \times 10^{-12}$ | $k_{25}$ | $1.8 \times 10^{-4}$ |
| $k_2$ | $0.982 \times 10^{-12}$ | $k_{26}$ | $1.1 \times 10^{-4}$ |
| $k_3$ | $2.51 \times 10^{-14}$ | $k_{27}$ | $1.8 \times 10^{-4}$ |
| $k_4$ | $3.48 \times 10^{-12}$ | $k_{28}$ | $1.3 \times 10^{-4}$ |
| $k_5$ | $3.72 \times 10^{-7}$ | $k_{29}$ | $0.0$ |
| $k_7$ | $1.78 \times 10^{-2}$ | $k_{30}$ | $1.7 \times 10^{-3}$ |
| $k_8$ | $3.74 \times 10^{-7}$ | $k_{31}$ | $1.7 \times 10^{-3}$ |
| $k_9$ | $63.2$ | $k_{32}$ | $4.87 \times 10^{-7}$ |
| $k_{10}$ | $3.24$ | $k_{33}$ | $2.0$ |
| $k_{11}$ | $3.24$ | $k_{34}$ | $2.7 \times 10^{-3}$ |
| $k_{13}$ | $364$ | $k_{35}$ | $2.7 \times 10^{-3}$ |
| $k_{15}$ | $2.92 \times 10^{-2}$ | $k_{36}$ | $6.8 \times 10^{-4}$ |
| $k_{16}$ | $6.87 \times 10^{-3}$ | $k_{37}$ | $1.1 \times 10^{-3}$ |
| $k_{17}$ | $365$ | $k_{38}$ | $2.0 \times 10^{-3}$ |
| $k_{18}$ | $20.1$ | $k_{39}$ | $5.0 \times 10^{-3}$ |
| $k_{19}$ | $4.56$ | $k_{40}$ | $5.0 \times 10^{-3}$ |
| $k_{21}$ | $0.0$ | $k_{41}$ | $1.3 \times 10^{-4}$ |
| $k_{22}$ | $1.1 \times 10^{-4}$ | $k_{42}$ | $1.3 \times 10^{-4}$ |
| $k_{23}$ | $1.1 \times 10^{-4}$ | $k_{56}$ | $1,500$ |
| $k_{24}$ | $1.8 \times 10^{-4}$ | $k_{57}$ | $0.05$ |

## 7.6. MODELS IN ECOTOXICOLOGY

### 7.6.1. Introduction

An increasing interest in toxic substance models has emerged during the last decade due to a rapidly growing concern for the related environmental problems.

Toxic substance models differ from other biogeochemical models by:

1) The need for parameters to cover all possible toxic substance models is great and general estimation methods are therefore used quite widely, see also section 2.7.
2) The safety margin should be high f.inst. expressed as the ratio between the actual concentration and the concentration, that gives undesired effects.
3) The possible inclusion of an effect component, which relates the output concentration to its effect.
4) The possibilities and needs to use simple models due to points 1 and 2 and due to our limited knowledge of process details, sublethal effects, antagonistic and synergistic effects.

A number of toxic substance models is reviewed in table 7.19 to give an impression of the types of model, which are available in this field of ecological modelling. Most models reflect that a good knowledge to the problem and ecosystem can be used to make reasonable simplifications. Therefore only a few models consider all trophic levels and all possible processes. Model characteristics indicated in the table are state variables and/or processes considered in the model. Note in the table the number of modelled toxic substances and the processes taken into, account.

## 7.6.2. Principles of Modelling the Distribution and Effects of Toxic Substances

The most difficult part of modelling the effect and distribution of toxic substances is to obtain the relevant knowledge about the behaviour of the toxic substances in the environment and to use this knowledge to make the feasible simplifications.

**TABLE 7.19**

**Examples of Toxic Substance Models.**

| Toxic Substance | Model Characteristics | Reference |
|---|---|---|
| Cadmium | Food chain similar to a eutrophication model | Thomann et al., 1974 |
| Mercury | 6 state variables: water, sediment, suspended matter, invertebrates, plant and fish | Miller (1979) |
| Vinyl chloride | chemical processes in water | Gillette et al.,1974 |
| Methyl parathion | chemical processes in water and benzothiophenemicrobial degradation, adsorption, 2-4 trophic levels | Lassiter (1978) |
| Methyl mercury | a single trophic level: food intake, excretion, metabolism growth | Fagerstroem and Aasell (1973) |
| Heavy metals | Concentration factor, excretion, bioaccumulation | Aoyama etal. (1978) |
| Pecticides in fish DDT and methoxychlor | Ingestion, concentration factor adsorption on body, defecation excretion, chemical decomposition, natural mortality | Leung (1978) |
| Zinc in algae | Concentration factor, secretion hydrodynamical distribution | Seip (1978) |
| Copper in sea | Complex formation, adsorption sublethal effect of ionic copper | Orlob et al. (1980) |
| Lead | Hydrodynamics, precipitation, toxic effects of free ionic lead on algae, invertebrates and fish | Lamm and Simons (1980) |
| Radionuclides | Hydrodynamics, decay, uptake and release by various aquatic surfaces | Gromiec and Gloyna (1973) |
| Radionuclides | Radionuclides in grass, grains, vegetables, milks, eggs, beef and poultry are state variables | Kirschner and Whicker (1984) |
| $SO_2$, $NO_x$ and heavy metals- | Threshold model for accumulation effect of pollutants. Air and soil. | Kohlmaier et al. (1984) |

on sprucefir forests

| | | |
|---|---|---|
| Toxic environmental chemicals in general | Hazard ranking and assessment from physicochemical data and a limited number of laboratory tests. | Bro-Rasmussen and Christiansen (1984) |
| Heavy metals | Adsorption, chemical reactions, ion exchange | |
| Polycyclic aromatic hydrocarbons | Transport, degradation, bioaccumulation | Bartell, Gardner and O'Neill (1984) |
| Persistent toxic organic substances | Groundwater movement, transport and accumulation of pol-groundwater lutants | Uchrin (1984) |
| Cadmium, PCB | Hydraulic overflow rate (settling), sediment interactions, steady state food chain submodel | Thomann (1984) |
| Hydrophobic organic compounds | Gas exchange, sorption/desorption, hydrolysis, photolysis, hydrodynamics | Schwarzenbach and Imboden (1984) |
| Mirex | Water-sediment exchange processes, adsorption, volatilization, bioaccumulation | Halfon (1984) |
| Toxins (aromatic hydrocarbons, Cd) | Hydrodynamics, deposition, resuspension, volatilization, photooxidation, decomposition, adsorption, complex formation, (humic acid) | Harris et al. (1984) |
| Heavy metals | Hydraulic submodel, adsorption | Nyholm, Nielsen and Pedersen (1984) |
| Oil slicks | Transport and spreading, influence of surface tension, gravity and weathering processes | Nihoul (1984) |
| Acidic rain (soil) | Aerodynamic, deposition | Kauppi (1984) |
| Acidic rain | C, N and S cycles and their influence on acidity. | Arp (1983) |

It is recommended to clarify several questions before entering the modelling procedure generally applied to ecological modelling, see section 2.2:

A. Obtain best possible knowledge about the possible processes of the considered toxic substances in the ecosystem. As far as possible knowledge about the quantitative role of the processes should be obtained.

B. Attempt to get parameters of the toxic substance processes in the environment.

C. Estimate all parameters by use of the methods described in section 2.7.

D. Compare the results from B and C and attempt to explain discrepancies, if present.

E. Estimate which processes and state variables it would be feasible and relevant to include into the model. If there is the slightest doubt then include at this stage too many processes and state variables rather than too few.

F. When the model is set up in accordance with the general procedure for ecological modelling, use a sensitivity analysis to evaluate the significance of the individual processes and state variables. In many cases this will lead to further simplification.

The description of the chemical, physical and biological processes will in general be in accordance with the equations presented in chapter 3. The processes involved in the interaction between an organism and a toxic substance are shown in fig. 7.19. The organism takes up toxic substances either from the feed or directly from the environment (air or water). The following equations can be developed (for symbols see Table 7.20)

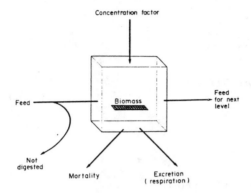

Fig.7.19: Principle for modelling the concentration of a toxic substance at a trophic level.

**TABLE 7.20**

**List of Symbols.**

| | |
|---|---|
| Bio (n) | Concentration of biomass, nth trophic level |
| CF | Concentration factor |
| EXC (n) | Rate of excretion, |
| MORT (n) | Mortality, nth trophic level |
| MY (n) | Growth rate of nth trophic level |
| n | Trophic level (n = 0 water) |
| RESP (n) | Respiration rate, nth trophic level |
| t | Time |
| TOX (O) | Concentration of toxicant in water (mg/l) |
| TOX (n) | Concentration of toxicant, nth trophic level (mg/1 water) |
| UT | Uptake of toxicant |
| YF (n) | Yield factor of feed, nth trophic level |
| YT (n) | Yield factor of toxicant, nth trophic level |
| $\Gamma$ (n) | Concentration of toxicant (mg/kg biomass = TOX(n)/BIO(n) |
| $\Gamma$ (s) | Concentration of toxicant in sediment (mg/kg dry matter) |

$$\frac{d\,\mathrm{BIO}(n)}{d\,t} = \mathrm{BIO}(n)\mathrm{YF}(n) - \mathrm{MORT}(n) - \mathrm{RESP}(n) - \mathrm{MY}(n-1) \qquad (7.129)$$

$$\frac{d\text{TOX}(N)}{dt} = \text{BIO}(n)\text{MY}(n)\text{YT}(n) \cdot \gamma(n-1) - \text{MORT}(n)\gamma(n) - \text{EXC}(n) \cdot \gamma(n) - \text{MY}(n+1) \cdot \gamma(n) + \text{UT}(n) \cdot \text{TOX}(o) \tag{7.130}$$

$$\text{As} \quad \gamma(n) = \frac{\text{TOX}(n)}{\text{BIO}(n)} \tag{7.131}$$

and

$$\frac{d\gamma(n)}{dt} = \frac{\text{TOX}'(n)\text{BIO}(n) - \text{BIO}'(n) \cdot \text{TOX}(n)}{(\text{BIO}(n))^2} \tag{7.132}$$

we have

$$\frac{d\gamma(n)}{dt} = \text{MY}(n)[\gamma(n-1) \cdot \text{YT}(n) - \gamma(n) \cdot \text{YF}(n)]$$
$$+ \gamma(n)[\text{RESP}(n) - \text{EXC}(n)] + \text{UT}(n)\text{TOX}(o) \tag{7.133}$$

Only few data, however, are available on the uptake rate, while many references give information about the concentration factor, CF, at steady state (Jørgensen et al., 1979). This means that

$$\frac{d\text{BIO}}{dt}, \frac{d\text{TOX}(n)}{dt} \text{ and } \frac{d\gamma(n)}{dt}$$

are all equal to zero, and under the circumstances of the experiments on which the CF-value is based, MORT(n), MY(n + 1) and $\gamma$(n-1) are zero as well. This means:

$$\text{MY(n) YF(n) - RESP(n)} = 0 \text{ and} \tag{7.134}$$

$$- \gamma\text{(n) EXC(n)} + \text{UT(n) TOX(o)} = 0 \text{ or} \tag{7.135}$$

$$\frac{\gamma(n)}{TOX(o)} = \frac{UT(n)}{EXC(n)} = CF \qquad (7.136)$$

$n = 0$ corresponds to the waterphase in these equations.

From equation (7.136), UT(n) or EXC(n) can be found if EXC(n) or UT(n), respectively, and CF are known. Estimation of relevant parameter has already been touched upon in section 2.7. A simpler approach to modelling accumulation of toxic substances in a single trophic level will be presented in section 7.7.2. The scope of modelling the distribution of toxic substances is often limited to show a relationship between the input of toxic substances to the aquatic ecosystem and the approximate concentrations at different trophic levels. However, the concentrations will show seasonal variations, probably with maximum concentration at maximum growth rate (summer) (see Betzer et al., 1974, and Gallegos et al., 1972). As the objectives of the model are to find the maximum concentration levels rather than to model the seasonal variations, it is suggested that the maximum growth situation is modelled, but also that the concentration levels for different growth rates are found.

### 7.6.3. Simplifications in Ecotoxicological Models.

It has been stated above that the key to ecotoxicological models is the relevant simplifications. The feasibility can be seen from the answers to questions A-F, shown above. Three illustrations are given below to demonstrate the introduction of simplifications.

### ILLUSTRATION 7.2

Modelling the Distribution and Effect of Copper Ions in Aquatic Ecosystems Free copper ions are very toxic to fish and zooplankton. The $LC_{50}$ value for Daphnia magna is as low as 10 $\mu g$ $l^{-1}$ and for Salmonoid species 100-200 $\mu g$ $l^{-1}$. A copper model should consequently focus on the concentration of free copper ions in the water, and include the processes determining this concentration.

In the case of copper a lethal concentration will be reached before the concentration in fish becomes toxic to human beings and as the uptake and excretion of copper by plants and animals are insignificant for the free copper ion concentration, these processes can be omitted. This means that a model as simple as the one illustrated in fig. 7.20 can be applied to give at least a first estimation of how much copper, and in what form, can be discharged to a lake.

The input of copper is the forcing function, and includes copper in rainwater, copper in tributaries and in waste water. The partition between free copper ions and copper adsorbed on suspended matter should also be included as an equation.

The equilibrium: copper ions + ligands = copper complexes can be described to a certain extent by the use of known equilibrium constants (Jørgensen et al., 1979).

The process copper ions converted to copper adsorbed on suspended matter requires laboratory investigation to find the adsorption capacity of the matter.

The release of copper from the sediment should also be studied, although some information is available (Lu and Chen, 1977). As seen from this case study a rather

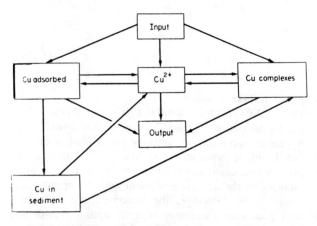

Fig.7.20: Simple copper model, conceptual diagram.

simple approach, which, however, is complex enough to require the application of a model can be used as a management tool, although the amount of data necessary to calibrate and validate the model is limited.

## ILLUSTRATION 7.3

Modelling the Distribution and Effect of DDT in Aquatic Ecosystems

A useful, but simple model for the distribution and effect of DDT in aquatic ecosystems can also be built. Here, the problem is the DDT concentration in fish at the highest trophic level in the lake, as DDT is accumulated mainly through the food chain. WHO has recommended the maximum permissible concentration of DDT in human food as 1-7 mg per kg net weight, which corresponds to a daily intake of 0.005 mg/kg body weight.

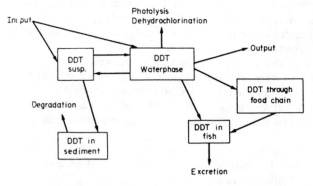

Fig.7.21: Simple model of DDT, conceptual diagram.

The management problem could be to keep the DDT concentration in all fish species below this value, divided by a safety factor of, say, 10. which means below a concentration of 0.1 mg per kg weight. The model shown in fig. 7.21 is suggested for this purpose.

The direct uptake rate from water and the excretion rate coefficients are known for some fish (Jørgensen et al., 1979) and are not significantly different from species to species. The DDT accumulation through the food chain and the rate of photolysis and dehydrochlorination is known with acceptable accuracy. The equilibrium between DDT in the water phase and adsorbel on suspended matter must be studied in the laboratory, while the degradation rate in the sediment can be appointed with data from wet soil.

As can be seen from these two cases of modelling the distribution of copper and DDT, it is possible to make simple, workable models, which provide the answer to specific management problems related to discharge of toxic materials into lakes.

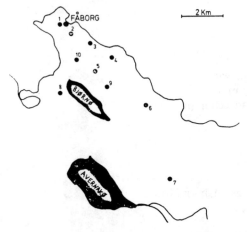

Fig.7.22: Faaborg Firth. The chromium profile was found at stations 2, 5, 6 and 7. The chromium concentration in the sediment surface sample was determined at the other stations.

## ILLUSTRATION 7.4

A case Study: Modelling the Distribution of Chromium in a Danish Firth (Faaborg Firth).

A tanning plant has for decades discharged waste water with a high concentration of chromium (III) into Faaborg Firth. 20 years ago the production was expanded significantly and gave a pronounced increase in the chromium concentration in the sediment (Mogensen, B.B. and Jørgensen, S.E. (1979).

The chromium profile has been found for sediment cores from four stations in the firth, and the chromium concentration in surface sediment samples was determined from six other stations (see fig. 7.22). (For details see B.B. Mogensen, 1978)

It is the scope of this investigation to set up a model for the distribution of chromium in the Firth on the basis of the sediment analysis. The concentration gradient in the water phase is much too small to get a picture of chromium distribution in the Firth, as a substantial part of the chromium is precipitated as chromium (III) hydroxide or other insoluble chromium compounds.

One model of the fate of chromium, including accumulation in the sediment and through the food chain, can be set up (Jørgensen, S.E., 1979 and Lu and Chen, 1977), provided that the concentration in the water is known as a function of the distance to the discharge point. Consequently, the results of the distribution model presented can be used as a forcing function in a model, considering the chromium concentration in the different trophic levels, in the water and in the sediment at a given station.

The distribution model is based on the following simple chromium transport equation: (see e.g. Rich (1973) and Chapter 3).

$$\frac{\partial C_t}{\partial t} = D\frac{\partial^2 C_t}{\partial X^2} - Q_w\frac{\partial C_t}{\partial 2} - \frac{K_s}{h}(C_t - C_o)$$

(7.137)

where

$C_t$  is the concentration of total chromium in water (g m$^{-3}$)
$C_o$  is the solubility of chromium in water (g m$^{-3}$)
$Q_w$  is the inflow to the firth = outflow by advection (m$^3$ day$^{-1}$)
$D$   is a mixing coefficient (m$^2$ day$^{-1}$)
$X$   is the distance from the discharge point
$K_s$  is the settling rate (m day$^{-1}$)
$h$   is the mean depth (m)

For a tidal firth without net advection such as Faaborg Firth,

$Q_w$ = 0, the stationary situation

$$\frac{\partial C_t}{\partial t} = 0$$

(7.138)

gives

$$D\frac{\partial^2 C_t}{\partial X^2} = \frac{K_s(C_t - C_o)}{h}$$

(7.139)

For the solution of this equation $C_u$ = total discharge of chromium (g day$^{-1}$) and F = cross sectional area (m$^2$), to state the boundary conditions. With a constant $C_u$ the following expression is obtained:

$$C_t - C_o = \frac{C_u}{FD^{1/2}}\left(\frac{h}{K_s}\right)\exp\left[-\left(\frac{K_s}{hD}\right)^{1/2} \cdot X\right] + C_1$$

(7.140)

260

F is known only approximately in this equation, due to the nonuniform geometry. $C_{tt}$ is known to be 22400 kg year$^{-1}$, based on the consumption and analytical determination of the waste water. About 10000 kg year-1 can be found in the sediment of the firth in accordance with the sediment examinations; $C_1$ is an integration constant.

8 m is used as an average of h, corresponding to an average of the inner and outer Firth.

From sediment analysis we have corresponding values between $Y = K_s(C_t - C_o)$ and X.

$$Y = K_s(C_t - C_0) = \frac{C_u}{F}\left(\frac{h \cdot K_s}{D}\right)^{1/2} \exp\left[-\left(\frac{K_s}{h \cdot d}\right)^{1/2}\right] \cdot X + C_1 \cdot K_s \quad (7.141)$$

The plot Y versus X confirms the exponential relation, see fig. 7.23.

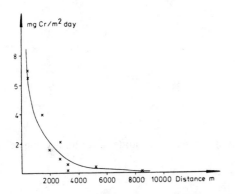

Fig.7.23: $Y = K_s(C_t - C_o)$ is plotted versus X.

Table 7.21 gives the values of Y and X found by the sediment investigation, taking natural mixing of the sediment (by means of a model developed by B. Larsen et al., 1981) and compression into consideration.

As a result of these considerations, the best possible fit to the equation:

$$Y = a.e^{-bx} + c \quad (7.142)$$

was found by application of a SAS-computer program (Marquardt method). The results are shown in table 7.22 including the standard error.

## TABLE 7.21

**Y versus X.**

| Station Number | g Cr/m² Year | (Y) g Cr/m² Day | (X) Distance from Discharge point (m) |
|---|---|---|---|
| 1 | 2.55 | 7.0 $10^{-3}$ | 500 |
| 2 | 2.39 | 6.5 $10^{-3}$ | 500 |
| 3 | 1.47 | 4.0 $10^{-3}$ | 1500 |
| 4 | 0.35 | 1.0 $10^{-3}$ | 2750 |
| 5 | 0.78 | 2.1 $10^{-3}$ | 2750 |
| 6 | 0.14 | 3.8 $10^{-4}$ | 5250 |
| 7 | 0.03 | 8.2 $10^{-5}$ | 8500 |
| 8 | 0.20 | 5.5 $10^{-4}$ | 3250 |
| 9 | 0.06 | 1.6 $10^{-4}$ | 3500 |
| 10 | 0.58 | 1.6 $10^{-3}$ | 2000 |

## TABLE 7.22

**Estimations of a,b and c.**

| | Estimate | Asymtotic St. error |
|---|---|---|
| a | 0.009909 | 0.00084 |
| b | 0.000723 | 0.00015 |
| c | -0.000081 | 0.00045 |

Table 7.23 shows the result of a variance analysis. As seen, the model gives an F-value as high as 114.5 compared with $F_{0.9995} = 30.4$.

## TABLE 7.23

**Statistical Analysis.**

| | Degree of Freedom | Sum of Squares | Mean Square |
|---|---|---|---|
| odel | 3 | 0.00011337 | 0.00003779 |
| sidual | 6 | 0.00000233 | 0.00000033 |
| ɔtal | 9 | | |
| | F = 114.5 | | |

From the regression analysis we have:

$$\frac{C_u}{F} \cdot \left(\frac{h \cdot K_s}{D}\right)^{1/2} = 0.00990 = a \qquad (7.143)$$

and

$$\left(\frac{K_s}{h \cdot D}\right)^{1/2} = 0.000723 = b \qquad (7.144)$$

which gives

$$\frac{C_u}{F} \cdot h = \frac{a}{b} = 13.7 \qquad (7.145)$$

$F = 35800$ m² which seems a reasonable average value of the cross sectional area.

From analysis of $C_t$ at stations 2, 5, 6, 7 and 8 (see Table 7.24) we get an estimation of $K_s$ since

$$Y = \frac{gCr}{m^2day} = K_s(C_t - C_o)(c \approx 0) \qquad (7.146)$$

$C_o$ is found to be 0.2 mg/m³.

As seen from Table 7.24 we get approximately the same $K_s$ values at three of the five stations. It would be expected that the settling rate would be lower, the greater the distance to the discharge point. In this context it should be stressed that the determinations of $C_t$ are not accurate, especially with respect to the concentration of particulate bound chromium.

**TABLE 7.24**

**Settling rates.**

| Station | mg $C_r$/m² day | $C_t$ - $C_o$ (mg m⁻³) | $K_s$ (m day⁻¹⁹) |
|---------|------------------|------------------------|--------------------|
| 2       | 6.5              | 2.5                    | 2.6                |
| 5       | 2.1              | 0.9                    | 2.3                |
| 6       | 0.4              | 0.6                    | 0.7                |
| 7       | 0.1              | 0.2                    | 0.5                |
| 8       | 0.6              | 0.3                    | 2.0                |

Average value of $K_s$ = 1.6

D can be found from $K_s$ = 1.6 m day⁻¹, since

$$b = \left(\frac{K_s}{h \cdot D}\right)^{1/2} = 0.000723 \qquad (7.147)$$

263

D = 3.8 . 105 m² day⁻¹, which corresponds to about 4.4 m² s⁻¹ , a quite reasonable value (Lerman 1971). From interpretation of sediment analysis it has, as illustrated, been possible to set up an equation for the distribution of chromium in a firth. It has been found that

$$C_t - C_o = \frac{a}{K_s}e^{-bX} + c = 0.00619 \cdot e^{-0.000723 \cdot X} \tag{7.148}$$

As seen for X = 0 $C_t$ = 0.0060 g m⁻³ or 6.0 ppb, which is reasonable. $K_s$ = 1.6 m day⁻¹ is found on the basis of analysis of the chromium concentration in the water (see Table 7.24), but as it leads to an acceptable D value (4.4 m²s-1) and three stations out of five give $K_s$ values close to the average, the estimation can be considered reasonably accurate. The cross sectional area has been determined on the basis of the distribution equation to 35.800 m², which is slightly more than the width of the inner firth, but as a weighted average for the inner - and outer - firth it seems acceptable.

With distribution set on the basis of the water analysis a much less reliable equation would be the result. The chromium concentration in sediment is much higher than in water, which gives a considerably more accurate determination of the concentrations in the sediment. Consequently, it is recommended, as demonstrated, to work out a distribution model on the basis of sediment analysis.

If the discharge of chromium is changed from its present level of 22.400 kg per year, equation (7.147) is still valid, only a` has another value as it is proportional to the annual emission (compare (7.141 and 7.142). Having set up a model for the distribution of chromium in the firth on the basis of sediment analysis, we turn to the question: what effect does a certain emission of chromium have on the aquatic life in the firth. Based on several surveys of the present condition of phytoplankton, zooplankton, fish and benthic animals, it can be concluded that:

1) Zooplankton and fish caught in the water body show only a slightly higher chromium concentration than those found in the open sea,

2) the chromium concentration of phytoplankton can hardly be determinded, as it is impossible to distinguish between chromium in the phytoplankton and suspended chromium hydroxide adhering to the phytoplankton,

3) the benthic animals show a pronounced chromium concentration in the firth than in open sea, probably due to the high concentration of chromium in the sediment. hus, this part of the aquatic ecosystem is affected by the discharge, as the major proportion of the metal settles and accumulates in the sediment.

Equation (7.133) can be applied to relate the concentration of chromium in the sediment with the concentration og chromium in benthic animals. The $n^{th}$ link in the food chain is the benthic animals and the $(n-1)^{th}$ link is the sediment. If a steady state is presumed, it means that $d\gamma(n)/dt = 0$, we get:

$$\gamma(n) = \frac{MY(n)\,\gamma(n-1)YT(n)}{MY(n) \cdot YF(n) - RESP(n) + EXC(n)} = K' \cdot \gamma(n-1)$$

For some of the species present in Faaborg Firth the parameter values in equation (11.32) are known. If we consider the mussel, Mytilus edulis, the following parameter values can be found in the literature (Jørgensen et al. 1979):

MY(n)   = 0.03 day$^{-1}$
YT(n)   = 0.07
YF(n)   = 0.06
RESP(n) = 0.001 day -1
EXC(n)  = 0.04

This equation implies that K' = 0.036 for Mytilus edulis. In other words the concentration of chromium in mussels should be expected to be 0.036 times the concentration in the sediment, which again can be related to the discharge of chromium.

21 mussels from Faaborg Firth have been analyzed and by a statistical analysis it was found that the relation between the concentration of chromium in the sediment and the mussel is:

$$\gamma(n) = \frac{MY(n) \cdot \gamma(n-1)YT(n)}{MY(n) \cdot YF(n) - RESP(n) + EXC(n)} = K' \cdot \gamma(n-1) \qquad (7.149)$$

where the standard deviation of the constant K' = 0.015 is 0.002 (see B. B. Mogensen, 1978).

The discrepancy between the observed K' value and the theoretical value can probably be explained by the uncertainty of biological parameters, which in many case s can only be considered as approximate values. The low standard deviation of the observed value confirms, however, the relation used.

The model presented can be used to assess environmentally acceptable chromium concentration in the sediment of about 70 mg per kg dry matter in the most polluted areas of the Firth. This would correspond to about 2 ppm chromium in the mussels, which is only twice the concentration found in the open sea.

### 7.7. MODELS IN TOXICOLOGY.

#### 7.7.1. Introduction.

Models of toxic substances in the environment might include effects on various organisms by use of a relationship between a concentration in water, air, soil or food and effects. In many problems it might however be necessary to go into more detail on the effect to be able to answer the following relevant questions:

1) Does the toxic substance accumulate in the organism? As an illustrative example see fig. 7.24.
2) What will the long term concentration in the organism be? Uptake rate, excretion rate and biochemical decomposition rate must be considered. see also 7.6.2.
3) What is the cronic effect of this concentration?

4) Does the toxic substance accumulate in one or more organs?
5) What is the transfer between various parts of the organism?
6) Will eventually decomposition products in the organism cause additional effects?

A detailed answer to those questions will require a model of the processes, that takes place in the organism and a translation of concentrations in various parts of the organism into effects. This implies of course that the intake = (uptake by organism). (efficency of uptake) is known. Intake might either be from water or air, which is often expressed by use of a concentration factor (ration of toxic substance in organism and in the environment or from the food. This part of the model has been touched in 7.6.2. and the related parameter estimation in 2.7.

### 7.7.2. Accumulation.

An accumulation as presented in fig. 7.24 might often be described more simple than in 7.6.2., when a single trophic level is considered.

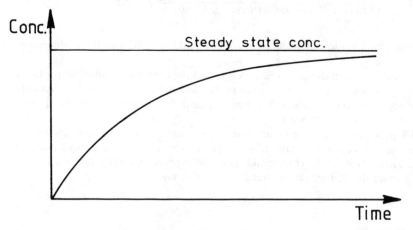

Fig. 7.24 Concentration of toxic substance in an organisation versus time.

If $C_t$ is the concentration of the toxic substance, ef is the efficiency of uptake through the food, $e_w$ the efficiency of uptake from water or air, $C_f$ the concentration of the toxic substance in the food, $C_w$ the concentration of the toxic substance in water or air, F the amount of food per day, V the volume of water or air which passes the gills or lungs per day, W is the body weight and EXC the excretion rate per day, we can set up the following differential equation:

$$\frac{dC_t}{dt} = (e_f \cdot C_f \cdot F + e_w \cdot C_w \cdot V) \, W' - EXC \cdot C_t \qquad (7.150)$$

At steady state, we get:

$$C_{t,max} = \frac{e_f \cdot C_p \cdot F + E_w \cdot C_w \cdot V}{W \cdot EXC} \qquad (7.151)$$

Equation (7.150) gives a curve for $C_t = f(t)$ similar to fig. 7.24. Not all the parameters in equation (7.150) are necessarily known; it is possible to compute one unknown parameter. If corresponding values of $C_t$ and t are known, more unknown parameters can be found by use of analytical solution to equation (7.150):

$$\frac{C_t}{C_{t,max}} = \frac{E_f \cdot C_p \cdot F + e_w \cdot C_w \cdot V}{W \cdot EXC}(1\text{-e}^{EXC.T}) \qquad (7.152)$$

The accumulation in an organ can be described by use of a similar equation as (7.152). If the percentage of the intaken toxic substance accumulates in the organ, p%, the weight of the organ, $W_0$, and the excretion rate from the organ EXC are known the equation will be:

$$\frac{dC_t}{dt} = \frac{(e_f \cdot C_f \cdot F + e_w \cdot C_w \cdot V)p}{W \cdot 100} - EXC \cdot C_t \qquad (7.153)$$

where $C_t$ is the concentration of the toxic substance in the considered organ. The steady state equation and the analytical soultion are parallel to equations (7.151) and (7.152).

W is in equations (7.150) and (7.152) considered constant. This a simplication, which is only valid for adult mammals. It is however not complicated to take into account W = f(t).

## EXAMPLE 7.3.

Man accumulates cadmium. It is known that

$e_f = 0.07, e_w = 0.5, C_f = 10\ \mu/kg, F = 1.2\ kg, C_w = 1.6\ \mu/m3\ air, V = 20\ m3/\ day$ W is 75 kg and a steady state concentration of 100 $\mu g/kg$ is upproximately obtained after 50 years. Estimate the excretion rate.

## Solution

By use of equation (7.151), we get:

$$EXC = \frac{e_f \cdot C_f \cdot F + e_w \cdot C_w \cdot V}{W \cdot C_{t,max}} = \frac{0.07 \cdot 10 \cdot 1.2 + 0.5 \cdot 1,6 \cdot 20}{75 \cdot 100} = 0.0022 day^1$$

50 years implies that $(1\text{-e}^{EXC.T})$

$$1 - e^{0.002 \cdot 50 \cdot 365} = 1 - 3.7 \cdot 10^{-18}$$

which means that $C_t$ is very closed to $C_{t,max}$

### 7.7.3. Multiple Compartment Models.

The models presented in 7.7.2 are in principle one-compartment models. Either the body or the most sensitive organ is used as the one-compartment. The toxicological problem might, however, require more detail i.e. more compartments to give a useful picture, which are the same considerations as we know from ecology. A compartment in toxicology is strictly defined as a mass of pollutants that has uniform kinetics of transformation and transport, and whose kinetics are different from those of other compartments.

Obviously, a multicompartment model in toxicology as in ecology requires more knowledge about the system and more data.

Fig. 7.25 shows a typical three compartment model for the distribution of a pollutant within an animal. Pollutant is absorbed into the blood (compartment 1) at a rate R. The peripheral compartments 2 and 3 - the liver and bones respectively - are linked to the central compartment, but not with each other. Most metabolism occurs in the liver, while assimilation might take place in the bones. Excretion takes place from the blood via the kidney. Rates of transfer between compartments are indicated by the rate constants (k). R corresponds to the intake and the transfers are often expressed either as first order reactions or by means of the Michaelis Menten kinetic. Total amounts are often used in such models - not concentrations.

## EXAMPLE 7.4.

Set up differential equations for the model in fig. 7.25, when R = 1.3 mg/day and the metabolism can be expressed by means of Michaelis Menten kinetic with a rate of 1.2 mg day$^{-1}$ and a half saturation constant of 1 mg in the liver. Transfer processes follow with good approximations first order kinetic. The rate constants are

$k_{12} = 0.5$ day$^{-1}$, $k_{13} = 0.08$ day$^{-1}$, $k_{21} = 0.8$ day$^{-1}$, $k_{31} = 0.1$ day$^{-1}$ and $k_{01} = 0.4$ day$^{-1}$. The amount of pollutants are denoted $P_1$, $P_2$ and $P_3$. At t = 0 $P_1 = 2$ mg, $P_2 = 4$ mg and $P_3 = 10$ mg (boundary conditions). Find steady state concentrations.

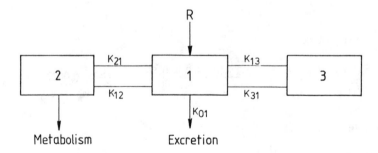

Fig. 7.25 A three-compartment model for distribution of a pollutant. 1 is the blood, 2 is the liver and 3 the bones. Pollutant is absorbed into the blood at the rate R. Rates of transfer between compartments are indicated by the rate constants. (K)

**Solution**

$$\frac{dP_1}{dt} = R - k_{31} \cdot P_1 - k_{21} \cdot P_1 - k_{01} \cdot P_1 + k_{12} \cdot P_2 + k_{13} \cdot P_3$$

$$\frac{dP_2}{dt} = k_{21} \cdot P_1 - k_{12} \cdot P_2 - \text{rate}\frac{P_2}{k_m + P_2}$$

$$\frac{dP_2}{dt} = k_{31} \cdot P_1 - k_{13} \cdot P_3$$

$$\frac{dP_1}{dt} = 1.3 - 0.1 \cdot P_1 - 0.8 \cdot P_1 - 0.4 \cdot P_1 + 0.5 \cdot P_2 + 0.08 \cdot P_3 \quad \left(\text{mgday}^{-1}\right)$$

$$\frac{dP_2}{dt} = 0.8 \cdot P_1 - 0.5 \cdot P_2 - 1.2\frac{P_2}{5 + P_2} \quad \text{mgday}^{-1}$$

$$\frac{dP_3}{dt} = 0.1 \cdot P_1 - 0.08 \cdot P_3 \text{mgday}^{-1}$$

$$\frac{dP_1}{dt} = \frac{dP_2}{dt} = \frac{dP_3}{dt} = 0$$

$$P_3 = \frac{0.1}{0.08}P_1 ; \quad P_2 = -2.6 + 1.2P_1 \quad \text{or} \quad P_1 = \frac{P_2 + 2.6}{2.4}$$

$$0.8 \cdot \frac{P_2 + 2.6}{2.4} - 0.5P_2 - 1.2\frac{P_2}{1 + P_2} = 0$$

$$P_2^2 + 1.7P_2 - 6.5 = 0$$

$$P_2 = -0.85 \pm 2.69$$

only $P_2 = 1.84$mg can be used.

$$P_1 = 2.92\text{mg} \quad \text{and} \quad P_3 = 23.39\text{mg}$$

Fig. 7.26 Illustrates a nine-compartment model describing the transport of cadmium in mammals. The model was proposed and investigated by Shank et al. (1977). The compartments in this model correspond to selected organs and residual carcass, liver, feces and blood, which is the central compartment. All processes were considered first order reactions. Models, which consider response of a certain concentration in the body or an organ, have already been touched in 5.3.

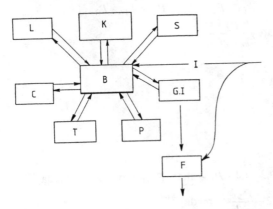

Fig.7.26: A nine-compartment models for transport of cadmium in mammals. The following abbreviations are used K for kidney, S for spleen, G for gastroin-testinal, P for pancreas, T for testes, C for residual carcass, L for liver, B for blood, F for feces and I for intake.

## 7.8. DISTRIBUTION OF AIR POLLUTANTS

### 7.8.1. General about Air Pollution Modelling.

Many air pollution models contain a huge system of equations to cover aerodynamic processes. In principle these equations are not different from hydrodynamic equations, but in pratical modelling situations the aerodynamic models are quite different from the hydrodynamic ones, as the considerations on possible simplifications will be different.

It would be wrong in this context to present an aerodynamic model in detail. The following topics have therefore been selected for presentation on the following pages:

1) Basic equations for long-range transport models of air pollution.
2) Some considerations on the application of such models on the distribution and effect on acidic rain.
3) Some considerations on inclusion of vertical transport in the models.
4) Models of airchemistry.
5) Models of plume dispersion.

By this presentation the reader will get an overview of the problems in air pollution modelling and also get an idea of the state of the art in the field. Furthermore, the reader will be able to develop simple air pollution models.

Dennis (1983) distinguish between three classes of models: analytical models, numerical models and statistical models.

If it is assumed that a single point source of air pollution exists, that there are no sinks and the source strength is Q, the mass conservation equation has the following solution:

$$C(x,y,z) = \frac{Q}{2\pi\delta_y\delta_x \cdot U} \exp\left[-\frac{1}{2}\left(\frac{y}{\delta_y}\right)^2 - \frac{1}{2}\left(\frac{z-H}{\delta_z}\right)^2 - \frac{1}{2}\left(\frac{z+H}{\delta_z}\right)^2\right] \quad (7.154)$$

where z is height above the ground, H is the source of the source $\delta_y$ and $\delta_z$ are dispersion variances, U average wind speed and C the pollutant concentration.

This equation is Gaussian in form, which means that the distribution of pollutant plume about the plume center is Gaussian. Therefore equation (7.154) is named the Gaussian plume model.

Models based upon this formulation are used extensively and will be demonstrated in 7.8.6.

The major strength of this type of model is that it is easy to use. It is well working in simple situations i.e. flat terrain and short travel time from the source ($\leqslant$ 10km).

The weakness of this type of models is that they cannot treat or only very poorly dispersion problems and chemical transformations beyond simple first order reactions. Because it is a steady state formulation they cannot account for build up of pollutants with time.

Numerical models attempt to overcome these weaknesses. They account for wind shear and eddy diffusivity shear and non-linear chemistry. Numerical models go back to the turbulent diffusion equation and specify the vertical and horizontal diffusivities. It results in partial differential equations, which must be solved numerically.

There are two major groups of numerical models: Eulerian multiple box models, which use a fixed coordinate system and Lagrangian trajectory models, which use a moving coordinate system. 7.8.2. will present the basic equations for the latter type, which is used in 7.8.3. The first type is the basis for 7.8.4.

The strength of numerical models are in first hand that they eliminate the weakness of analytical models furthermore they can be used rather generally for urban air quality and for long-range transport of pollutants, as they are easy to adapt to different areas. The numerical models require significant computer time - a disadvantage which is steadily reduced as the computer technology develops. At the more theoretical level, the gradient tranfer hypothesis is inconsistence with observations (this affects, of course, also the analytical models), Numerical models do not consider for point sources that the diffusivity is a function of travel time. Further development of the numerical models should be expected in the coming years to eliminate these theoretical disadvantages.

Analytical and numerical models are deterministic and they will predict concentrations, that are ensemble averages at a given time and place. The stochastic nature of the atmospherie processes has been subsumed in the ensemble average concentration. The statistical models recognize and accommodate the description of the stochastic part of the atmospheric diffusion processes. They treat the data as a time series and set up auto correlations. The observed time series is taken as the realization of some underlying stochastic processes. This realization is used to build a model of the process, that generated the time series. The model is built as a single time series model, a multiple time series model with or without deterministic elements.

Statistical models are cost-effective and can as discussed account to a certain extent for stochastic processes, but they are non-causal and might therefore have a large uncertainty of predictions associated with them. This type of models will not be treated in this chapter.

### 7.8.2. Basic Equations for Long -Range Transport Models of Air Pollution.

Lagrangian models can be used to predict long-range transport of air pollutants. A parcel of air is followed as it blows with the wind and the models keep track of the pollutant content of the parcel. This is in contrast to Eulerian models, where the integration of the mass balance equation is performed in a geographically fixed grid. It implies that Lagrangian models avoid the problems associated with the advection terms.

The mass balance equation for the considered pollutant within an air parcel is:

$$\frac{d(AhC)}{dt} = V_d \cdot A \cdot C - k \cdot A \cdot h \cdot C + Q \cdot A \qquad (7.155)$$

where C is pollutant concentration, $V_d$ the deposition velocity, k the rate constant of the first order decay, Q is the emission per unit area and time, A the base area and B the heigth of the air parcel.

The Lagrangian models fall into two main types: source oriented and receptor oriented models. In the first type the positions of emissions from each source are traced as a function of time. In the latter type the pollutant content of an air parcel is followed until the air parcel arrives at one of the selected receptor points. During its travel the air parcel receives emitted material from the sources it passes over.

For the source oriented models equation (7.155) takes the form:

$$\frac{d(A \cdot h \cdot C)}{dt} = -V_d \cdot A \cdot C - k \cdot A \cdot h \cdot C$$

By introduction of the pollutant mass of the puff, M = A·h C we obtain:

$$\frac{dM}{dt} = -\frac{V_d}{h}M - k \cdot M \qquad (7.156)$$

where $V_d$,h and k can be functions of t. If equation (7.156) is integrated to give M, C can be found, provided A(t) and h(t) are known.

A good example of a model of this type is the EURMAP/ENAMAP model (Johnson et al 1978, Bhumralkar et al 1979). It has been applied to calculate the transport and deposition of air-borne sulphur pollution over Europe and North America. It will be mentioned further in 7.8.3.

Equation (7.155) may be written for a receptor oriented model as follows:

$$A \cdot h\frac{dC}{dt} = -C\frac{d}{dt}(A \cdot h)_{turb} V_d A \cdot C - k \cdot A \cdot h \cdot C + Q \cdot A \qquad (7.157)$$

where d/dt(Ah)$_{turb}$ is due only to turbulence. Good example of the use of such models are given in Voldner et al (1980), Olson and Voldner (1981), OECD (1977) and Eliassen (1978).

### 7.8.3. Models of the Distribution and Effect of Acidic Rain.

A few models must be coupled to relate emission of sulphur and nitrogen with the effect of acidic rain on the ecosystems. Fig. 7.25 shows how a chain of models can be built. A given energy- and enviromental protection policy is related with emission of sulphur and nitrogen compounds, which again determine the air chemical processes and thereby the concentration of various sulphur - and nitrogen compounds. The next model step in the chain is a long-range transport model (see also 7.8.2), which gives the emission of acidic components. A soil model uses this as input and describe changes in soil water composition including pH-changes. The result of this model could be used to direct the energy- and environmental protection policy, as we can set a minimum pH-value of f.inst. 4.2. But the results of the soil model can also be used as input to a model concerned with the effect on lakes, streams, forests and agriculture, as the composition of soil water influences these ecosystems significantly. The output from the effects models can be used as feedbacks, see fig. 7.27 to the political decisions.

The total S-emission per unit of time, $S_r$ can be found from the following simple equation:

$$(7.158)$$

$$S_r = E_k \times s_k \times P_k(1\text{-}r_k)$$

where $E_k$ is energy consumption, $S_r$ the sulphur concentration in in fossil fuel, $P_k$ is the fraction of sulphur emitted in accordance with the applied environmental technology and $r_k$ the fraction of sulphur removed by ash and slags. $E_k$ and $s_k$ are political decisions. $r_k$ is known for most industrial processes and $P_k$ is dependent on the legislation i.e. which equipment is the industry forced to buy to reduce the sulphur emission.

In the long-range transport model considers a grid of the geographical area modelled. In fig. 7.28 is shown how the so-called EMEP-model (see Eliassen and Saltbones (1983), Fischer (1984) and Lamp (1984) has divided Europe in parcels of 150 x 150 km². The following equations are used to account for the emission, the chemistry and the deposition:

$$\frac{dC_{SO_2}}{dt} = Q_{SO_2} - D_{SO_2} - Ox \qquad (7.159)$$

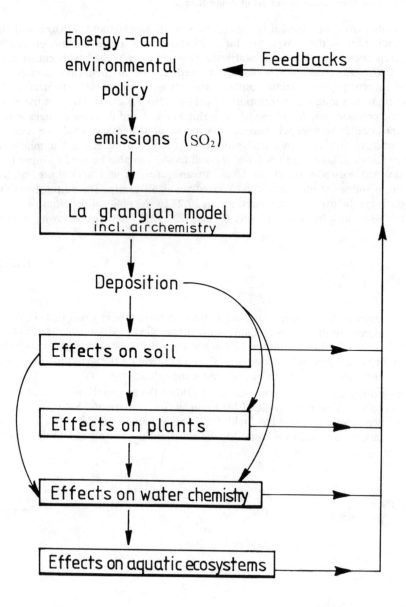

Fig. 7.27 A series of models must be used to relate energy - and environmental protection policy to effects on soil, plants, water chemistry and aquatic ecosystems, which serve as feedback to political issues.

$$\frac{dC_{SO_4^{2-}}}{dt} = Q_{SO_4^{2-}} - D_{SO_4^{2-}} + Ox \qquad (7.160)$$

$$Ox = k \cdot C_{SO_2} \qquad (7.161)$$

$$Q_{SO_2} + Q_{SO_4^{2-}} = S_k \qquad (7.162)$$

$$D_{SO_2} = \delta_{SO_2} C_{SO_2} \qquad D_{SO_4^{2-}} = S_{SO_4^{2-}} \cdot C_{SO_4^{2-}}$$

where Q are sources, D depositions, k and $\delta$ rate constants Ox represents the rate of oxidation $SO_2 \rightarrow SO_3$

At this stage of the model chain it is possible to predict sulphur deposition as g/m² yr, see fig. 7.29 for a given set of political decisions.

Fig. 7.29 is based upon an average metereological year and a total sulphur emission in Europe equal to the emission in 1970, but with 20% higher energy consumption, which implies that $s_k \cdot P_k$ must be reduced with a factor 1/1.2. The emission used in fig. 7.29 is unacceptable, however, because it is desirable to reach an emission of 0.5 gS/m² year for the major part of Europe to assure that pH in soil is $\geqslant 4.2$.

A number of models have been developed to translate the emission of sulphur- and nitrogen compounds into changes in soil chemistry in the first hand to changes in soilwater-pH. Kauppi et al (1984) has used knowledge about the buffer-capacity and -velocity to relate the emission with soilwater-pH. The results of this model is shown in fig. 7.30, where the same conditions as in fig. 7.29 have been used. A critical pH-value of 4.2 is used to interprete the results.

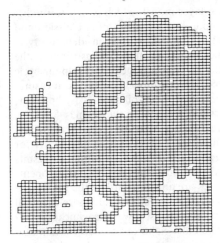

Fig. 7.28 Parcels in the EMEP-model.

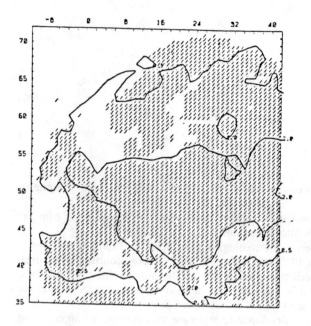

Fig. 7.29  Prognosis of sulphur deposition (g/m² year) in assistance with EMEP.

Christophersen and Wright have constructed a two layer model, which is able to predict the acidity of the tributaries to a stream or lake. Input data are the sulphur emission and composition of soil. Arp (1983) uses a more complex model, which considers n- layer. This model takes into account a number of chemical reactions in the soil as result of the acidic rain. It takes, furthermore, into account the carbon- nitrogen- and sulphur cycle in soil, see fig. 7.31. Arp obtains good accordance between model results and measurements for a number of case studies. Further development of soil models is, however, needed as the soil model must be considered the weakest model in the chain.

Several simple models are able to relate the total loading of acidic components to pH of a stream or a lake. Fig. 7.32 shows a relationship between total S-loading and pH for two lakes. The curves are in principle a simple titration curve and based upon the results from the two previous models, it is possible to calculate the input of acidic components from the precipitation and the tributaries. The curves are able to translate the total input of acidic components into a pH-value. Henriksen and Seip (1982) include more chemical processes into their model and several other more comprehensive water chemical models for the prediction of pH in water are available. The most elaborate models contain also the influence of pH on the eutrophication, which again is determining pH of the water.

Muniz and Seip (1982) use a simple statistical model to relate the pH of water in streams

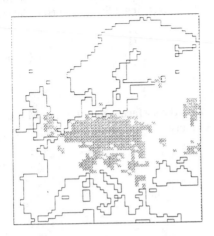

Fig.7.30: Prognosis of pH in soil. Areas with soil-pH < 4.2 are shown.

and lakes with the effect on the fish population see fig. 7.33. Chester (1982) critisises this method, as he means that the calcium concentration of the water influences the pH-effect.

Chen et al (1982) have developed a more comprehensive pH-effect model. They consider the effect on all levels in the food chain and the total effect on the ecosystem.

Kohlmaier et al. (1984) have developed a model, which relates the atmospheric composition and the soil water composition, including pH and the concentration of aluminium ions, with the effect on trees. The model is a black box approach as it is based upon a statistical analysis of these relations.

### 7.8.4. Inclusion of Vertical Transport in Air Pollution Models.

It is often needed to include vertical transport processes in pollution models of urban areas. It is possible by use of Eulerian models, which have found their widest application, when wind shear and vertical eddy diffusivity profiles must be considered.

Characterization of the wind field requires knowledge about the wind shear, i.e. the increase of wind speed with height, and about the wind direction in the x-y plane throughout the region.

For a neutral atmosphere in the surface layer (the first tens of meters) is:

$$U = \frac{U^+}{k} \ln \frac{Z + Z_o}{Z_o} \qquad (7.163)$$

where U is the wind speed at height Z, U+ is the friction velocity, $Z_0$ is the surface roughness and k is Von Karman's constant.

This relationship do not hold throughout the boundary layer. Many models use the emperical power relationship for the entire boundary layer, including the surface layer. This relationship is

$$U = U_R \cdot \left(\frac{Z}{Z_0}\right)^M \tag{7.164}$$

where $U_r$ is the wind speed at a reference height, $Z_0$ is the surface roughness and M is a function of surface roughness and atmospheric stability.

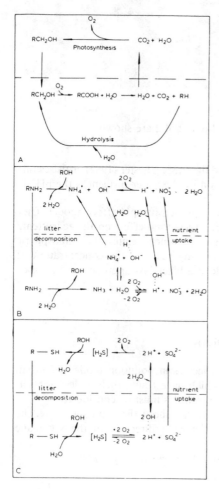

Fig. 7.31  C, N and S cycles considered in Arp's model.

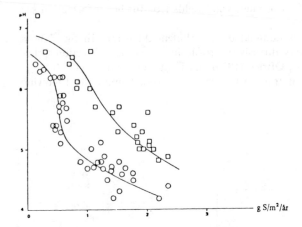

Fig. 7.32 Relationship between S- emission and pH. □-corresponds to high calcium
concentration and o-low calcium concentration.

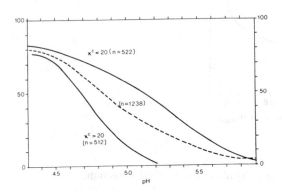

Fig. 7.33 Mortality % versus pH of fish population. Dotted line represents the total
statistical material. The other lines corresponds to conductirety of water
below and above 20 ms/cm. N indicates number of observations.

For some urban areas the terrain is simple enough that it can be assumed that the wind direction in the x-y plane is the same across the urban area. For other urban areas, however, the convergence and divergence of wind fields are important. Interpolation schemes have had to be developed to produce wind fields (for further information see Reynolds et al 1976, Killus et al 1980).

Vertical transport of pollutants is dominated by turbulent diffusion. In fig. 7.34 is shown a possible profile in the eddy diffusivity profile in the vertical direction Other formulations exist, see Lamb (1977), OBrien (1970) and Lamb and Durran (1977).

A model considering the vertical transport processes is based upon the following governing equations:

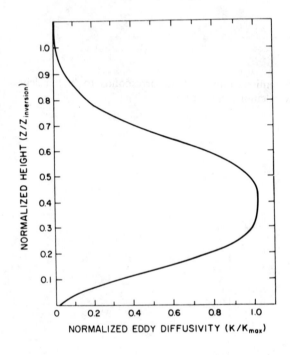

Fig. 7.34 Characteristic Eddy Diffusivity Profile in the Vertical Direction.

$$\underbrace{\frac{\partial C_i}{\partial t}}_{\text{Time Dependence}} + \underbrace{\frac{\partial(\bar{u}c_i)}{\partial x} + \frac{\partial(\bar{v}c_i)}{\partial y} \frac{\partial(\bar{w}c_i)}{\partial z}}_{\text{Advection}}$$

$$= \underbrace{\frac{\partial}{\partial x}\left(K_H\frac{\partial c_i}{\partial x}\right) + \frac{\partial}{\partial y}\left(K_H\frac{\partial C_i}{\partial y}\right) + \frac{\partial}{\partial z}\left(K_V\frac{\partial C_i}{\partial z}\right)}_{\text{Diffusion}} \qquad (7.165)$$

$$+ \underbrace{R_i}_{\substack{\text{Chemical} \\ \text{Reaction}}} + \underbrace{L_i}_{\text{Removal}} + \underbrace{S_i}_{\text{Emission}}$$

where:

$c_i$ = concentration of species i.
$\bar{u}, \bar{v}$ = horizontal components of the wind.
$\bar{w}$ = vertical component of the wind.
$K_H$ = horizontal turbulent diffusivity.
$K_V$ = vertical turbulent diffusivity.
$R_i$ = rate of formation of species i by chemical reactions.
$N_i$ = rate of removal of species i.
$S_i$ = rate of emission of species i.

### 7.8.5. Models of Air Chemistry.

The chemical reactions in the atmosphere are mentioned in 7.8.3 and in 7.8.4. They are included in the models as rate of production or destruction of pollutant species i. (see equation (7.154)).

The fate of any one species is related to the fate of several other species. The model would be very comprehensive, if it should include all possible pollutants and it is therefore often necessary to consider the possibilities of setting up a more generalized scheme. Carbon-Bond mechanism (CMB·I) will be presented here. Table 7.25 overviews the reactions considered (Whitten et al 1980).

CBM-I is based on the concept of grouping carbon atoms with similar chemical bonding. Four groups are used: 1) single-bonded carbon atoms (PAR), 2) very reactive double-bonded carbon atoms (OLE) 3) moderately reactive double-bonded carbon atoms (ARO) and 4) carbonyl-bonded carbon atoms (CAR). PAR includes not only alkanes, but also the single-bonded carbon atoms of alkenes, aromatics and aldehydes. F.inst. 1 ppm of propylene would give 1 ppm OLE and 1 ppm of PAR.

The CMB-I system has no adjustable parameters (see table 7.25). It must be considered an advantage in the first hand as the model otherwise would contain too many parameters to be calibrated, see also the discussion in 2.7.

In spite of the reduction in number of state variables, which is possible by use of CMB-I, further reduction is often needed to avoid a too complex model. This is discussed and demonstrated by Dennis (1983)

**TABLE 7.25**

**The Carbon-Bond Mechanism**

| No. | Reaction | Rate Constant[a] |
|-----|----------|-------------|
| 1 | $NO_2 + h\nu \rightarrow NO + O$ | $K_1$[b] |
| 2 | $O + O_2 (+M) \rightarrow O_3 (+M)$ | $2.08 \times 10^{-5}$ |
| 3 | $O_3 + NO \rightarrow NO_2 + O_2$ | $2.52 \times 10$ |
| 4 | $O + NO_2 \rightarrow NO + O_2$ | $1.34 \times 10^4$ |
| 5 | $O_3 + NO_2 \rightarrow NO_3 + O_2$ | $5 \times 10^{-2}$ |
| 6 | $NO_3 + NO \rightarrow NO_2 + NO_2$ | $2.5 \times 10^4$ |
| 7 | $NO_3 + NO_2 + H_2O \rightarrow 2HNO_3$ | $2.0 \times 10^{-3}$ |
| 8 | $HO_2\cdot + NO_2 \rightarrow HNO_2$ | $2.0 \times 10$ |
| 9 | $NO_2 + OH\cdot \rightarrow HNO_3$ | $1.4 \times 10^4$ |
| 10 | $HNO_2 + h\nu \rightarrow NO + OH\cdot$ | $1.9 \times 10^{-1} K_1$ |
| 11 | $NO + OH\cdot \rightarrow HNO_2$ | $1.4 \times 10^4$ |
| 12 | $CO + OH\cdot \xrightarrow{O_2} CO_2 + HO_2\cdot$ | $4.5 \times 10^2$ |
| 13 | $HO_2\cdot + NO \rightarrow OH\cdot + NO_2$ | $1.2 \times 10^4$ |
| 14 | $HO_2\cdot + HO_2\cdot \rightarrow H_2O_2 + O_2$ | $1.5 \times 10^4$ |
| 15 | $PAN \rightarrow HC(O)O_2\cdot + NO_2$ | $2.0 \times 10^{-2}$ |
| 16 | $H_2O_2 + h\nu \rightarrow OH\cdot + OH\cdot$ | $7.0 \times 10^{-4} K_1$ |
| 17 | $OLE + OH\cdot \xrightarrow{O_2} CAR + CH_3O_2\cdot$ | $3.8 \times 10^4$ |
| 18 | $OLE + O \xrightarrow{2O_2} HC(O)O_2\cdot + CH_3O_2\cdot$ | $5.3 \times 10^3$ |
| 19[c] | $OLE + O_3 \xrightarrow{O_2} \gamma(HC(O)O_2\cdot + HCHO + OH\cdot)$ | $1.5 \times 10^{-2}$ |
| 20 | $PAR + OH\cdot \xrightarrow{O_2} CH_3O_2\cdot + H_2O$ | $1.3 \times 10^3$ |
| 21 | $PAR + O \xrightarrow{O_2} CH_3O_2\cdot + OH\cdot$ | $2.0 \times 10$ |
| 22 | $CAR + OH\cdot \xrightarrow{O_2} HC(O)O_2\cdot + H_2O$ | $1.0 \times 10^4$ |
| 23[c] | $CAR + h\nu \xrightarrow{2O_2} \alpha HC(O)O_2\cdot + \alpha HO_2\cdot + (1-\alpha)CO$ | $6.0 \times 10^{-3} K_1$ |
| 24 | $ARO + OH\cdot \xrightarrow{O_2} CAR + CH_3O_2\cdot$ | $8.0 \times 10^3$ |
| 25 | $ARO + O \xrightarrow{2O_2} HC(O)O_2\cdot + CH_3O_2\cdot$ | $3.7 \times 10$ |
| 26 | $ARO + O_3 \xrightarrow{O_2} HC(O)O_2\cdot + CAR + OH\cdot$ | $2.0 \times 10^{-3}$ |
| 27 | $ARO + NO_3 \rightarrow products \ (aerosols)$ | $1.0 \times 10^2$ |
| 28 | $CH_3O_2\cdot + NO \rightarrow NO_2 + CAR + HO_2\cdot$ | $1.2 \times 10^4$ |
| 29 | $HC(O)O_2\cdot + NO \rightarrow NO_2 + CO_2 + HO_2\cdot$ | $3.8 \times 10^3$ |
| 30 | $HC(O)O_2\cdot + NO_2 \rightarrow PAN$ | $6.0 \times 10^2$ |
| 31 | $CH_3O_2\cdot + HO_2\cdot \rightarrow CH_3OOH + O_2$ | $4.0 \times 10^3$ |
| 32 | $HC(O)O_2\cdot + HO_2\cdot \rightarrow HC(O)OOH + O_2$ | $4.0 \times 10^3$ |

a - in units of $ppm^{-1} min^{-1}$ except for photolysis reactions
b - Photolysis rate constant - depends on light intensity
c - $\alpha = 0.5$ and $\gamma = 0.67$
Source: Whitten, et al., 1980.

### 7.8.6. Plume Dispersion.

Determination of the atmospheric concentration of pollutants emitted from point sources is an important example of the use of the mass conservation principle, see equation(7.154)in 7.8.1.

Fig. 7.35 illustrates a plume from a source at x = 0, y = 0. As the plume moves downwind it grows through the action of turbulent eddies. The instantaneous plume has a high concentration over a narrow width. Over, for instance, ten minutes the plume will touch a much broader area but the concentration will, of course, be correspondingly lower. Likewise in two hours the plume will stretch out in the cross wind direction and its concentration will be further reduced. The concentration distribution perpendicular to the axis of wind appears to be Gaussian or, rather the Gaussian distribution seems to be a useful model for the calculation of plume concentration. The shortcomings of this model will be discussed later and have been touched in 7.8.1.

The scope of the plume dispersion model is to determine, C, the concentration of the pollutant, as a function of its position downwind from the source. Equation (7.154) is used. Plume from a source (1) corresponds to the instantaneous plume. (2) is 10 minutes average plume and (3) is the 2 hour average plume. The diagram represents the cross-plume distribution patterns.

Fig. 7.36 shows the co-ordinate system applied in the model. Notice that z - H accounts for the reflection of the plume, corresponding to an imaginary source below the ground. H is the effective stack height. Below is discussed how it can be calculated.

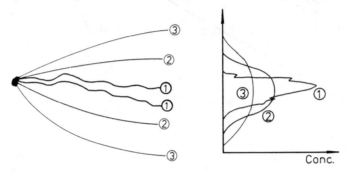

Fig. 7.35  Plume from af source (1) corresponds to the instantaneous plume. (2) is 10 minutes average plume and (3) is the 2 hour average plume. The diagram represents the cross-plume distribution patterns.

For z = 0, at ground level, we find:

$$C(x, y, 0) = \frac{Q}{\pi \delta_y \delta_z \cdot U} \exp\left[-\left(\frac{y^2}{2\delta_y^2} + \frac{H^2}{2\delta_z^2}\right)\right] \tag{7.166}$$

For the concentration along the centerline (y = 0) and at ground level, we have:

$$C(x, 0, 0) = \frac{Q}{\pi \delta_y \delta_z \cdot U} \exp\left[-\frac{H^2}{2\delta_z^2}\right] \tag{7.167}$$

U is function of z but usually the average wind speed at the effective stack height is used. If the wind speed $U_1$ at level $z_1$ is known, U can be estimated by the following equation:

$$U = U_1\left(\frac{H}{z_1}\right)^n \tag{7.168}$$

Smith (1968) recommends n = 0.25 for unstable and n = 0.50 for stable conditions. For further discussion on this subject see Turner (1970). Compare equation (7.168) with (7.164).

$\delta_y$ and $\delta_z$ can be considered as diffusion coefficients and can be found as functions of

283

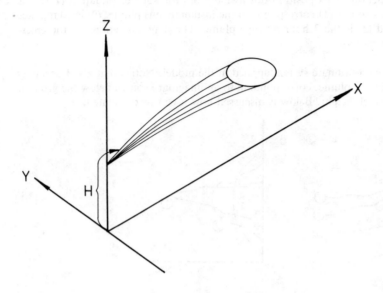

Fig. 7.36 Co-ordinate system applied in the plume model.

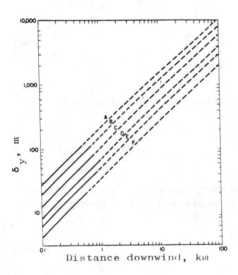

Fig. 7.37 Horizontal dispersion coefficient as a function of downwind distance from the source. (Turner, 1970).

atmospheric conditions (Smith, 1968). These relationships are demonstrated in Table 7.26 and Fig. 7.37 and 7.38.

**TABLE 7.26 Key to stability categories**

| Surface Wind speed (at 10m) m set⁻ | Day | | | Night | |
|---|---|---|---|---|---|
| | Incoming solar radiation | | | Thinly overcast or | |
| | Strong | Moderate | Slight | >4/8 low cloud | <3/8 cloud |
| <2 | A | A-B | B | | |
| 2-3 | A-B | B | C | E | F |
| 3-5 | B | B-C | C | D | E |
| 5-6 | C | C-D | D | D | D |
| >6 | C | D | D | D | D |

Uncertainties in the estimation of $\delta_y$ are generally fewer than those of $\delta_z$. However, wide errors in the estimate of $\delta_z$ can occur over longer distances. In some cases $\delta_z$ may be expected to be correct within a factor of 2. These cases are 1) stability for distance of travel to a few hundred metres, 2) neutral to moderately unstable conditions for distances to a few kilometres, 3) unstable conditions in the lower 1000 metres of the atmosphere with a marked inversion above for a distance of 10 km or more.

Turner (1970) discusses a procedure for handling diffusion, when the plume expansion is limited by an upper level inversion. The principle of the method is to calculate the concentrations as if they were distributed uniformly throughout the layer of height $H_i$ - the distance from ground level to the inversion. In this case the expression for C at ground level becomes:

$$C = \frac{Q}{\sqrt{2\pi}\,UH_i\delta_y} \exp\left[-\frac{1}{2}\left(\frac{y}{\delta_y}\right)^2\right] \tag{7.169}$$

This equation can be applied downwind, when the plume first reaches the $H_i$ elevation, which can be estimated when $2.15\delta_z = H_i\text{-}H$. Turner (1970) recommends using equation (7.169) for $x > 2x_i$, where $x_i$, is the point at which the plume reaches the inversion. Before $x_i$ the regular diffusion result applies and between $x_i$ and $2x_i$ an interpolation between the results at $x_i$ and $2x_i$ is recommended.

The same equation can be applied under fumigation conditions, in which case $H_i$ is the height to which the unstable air risen. For an explanation of inversion and fumigation, see Fig.7.39. However, Q must then be corrected to the fraction of the plume which can be carried down to the ground. Note that the stable classifications used to calculate $\delta_x$ and $\delta_z$ are also used to predict fumigation conditions. Turner, however, uses a correction factor to account for the plume spreading in the y direction:

$$\delta_{y,fum} = \delta_{y,stable} + \frac{H}{8} \tag{7.170}$$

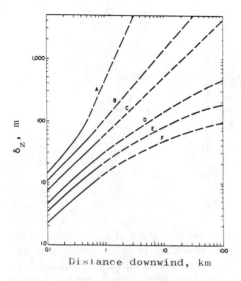

Fig. 7.38 Vertical dispersion coefficient as a function of downwind distance from the source. (Turner, 1970).

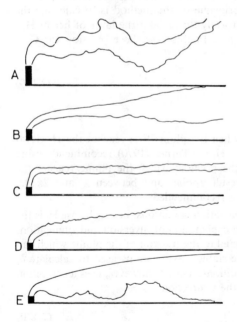

Fig. 7.39 Stack gas behaviour under various conditions. A) Strong lapse (looping), B) Weak lapse (coning), C) Inversion (fanning), D) Inversion below, lapse aloft (lofting), E) Lapse below, inversion aloft (fumigation).

**TABLE 7.27 Various atmospheric conditions**

| | |
|---|---|
| strong lapse (looping) | Enviromental lapse rate > adiabatic lapse rate |
| Weak lapse (conning) | Environmental lapse rate < adiabatic lapse rate |
| Inversion (fanning) | Increasing temperature with height |
| Inversion below lapse aloft (lofting) | Increasing temperature below, app. adiabatic lapse rate aloft |
| Lapse below, inversion aloft (fumigation) | App. adiabatic lapse rate below, increasing temperature aloft |

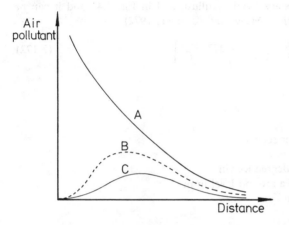

Fig. 7.40 Distribution of emission from stack of different heights under coning conditions. A) height 0 m, B) height 50 m, C) height 75 m.

As the example below demonstrates it is easy to use the plume model in practice, but, of course, such a relatively simple approach has some limits:

1) the diffusion coefficients applied are not very accurate,
2) the turning of the wind height owing to friction effects is neglected,
3) adsorption or deposition of pollutants is neglected, but could easily be included, if required,
4) chemical reactions along the plume path have been omitted, but might be included if the necessary knowledge were available,
5) shifts in wind direction are not taken into consideration.

In view of these limitations, the plume model should only be used as a first estimate of pollutant concentration (see Scorer, 1968).

The maximum concentration, $C_{max}$, at ground level can be shown to be approximately proportional to the emission and to follow approximately this expression:

$$C_{max} = k\frac{Q}{H^2}$$

(7.171)

where Q is the emission (expressed as g particulate matter per unit of time), H is the effective stack height and k is a constant.

It can, furthermore, be shown that the maximum ground level concentration occurs where $\delta_z = 0.707\, H$, provided that the ration $\delta_z/\delta_y$ is constant with downwind distance x.

The definition of the effective stack height is illustrated in fig. 7.41 and it can be calculated from the following equation (Moses and Kraimer, 1972):

$$H = h + 0.28 \cdot V_s \cdot D_s\left[1.5 + 2.7\frac{T_s - 273}{T_s} \cdot D_s\right]$$

(7.172)

where

$V_s$    = stack exit velocity in m per second
$D_s$    = stack exit inside diameter in m
$T_s$    = stack exit temperature in degree Kelvin
h     = physical stack height above ground level in m
H    = effective stack height in m

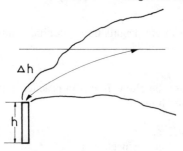

Fig.7.41: Effective stack height H = h + Δh.

These equations explain why a lower ground-level concentration is obtained when many small stacks are replaced by one very high stack. In addition to this effect, it is always easier to reduce and control one large emission than many small emissions, and it is more feasible to install and apply the necessary environmental technology in one big installation.

As the example below demonstrates it is easy to use the plume model in practice, but, of course, such a relatively simple approach has some limits:

1) the diffusion coefficients applied are not very accurate,
2) the turning of the wind with height owing to friction effects is neglected,

3) adsorption or deposition of pollutants is neglected, but could easily be included, if required,
4) chemical reactions along the plume path have been omitted, but might be included if the necessary knowledge were available,
5) shifts in wind direction are not taken into consideration.

In view of these limitations, the plume model should only be used as a first estimate of pollutant concentration (see Scorer, 1968).

## EXAMPLE 7.5

Estimate the concentration of $SO_2$ downwind of a power plant burning 12,000 tons of 1.5% sulphur coal per day. The effective stack height is 200 m. The wind speed has been measured on a clear sunny day as 4 m/sec at the top of a 10 m tower. Find the concentration at ground level at x = 1 km and at 5 km, if it is estimated that 25% of the sulphur remains in the ash and is collected. At what height would an inversion affects the ground level concentration at x = 10 km?
Determine the maximum ground level concentration of $SO_2$ and the distance of the stack, at which the maximum occurs.

**Solution:**

From Table 7.26: stability category B

From Fig. 7.37 and 7.38

| x | $\delta_y$ | $\delta_z$ |
|---|---|---|
| 1 km | 150 | 100 m |
| 5 km | 700 m | 700 m |
| 10 km | 1000 m | 1200 m |

From equation (7.168):

$$U = U_1\left(\frac{H}{z_1}\right)^n = 4\left(\frac{200}{400}\right)^{0.5} = 18 \ m \, sec^{-1} \qquad (7.173)$$

S-emission: $\frac{0.75 \cdot 12 \cdot 10^9 \cdot 1.5}{100 \cdot 24 \cdot 3600} = 1562.5 \ g \ sec^{-1}$

$SO_2$-emmision: $1562.5 \cdot \frac{64}{32} = 3125 \ g \ sec^{-1}$

$$x = 1 \text{ km}: C_{SO_2} = \frac{3125}{\pi \cdot 150 \cdot 100 \cdot 18} \exp\left(-\frac{200}{2 \cdot 100^2}\right) = 500 \ \mu\text{g m}^{-3}$$

$$x = 5 \text{ km}: C_{SO_2} = \frac{3125}{\pi \cdot 700 \cdot 700 \cdot 18} \exp\left(-\frac{200^2}{2 \cdot 700^2}\right) = 108 \ \mu\text{g m}^{-3}$$

$$2.15 \cdot 1200 = H_i - H = H_i - 200$$

$$H_i = 2780 \text{ m}$$

$$\delta_z = 0.707 \cdot 200 = 141 \text{ m}$$

which is obtained at $x = 1.1$ km.

$$C_{SO_2} = \frac{3125}{\pi \cdot 200 \cdot 100 \cdot 18} \exp\left(-\frac{200^2}{2 \cdot 141^2}\right) = 1000 \ \mu\text{g m}^{-3}$$

## 7.9. MODELS OF SOIL PROCESSES, PLANT GROWTH AND CROP PRODUCTION

### 7.9.1. Introduction

These models have found wide application: 1) to control groundwater pollution by models of soil transport processes 2) to understand the nitrogen cycle in soil by models of transport processes in soil combined with models of the chemical processes of nitrogen in soil, 3) to simulate plant growth in general, 4) to manage such ecosystems as grassland, forestland, etc. on the system level, 5) to optimize crop yields 6) to optimize use of irrigation 7) to control pollution caused by use of pesticides, 8) to optimize use of fertilizers from an economical-ecological viewpoint.

These models are obviously applicable to a spectrum of agriculture management problems. An increased use of models for the control and correct management of these man controlled ecosystems is foreseen. The combination of the many interacting ecological processes, the need for management and the peril of great pollution problems make the use of models self-evident.

The wide use of these various types of models has given a good basis of experience with model construction in the field. Generally, good data are available and a good basis of parameter estimations can be found in the literature.

The area is therefore one of the most developed fields within ecological modelling.

As for other areas of ecological modelling, many different models are available to solve the same problem, and the selection of model including the complexity of the model is dependent on the system, the problem and the data. It is therefore not possible to review all models, but a few characteristic ones will be presented in more or less detail. The most widely used equations applied in many models are presented, since they can be considered as core equations in the field.

A model of mass transfer in soil is presented. Only diffusion and vertical water flux are considered in its simplest version, which can be solved analytically. The model predicts the flux of water and dissolved substances. More complex versions include bound water, adsorption, ion exchange and decomposition during the transport. Also a model of heat transfer in soil is presented.

As plants and soil transpiration influences the water balance, a model, which accounts for these contributions, is included. Furthermore, the climate influences the plant growth and the water balance, which is considered in another submodel.

Plant growth is, however, not only dependent on the climatic factors i.e. temperature and radiation, but also on the atmospheric composition, water availability and concentrations of nutrients and toxic substances. A model of the influence of all these factors is therefore presented.

Finally a model is presented of the nitrogen processes in soil; these again influence the plant growth and the transfer of nitrogen in soil. Other components such as phosphorus and pesticides might also participate in other processes than those already included in the mass transfer mentioned above. Consequently similar models must be developed for these substances.

The presented models are related and the relationships must be included in a total model. In fig. 7.42 it is shown how the submodels are related in a total model of crop production. Similar diagrams can be set up for other applications (see above) of submodels to form total models.

### 7.9.2. Mass and Heat Transfer in Soil.

A differential equation in accordance with the mass conservation principle, as presented in 3.2.1, is valid for mass transfer in soil:

$$\frac{\partial(\Theta C)}{\partial t} = \frac{\partial}{\partial z}\left(\Theta D_{zz}\frac{\partial C}{\partial z}\right) + \frac{\partial}{\partial x}\left(\Theta D_{xx}\frac{\partial C}{\partial x}\right) - \frac{\partial}{\partial z}(qC) + F \tag{7.174}$$

where

| | | |
|---|---|---|
| C | = | concentration of dissolved substance |
| $\Theta$ | = | water content |
| t | = | time |
| F | = | sources and/or sinks |
| z | = | vertical coordinate |
| q | = | water flux (vertical) |
| $D_{ii}$ | = | hydrodynamic dispersion coefficient + (incl. mechanical molecular diffusion). |

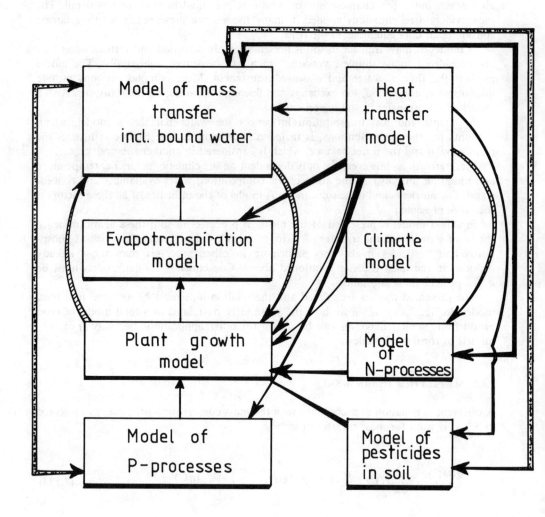

Fig.7.42: A total crop production model might require a total model, which consists of the shown 8 submodels. The relations are indicated by arrows. The arrow direction shows that output from one model is used in another model.

This equation takes into consideration convection in one dimension (Darcys equation) and diffusion in two dimensions.

If diffusion only takes place in one dimension the equation is reduced to:

$$\frac{\partial(\Theta C)}{\partial t} = \frac{\partial}{\partial z}\left(\Theta D_{zz}\frac{\partial C}{\partial z}\right) - \frac{\partial}{\partial z}(qC) + F \qquad (7.175)$$

This equation has an analytical solution. The concentration profile for the addition of the amount M in the depth $z = 0$ at time $t = 0$ can be found from:

$$C = \frac{M}{\Theta}\left(\frac{1}{(4\pi D \cdot t)^{\frac{1}{2}}} \exp\left(-\frac{(z - qt/\Theta)^2}{4Dt}\right)\right) \qquad (7.176)$$

For diffusion in two dimensions the analytical solution is:

$$C = \frac{M}{\Theta}\left(\frac{1}{\left(4\pi t(D_x \cdot D_z)^{\frac{1}{2}}\right)} \exp\left(-\frac{(z - qt/\Theta)^2}{4D_2 \cdot t} - \frac{x^2}{4D_x \cdot t}\right)\right) \qquad (7.177)$$

As in hydrodynamics F defines the boundary conditions.

Equation (7.174) does not account for the water bound to the soil in pores and capillars. It is possible by use of the following equation:

$$\frac{\partial(\Theta_m C_m)}{\partial t} + \frac{\partial(\Theta_{im}C_{in})}{\partial t} = \frac{\partial}{\partial z}\left(\Theta_m D_{zz}\frac{\partial C_m}{\partial z}\right)$$

$$+ \frac{\partial}{\partial x}\left(\Theta_m D_{xx} \cdot \frac{\partial C_m}{\partial x}\right) - \frac{\partial}{\partial z}(qC) + F \qquad (7.178)$$

m indicates mobile water and im immobile or bound water.

The exchange of water between the two phases can in accordance with Coats and Smith (1964) be described by:

$$\Theta_{im}\frac{\partial C_{im}}{\partial t} = \beta(C_m - C_{im}) \tag{7.179}$$

If the amount of bound or immobile water is constant, equation (7.179)

can be changed to:

$$\frac{\partial(\Theta_{im}C_{im})}{\partial t} = \beta(C_m - C_{im}) \tag{7.180}$$

Equation (7.178) can now be solved numerically, as it can be reformulated as follows:

$$\frac{\partial(\Theta_m C_m)}{\partial t} = \frac{\partial}{\partial z}\left(\Theta_m \cdot D_{zz}\frac{\partial C_m}{\partial z}\right) + \frac{\partial}{\partial x}\left(\Theta_m \cdot D_{xx}\frac{\partial C_m}{\partial x}\right)$$

$$-\frac{\partial}{\partial z}(qC) + F - \beta(C_m - C_{im}) \tag{7.181}$$

The analytical solution is as seen similar to (7.177).

If adsorption/desorption takes place the basic equation is changed to:

$$\frac{\partial(\Theta C)}{\partial t} + \varrho\frac{\partial S}{\partial t} = \frac{\partial}{\partial z}\left(\Theta D_{zz}\frac{\partial C}{\partial z}\right) + \frac{\partial}{\partial x}\left(\Theta D_{xx}\frac{\partial C}{\partial x}\right) - \frac{\partial}{\partial z}(q \cdot C) + F \tag{7.182}$$

where

S is the concentration of the adsorbed substance and $\varrho$ is the volume weight of soil.

If the adsorption can be considered a fast process relatively to other processes, it will be possible to describe the adsorption by means of an equilibrium isotherm such as:

$S = k_1 . C + k_2$, equation (3.22) or equation (3.23)

The results of the transport model applied on the case that an addition of the amount M at the surface ($z = 0$) takes place at time $z = 0$ is shown on fig. 7.43. The following equilibrium equation is applied in this case:

$$ (7.183) $$

$$ S = 0.05 . C $$

The adsorption causes a delay of the mass transport.

In case the adsorption rate is comparable to the rate of other processes equation (3.24) or the following equations can be applied:

$$ \frac{\partial S}{\partial t} = k_1 \cdot C - k_2 \cdot S + k_3 \quad \text{linear case} \qquad (7.184) $$

$$ \frac{\partial S}{\partial t} = k_r \left( \frac{k_1 \cdot C}{1 + k_2 \cdot C} - S \right) \quad \text{Langmuir adsorption isotherm} \qquad (7.185) $$

$$ \frac{\partial S}{\partial t} = k_r \left( k_1 \cdot C^{k_2} - S \right) \quad \text{Freundlich adsorption isotherm} \qquad (7.186) $$

Ion exchange is a third process not included in equation (7.174) and which might be of significance in some cases. A useful equation for the ion exchange equilibrium is:

$$ (7.187) $$

$$ K_{ij} = \left( S_j / C_j \right)^{v_i} (C_i / S_i)^{v_j} $$

$K_{ij}$ is the selectivity coefficient, C equivalent concentration in solution, S equivalent fraction in soil, $v_i$ and $v_j$ valences.

Table 7.28 gives the selectivity of relevant ions for various types of soil. Furthermore, CEC the Cation Exchange Capacity is used and considered a constant for a given type of soil. CEC is strongly dependent on the composition of soil and can in most cases easily be found in the literature, when the type of soil is given

$S_i$ is the fraction of CEC occupied by ion i.

It might also be necessary to account for decomposition of considered components during the transport. It can be done by use of the following equation for the source/sink:

$$F = \lambda(\Theta_m \cdot C_m + \Theta_{im} \cdot C_{im} + \varrho \; f \; S_m + \varrho(1\text{-}f) \; S_{im}) \quad \text{where}$$

(7.188)

F = part of adsorption surface in contact with mobile water

$\lambda$ = decomposition rate constant.

## TABLE 7.28

**Ion in Soil/Ion in Soil-Water.**

| Exchange | Value | Reference | Soil properties |
|---|---|---|---|
| Na + /NM$_4$+ | 4.5 | Dutt et al., (1972) | Not indicated |
| Na + /K + | 4.5-6.3 | Deist and Talibudeen (1967) | Not indicated |
| Na + /Ca$^2$ + | 1.8-7.2 | Bower (1959) | Organic soil |
| K + /Ca$^2$ + | 0.07-0.35 | Andre' (1970) | Alluvial soil |
| Ca$^{2+}$/Al$^{3+}$ | 215-435 | Coulten and Talibudeen | Acidic soil |

The vertical water flow is computed by use of the following differential equation, valid for unsaturated soil:

$$C \cdot \frac{\partial \psi}{\partial t} = \frac{\partial}{\partial z}\left(K \frac{\partial \psi}{\partial z}\right) - \frac{\partial K}{\partial z} - S$$

(7.189)

where

$\psi$ is the tension of soil water(m), t is time(h), z vertical coordinate(m), $C(\psi)$ specific water capacity(-1), $K(\psi)$ hydraulic conductivity (m/h) and S(z) is withdraw of water (h-1) due to f.inst. transpiration of plants. As seen from the equation two relations are needed to make use of this equation: the relation between the water content and the tension of soil water, the so-called retention curve, see fig. 7.44 and the relation between the hydraulic conductivity and the water content, see fig. 7.45.

The temperature profile is found from the following equations based upon the principle of energy conservation:

$$(7.190)$$

$$\frac{\partial (H \cdot T)}{\partial t} = \frac{\partial}{\partial Z}\left(C_T \cdot \frac{\partial T}{\partial Z}\right)$$

H is the volumetric heat capacity, t is time, T the temperature, $Z$ the vertical coordinate and $C_T$ the thermal conductivity.

The models described above can all easily be solved numericallly on computers. If we

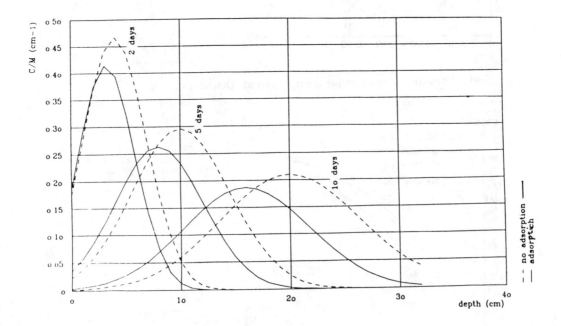

Fig.7.43: Concentration profiles at t = 2 days, t = 5 days and t = 10 days without adsorption.

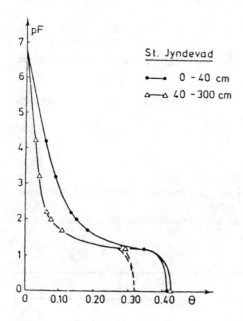

Fig.7.44: Retension curve (example from Jyndevad, Denmark).

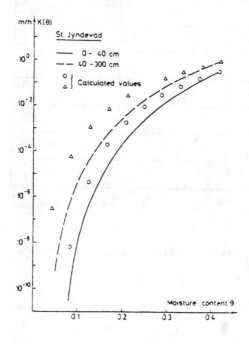

Fig.7.45: Conductivity as function of water content, (example from Jyndevad, Denmark).

consider the last equation we divide the soil column into N equal compartments with the thickness of COM. The heat flow from compartment N-1 to compartment N is represented by FLO(N), temperatures are symbolized by TEM and conductivity by CON. This means that we get the following equation:

$$(7.191)$$

$$FLO(N) = (TEM(N-1) - TEM(N)) \cdot CON/COM$$

The net flow to compartment N is now:

$$(7.192)$$

$$NFLO = FLO(N) - FLO(N + 1)$$

Correspondingly will the volumetric heat content, VHT $= H \cdot T$, of the $N^{th}$ compartment at time t later be given by:

$$(7.193)$$

$$VHT(N)_{t+\Delta t} \quad VHT(N)_t + NFLO(N) \cdot \Delta t$$

### 7.9.3. The Influx of Plants on the Water Balance.

The maximum interception capacity varies in accordance with the following expression

$$(7.194)$$

$$IC_{max} = 0.05 \cdot LAI \text{ where}$$

LAI is the leaf area index. This amount of water evaporates in accordance with the transpiration velocity given below as equation (7.195). The interception water loss is insignificant for agriculture systems but plays a certain role in forestry modelling.

The transpiration from plants, $E_{pt}$, is given by the Penman-Monteiths equation (Monteith, 1975):

$$E_{pt} = \frac{\delta R_{nc} + \varrho C_p \cdot \partial e / ra}{\lambda (\delta + j)} \qquad (7.195)$$

where

| | |
|---|---|
| $E_{pt}$ | = potential transpiration (mmh$^{-1}$) |
| $R_{nc}$ | = net radiation on plants (w m$^{-2}$) |
| $\varrho$ | = specific gravity of air (kg m$^{-3}$) |
| $C_p$ | = heat capacity of air (J kg$^{-1o}$C$^{-1}$) |
| $\partial e$ | = saturation deficit |
| $ra$ | = aerodynamical resistance (s m$^{-1}$) |

$\lambda$ = heat of evaporation (J kg$^{-1}$)
$j$ = psykometer constant 0.667 mb $^{\circ}$C$^{-1}$
$\delta$ = pressure of water vapour versus temperature (mb $^{\circ}$C$^{-1}$)

$R_{nc}$ is found from

$$R_{nc} = R_n (1 - \exp(-0.4 \cdot LAI)) \text{ where}$$

(7.196)

$R_n$ = net radiation
$R_{ns}$ = $R_n - R_{nc}$ is net radiation on soil.

$R_{ns}$ = $R_n - R_{nc}$ is used for the potential soil evaporation, computed by:

$$E_{ps} = \frac{\delta}{\lambda(\delta + j)}(R_{ns} - Q)$$

(7.197)

where

Q is the soil heat flux.

Equation (7.195) finds the potential transpiration, but the actual transpiration is dependent on the amount of water assessible by the plants. Fig. 7.46 gives the relation between the actual transpiration, $E_{AT}$, and potentiel transpiration, $E_{PT}$, as function of $(\theta - \theta_{WP})/(\theta_{FC} - \theta_{WP})$ where $\theta$ is the water content at depth z, $\theta_{FC}$ is the upper water capacity (field capacity) and $\theta_{WP}$ is the lower water capacity at withering.

$E_{AT}$ is found at various depths in the rootzone. Also the actual soil transpiration is different from the potential soil evapotranspiration as shown in fig. 7.47. in which ratio $E_{AS}/E_{PS}$ is plotted versus the actual water content at the soil surface.

The total evapotranspiration reduces the water content in equation (7.174) correspondingly and is incorporated in the sink, S(z). Equation (7.189) can be solved on a computer by use of the finite difference method. The parameters are dependent on $\psi$. However, to facilitate the computations, S(z) is found on the basis on the values at t-1. A discretization of the two independent variables Z and t is illustrated in fig. 7.48, in which the symbol $\psi_j^n$ with indices j and n is used to cover the value of for z = z$_j$ at time t = tn. Fig. 7.49 shows the solution procedure. The following formulation is used for the water flow equation:

$$C \cdot \frac{\partial \psi}{\partial t} = -\frac{\partial q}{\partial Z} - S$$

(7.198)

where

$$q = -K\frac{\partial h}{\partial Z} = -K\frac{\partial \psi}{\partial Z} + K$$

(7.199)

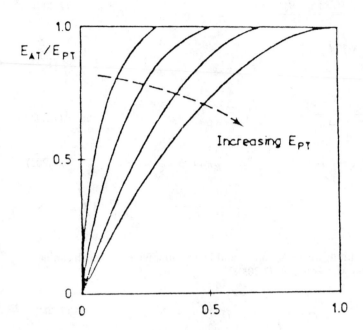

Fig.7.46: $E_{AT}/E_{PT}$ versus $\theta\text{-}\theta_{WP}/\theta_{FF}\text{-}\theta_{WP}$.

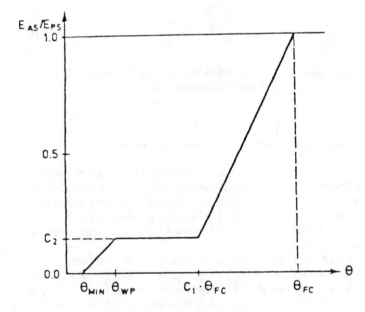

Fig. 7.47: $E_{AS}/E_{PS}$ versus $\theta$ C, and $C_2$ are constants.

The equations corresponding to the discretization are:

$$C_j^{n+1/2} \frac{\psi_j^{n+1} - \psi_j^n}{\Delta t^{n+1}} = \frac{q_{j+1/2}^{(n+1)} - q_{j-1/2}^{(n+1)}}{\Delta z_j} - S_j^n \tag{7.200}$$

$$q_{j+1-2}^{(n+1)} = -K_{j+1/2}^{n+1/2} \frac{\psi_{j+1}^{n+1} - \psi_j^{n+1}}{\Delta z_{j+}} + K_{j+1/2}^{n+1/2} \tag{7.201}$$

$$q_{j-1/2}^{(n+1)} = -K_{j-1/2}^{n+1/2} \frac{\psi_j^{n+1} - \psi_{j-1}^{n+1}}{\Delta z_{j-}} + K_{j-1/2}^{n+1/2} \tag{7.202}$$

$$C_j^{n+1/2} \frac{\psi_j^{n+1} - \psi_j^n}{\Delta t^{n+1}} = [K_{j+1/2}^{n+1/2} \frac{\psi_{j+1}^{n+1} - \psi_j^{n+1}}{\Delta z_{j+}} - K_{j+1/2}^{n+1/2} - K_{j-1/2}^{n+1/2} \tag{7.203}$$
$$- K_{j-1/2}^{n+1/2} \frac{\psi_j^{n+1} - \psi_{j-1}^{n+1}}{\Delta z_{j-}} + K_{j-1/2}^{n+1/2}] \Big/ \Delta z_j - S_j^n$$

or in a more simple formulation, where $A_j$, $B_j$, $C_j$ and $D_j$ are abbreviations which can be easily found by comparison with equation (7.203):

$$A_j \psi_{j-1}^{n+1} + B_j \psi_j^{n+1} + C_j \cdot \psi_{j+1}^{n+1} = D_j \tag{7.204}$$

By this method the equations can now be considered as N linear equations of N unknown. In matrix formulation, we have

$$[H] \bar{\psi} = \bar{D} \tag{7.205}$$

The boundary conditions are included in $B_1$, $C_1$ and $D_1$ respectively $A_n$, $B_n$ and $D_n$.
A solution of the system of equations is easily found on a computer.


### 7.9.4. How to consider the Climatic Influence on Plant Growth and Water Balance.

The weather forcing functions used in models of plant growth and water balance show a distinct daily and annual course. As many of the processes included in these models are non-linear and interacting, it might be of importance to account for the daily and annual course in simulation programs. The annual course is often considered by use of average day to day climatic data by the use of tables for the relevant forcing functions: temperature, radiance, precipitation etc.

It is usually more difficult to consider the daily course of the weather forcing functions. The incoming radiation flux during a time interval is computed from the radiation fluxes with a clear and overcast skies by assuming that the sky is overcast for the time fraction f and clear for the time fraction 1-f. The radiation flux in the 400-700 nm wave band in $J_m^{-2}S^{-1}$ depends on the height of the sun. Fig. 7.50 shows the radiation flux as function of the height of the sun for A) direct flux on a standard clear day B) diffuse sky radiation on a standard clear day and C) diffuse radiation on a standard overcast day. The value of f can be found from:

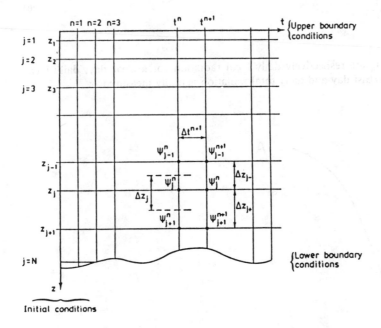

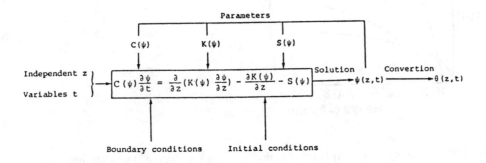

Fig.7.48: Discretization after the "finite difference" method.

Fig.7.49: Solution procedure.

$$f = \frac{\int I_c - \int I}{\int I_c - \int I_0} \qquad\qquad (7.206)$$

where $\int I_c$, $\int I_0$ and $\int I_0$ are respectively daily local radiation for a clear day, daily local radiation for an overcast day and daily total radiation acutally measured.

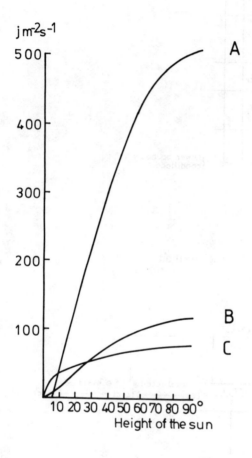

Fig.7.50: Incoming visible (photosynthetically active) radiation (400-700 nm) as a function of the solar height.

A is direct flux on a standard clear day, B is diffuse sky radiation on a standard clear day and C is diffuse radiation on a standard overcast day.

The current total short wave radiation at any moment of the day is estimated according to:

$$I = f \, Io + (1 - f)Ic \qquad (7.207)$$

The standard fluxes of short wave radiation for a clear and an overcast day can be calculated from the sine of the height of the sun. This sine is dependent on the sine and cosine of the declination and latitude and the cosine of the hour angle of the sun according to:

$$\sin \beta = \sin \Gamma \cdot \sin \alpha + \cos \Gamma \cdot \cos \alpha \cos 2 \, \pi(t_h + 12)/24 \text{ where} \qquad (7.208)$$

$\beta$ is the height of the sun, $\Gamma$ the latitude of the site, $\alpha$ the declination of the sun and $t_h$ the hour of the day.

Long wave radiation is estimated by use of Brunts formula (1932):

$$Bn = \sigma \cdot T^4(0.56-0.0923(0.75 \cdot PV)^{0.5}(1-0.9 \cdot f) \text{ where} \qquad (7.209)$$

T is the absolute air temperature, $\sigma$ is the Stefan-Boltzmann constant, PV is the vapour pressure in mm Hg and f is defined above.

As for short wave radiation Bn can be calculated separately for a clear or an overcast sky. f found in accordance with equation (7.206) is used. The main problem in using Brunts formula is the assumption that cloudiness during the night is the same as that during the day, which is unlikely. However, by measuring sky temperature, the calculation of loss of long wave radiation, can be improved.

Usually the maximum and minimum temperature are available for each day and these are used for computation of the daily courses. It can be assumed that the maximum occurs at 14hour and the minimum at sunrise. The daily course is now described by use of a sine curve for the period between sunrise and 14hour and another sine curve for the period 14hour to sunrise. The same procedure can be followed for the dew point.

The turbulence diffusion resistance in s m$^{-1}$ $r_n$, might be computed by use of:

$$r_n = \ln\left(\frac{z_r - d}{z_o}\right) \ln\left(\frac{z_r - d}{z_c - d}\right)/\left(k^2 \cdot v_r\right) \qquad (7.210)$$

$z_r$ is the height above surface, where the wind speed is $v_r$, k is the Von Karman constant (0.4) and $z_c$ is the height of the crop. The heights d and $z_o$ are the zero plane displacement and the roughness length of the crop, as it is presumed that the wind speed is zero at height $d + z_o$. It is often assumed that $d = 0.63 \, z_c$ and $z_o = 0.13 \, z_c$.

The incoming and outgoing fluxes of radiation after passing the first layer with leaf area index LAI can be calculated by use of the following equations:

$$I_2 = r_c \cdot I_1 \tag{7.211}$$

$$I_3 = I_1 \cdot \exp\left(\text{Ext}_{\text{dir}}\sqrt{(1-\sigma)} \cdot \text{LAI}_1\right) \tag{7.212}$$

$r_c$ is the reflection coefficient and $I_1$, $I_2$ and $I_3$ are explained in fig. 7.51. $\sigma$ is the scattering coefficient.

Because of the lower levels of leaves, there is also a reflected flux:

$$I_4 = r_c \cdot I_3$$

A flux balance gives us the flux adsorbed A, in the layer considered :

$$A_1 = I_1 + I_4 - I_3 - I_2 = (1 - r_c)S_1(1 - \exp\left(\text{Ext}_{\text{dir}}\sqrt{(1-\sigma)} \cdot \text{LAI}_1\right) \tag{7.213}$$

It is now possible to find the radiation at a given level and thereby the photosynthesis (see next section). Dependent on the scope of the model a computation at each level of an average radiation or of an average photosynthesis is carried out.

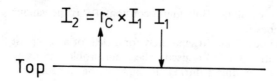

Top ————————————————

one leaf layer

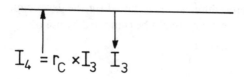

Fig. 7.51 The incoming and outgoing fluxes of radiation after passing the first leaf layer.

The wind speed within the canopy can be computed by use of the assumption that the wind profile is logarithmic:

$$u_c = u_r \ln \left( \frac{z_r - d}{z_0} \right) / \left( \ln \left( \frac{z_r - d}{z_0} \right) \right) \tag{7.214}$$

where

$u_c$ is the wind speed at the top of the canopy.

To compute the photosynthesis an extinction coefficient of the plant cover is needed. The incoming radiation, $I_i$, is found from:

$$I_i = I \cdot e^{-Ext \cdot LAI} \tag{7.215}$$

Ext can be found from the following expression: (for direct light)

$$Ext_{dir} = 0.5(1 - \sigma)^{0.5} / \sin \beta \tag{7.216}$$

where

$\sigma$ is the scattering coefficient and $\beta$ the solar height.

A good accordance between the extinction coefficient measured and computed by use of equation (7.216) has been found (de Wit et al., 1978).

For diffuse light the extinction coefficient is found by:

$$Ext_{dif} = 0.8(1 - \sigma)^{0.5} \tag{7.217}$$

Averaged over the wave length bands the scattering coefficient is 0.2 for green leaves and visible radiation. In the near infrared area the green leaves give a scattering coefficient of 0.8.

When no measurements are available the following equation can be used to estimate the fraction of the direct irradiation out of the total irradiation:

$$\frac{I_{dir}}{I_{tot}} = \exp \left( \frac{-0.15}{\sin \beta} \right) \tag{7.218}$$

where

$I_{dir}$ is the direct irradiation and $I_{tot}$ is the total irradiation.

### 7.9.5. Models of Plant growth and Crop Production

Plant growth is dependent on a number of factors:

1) Radiance
2) Respiration
3) Conversion of $CO_2$ to tissue
4) Water availability
5) Long-term water stress
6) Temperature
7) Nutrient concentrations
8) Concentrations of various substances.

A plant growth model will therefore be a relatively complex one, if it has to consider all these factors simultaneously. Furthermore, many of the factors are variables and it requires a submodel to compute them including a possible influence of the plant growth on the factors. Fig.7.52 illustrates the interactions of the submodels that are needed to set up a total model of plant growth. The relation between radiation and photosynthesis can be obtained from:

$$F_g = \frac{F_{mm} \cdot \Sigma \cdot A \cdot}{(F_{mm} \cdot \Sigma A r_x)/\epsilon + \Sigma \cdot A + F_{mm}}$$ (7.219)

where

$F_g$ is the gross assimilation (kg ha$^{-1}$ h$^{-1}$), $F_{mm}$ the maximum assimilation rate, $\Sigma$ is an efficiency coefficient, $r_x$ is the carboxylation resistance, $\epsilon$ is the carbon dioxide concentration and A the adsorbed radiant flux in the 400-700 nm range in Jm$^{-2}$(leaf) s$^{-1}$ = Wm$^{-2}$.

Other possible formulations are:

$$F_g = (F_{mm} + R_d)(1 - \exp(-A \cdot \epsilon/(F_{mm} + R_d))) \text{ where}$$ (7.220)

$R_d$ is the dark respiration in kg ha$^-$ h$^{-1}$

The dark respiration can be subtracted from the gross assimilation in equation (7.217).
    In $C_3$ plants (see Table 7.29) photorespiration occurs as well as dark respiration, so that the net assimilation is lower than in $C_4$ plants. Some $C_3$ and $C_4$ type species are listed Table 7.29. The photorespiration takes place during the assimilation process only and it does not use sugar from the reserve pool. Normally it is about 0.2-0.3 of the gross assimilation. According to Laing et al.(1974) the ratio of photorespiration, Rg, to gross assimilation, $F_g$ , can be found by:

$$\frac{R_f}{F_g} = \frac{t \cdot V_o \cdot 0 \cdot r_x}{K_o \cdot C} \qquad (7.221)$$

where t is the fraction of glycolate-carbon released (0.25), $V_o$ the maximum rate fo oxygenation, $K_o$ the Michaelis-Menten constant for the oxygen concentration 0. The gross rate $F_g$ itself is also reduced by the competing effect of oxygen, as shown in the following extended Michaelis-Menten equation:

$$F_g = \frac{\Sigma \cdot A \cdot C}{\Sigma A \cdot r_c (1 + 0/K_o) + C} \qquad (7.222)$$

where all symbols are defined above.

Growth implies the conversion of primary photosynthates into plant material. The efficiency of the conversion depends on the chemical composition of the dry matter formed. Some conversion factors are given in Table 7.30. As an average 0.7 may be used as conversion efficiency factor.

Growth is reduced by water shortage in accordance with the following equation:

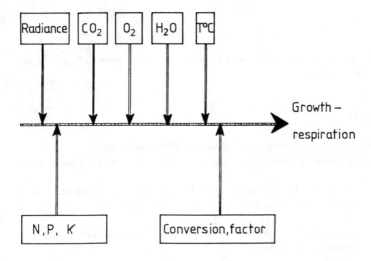

Fig,7.52: Plant growth is described as function of radiance, $CO_2, O_2$, water, temperature, nitrogen, phosphorus, pottassium and other elements and a conversion
factor determined by the composition of plant tissue.

$$G_{r,w} = G \cdot E_{at}/E_{pt} \text{ where} \tag{7.223}$$

$G_{r,w}$ is the reduced growth, while G is found from equation (7.222) after multiplication with the selected conversion efficiency factor and the ratio $E_{at}/E_{pt}$ is found in accordance with section 7.9.3.

However, prolonged water stress influences some of the basic plant properties. This effect is found by introduction of the concept relative transpiration deficit, TD, defined as:

$$TD = \frac{E_{pt} - E_{at}}{E_{pt}} \tag{7.224}$$

and the concept cumulative relative transpiration deficit, CTD, defined as:

$$CTD = \int_0^t \frac{TD}{t_c} dt \tag{7.225}$$

where

$t_c$ is a time coefficient often chosen to 10 days. The influence on the growth rate is found by use of the following equation:

$$G_{rw} = G (1 - CTD) \tag{7.226}$$

It is most often assumed that a mild water stress has no effect on growth.

Therefore TD is only included in (7.225), when it exceeds the arbitrary chosen value of 0.4.

When moisture becomes available again, the effect gradually disappears. It is described by an exponential extinction of CTD at a rate of 0.1 day-1, when the value of TD is below 0.4.

The influence of temperature on growth rate and respiration can be taken into account by use of the equations presented.

**TABLE 7.29**

**Typical C$_3$- and C$_4$-Plants.**

| C$_3$ | C$_4$ |
|---|---|
| Small grains (wheat, barley, oats,rye rice). Temperate grasses. Sugar-beet, potato, sunflower, cotton. All leguminous species with nitrogen fixation. Almost all trees (except Mangrove). | Tropical grasses as maize, sorghum millet, Cenchrus biflorus, sugar cane, Rhodes grass. Some halophytes as Spartina townsedii. Salsola kali, Atriplex rosea, Mangrove. |

**TABLE 7.30**

**Conversion Characteristics for Synthesis of 5 Categories of Plant Substances from Glucose.**

|  | production value gg$^{-1}$ | CO2 production factor gg$^{-1}$ | oxygen requirement factor gg$^{-1}$ |
|---|---|---|---|
| Carbohydrates | 0.860 | 0.7 | 0.051 |
| Lipids | 0.36 | 0.47 | 0.035 |
| Lignin | 0.46 | 0.27 | 0.090 |
| Organic acids | 1.43 | -0.25 | 0.13 |
| Organic N-compounds with NO$_3$ | 0.47 | 0.58 | 0.030 |
| Organic N-compounds with NH$_3$ | 0.70 | 0.15 | 0.74 |

Most often is simply used a $Q_{10}$ value. Usually this is set at 2.0.

The influence of nitrogen on the plant growth is simulated by use of the following expression:

$$G_{r,N} = G \cdot R_N \text{ where} \tag{7.227}$$

$G_{r,N}$ is the growth rate considering N-influence
$G$ is the growth rate
$R_N$ is found from fig. 7.53

Fig. 7.53 shows the reduction in growth by reduced nitrogen concentration in the plant. $N_{max,p}$ represents the maximum concentration in plants, $N_p$, the actual concentration of nitrogen in plants and Nmin,p the minimum concentration of nitrogen in plants. $N_{max,p}$ and $N_{min,p}$ can be found from fig. 7.54. When the distribution of biomass among the parts of the plant is known, an average value can be found and used, otherwise it is necessary to simulate the nitrogen concentrations in the different parts of the plant. It is done in the same way as illustrated below for the total plant.

$N_p$, see above, is a state variable which is included in the model by use of the following equations:

$$\frac{dN_B}{dt} = N_u = G \cdot N_{max,p} \cdot \frac{N_{max,p} - N_p}{N_{min,p}} \cdot \frac{N_s - N_{min,s}}{N_{min,s}} \tag{7.228}$$

where

$N_U$ is N uptake (kg N/(day . ha)), $N_B$ is N in plants as kg N/ha, $N_{max,s}$ is the maximum N-concentration in soil, $N_S$ the actual N-concentration in soil, $N_{min,s}$ the minimum N-concentration in soil and $N_p$ (defined above) $=N_B/B$, where B is the biomass in kg/ha (see above). $N_{max,s}$ is included in this context to indicate the maximum value of $N_s$. $N_s$-values above $N_{max,s}$ should be avoided by use of the nitrogen submodel presented in the next paragraph.

Having formulated a model of plant growth, the next step would be to model the crop production. An approximate but simple method that often works well relates the grain production to the total amount of N in plants. The usual range of the ratio of N to the total N in plants is 0.74-0.82. An average value of 0.78 can be used to translate the total N in plants to the N in grain. The total grain production is found by application of the nitrogen or protein concentration in grains. This approach is, however, not valid in situations of drastic water shortage.

**7.9.6. Models of the Nitrogen Processes in Soil.**

Fig. 7.55 illustrates a model, which considers the most important nitrogen processes in soil. Many modifications of this model exists, but the differences are minor. The processes can be described quantitatively in many different ways, but the equations presented below are of a rather general nature and can easily be substituted by alternative descriptions, if it is found relevant for the modeller.

Process (1), mineralization can be described by use of a first order expression:

$$\frac{dN_s}{dt} = k_n \cdot DN \tag{7.229}$$

where

$N_S$ is soluble nitrogen (ammonium), $k_N$ is the mineralization rate constant and DN the concentration of detritus nitrogen.

The rate is dependent on the temperature and an Arrhenius expression is used (see chapter 3):

If the soil temperature, $T_s$, is not available the following relation between soil and air temperature can be used at least for depths below 15 cm:

$$T_s = 0.66 + 0.93 \, T_a \text{ where} \tag{7.230}$$

$T_a$ is the average air temperature at 2 m height of the previous 7 days and $T_s$ the soil temperature.

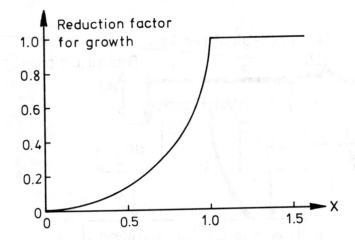

Fig.7.53: The reduction factor for dry matter accumulation as a function of a normalized nitrogen concentration. X computed as $(N_p - LNCL)/(N_{min,p} - LNCL)$. LNCL is shown fig. 7.54b.

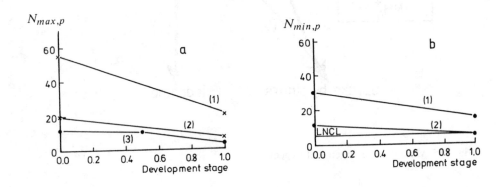

Fig.7.54: The maximum $N_{max,p}$ (a) and minimum $N_{min,p}$ (b) nitrogen concentration for various organs of natural grassland vegetation as a function of phenological age (development stage, here defined as being 1.00 at maturity). (1) are leaves, (2) is non leaf material and (3) are roots.

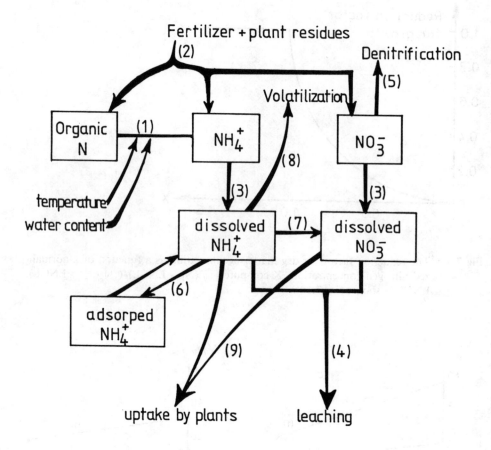

Fig.7.55: Models of nitrogen soil processes.

The rate is also dependent on the moisture content from the computation presented section 7.9.2.

The rate is regulated with a factor $F_W$ found from:

$$F_W = \frac{\theta}{\theta_{fc}}$$

(7.231)

where

$\theta$ is the actual water content and $\theta_{fc}$ the maximum water content. The moisture content at saturation and at field capacity for different soil types are shown in Table 7.31.

Process (2) accounts for the input of fertilizer and plant residue nitrogen. The former contribution is a forcing function, while the latter can be found from Table 7.32.

Process (3) corresponds to dissolution of the inorganic nitrogen. The dissolved nitrogen is transfer into the model for mass transfer in soil, presented in section 7.9.2. It is assumed that dissolution takes place if one of the following three criteria are satisfied:

$$I: \quad r = \frac{\theta - \theta_{WP}}{\theta_{FC} - \theta_{WP}} \geq 0.8$$

(7.232)

**II:Precipitation P of the present and previous 2 days:**

$$P = \sum_{i=j-2}^{j} P_i < 7mm(1 + 2(0.9 - r))$$

(7.233)

**III: Precipitation P of the present and the previous 6 days:**

$$P = \sum_{i=j-6}^{j} P_i > 12mm(1 + 2(0.9 - r))$$

(7.234)

**TABLE 7.31**

**Water content per 25 cm Soil Layer (Free space = Water-filled minus FC at porosity 44 vol.percentage.)**

| Soil | Water | 0-25 | 25-50 | 50-75 | 75-100 | Sum |
|------|-------|------|-------|-------|--------|-----|
| 1 | Unavailable | 15 | 15 | | | 30 |
| | Available | 40 | 20 | | | 60 |
| | Field cap. | 55 | 35 | | | 90 |
| | Free space | 55 | 75 | | | 130 |
| 2 | Unavailable | 20 | 15 | 15 | | 50 |
| | Available | 50 | 45 | 45 | | 140 |
| | Field cap. | 70 | 60 | 60 | | 190 |
| | Free space | 40 | 50 | 50 | | 140 |
| 3 | Unavailable | 15 | 15 | | | 30 |
| | Available | 50 | 30 | | | 80 |
| | Field cap. | 65 | 45 | | | 110 |
| | Free space | 45 | 65 | | | 110 |
| 4 | Unavailable | 20 | 15 | 15 | 15 | 65 |
| | Available | 50 | 35 | 25 | 25 | 135 |
| | Field cap. | 70 | 50 | 40 | 40 | 200 |
| | Free space | 40 | 60 | 70 | 70 | 240 |
| 5 | Unavailable | 25 | 25 | 25 | 25 | 100 |
| | Available | 50 | 40 | 30 | 30 | 150 |
| | Field cap. | 75 | 65 | 55 | 55 | 250 |
| | Free space | 35 | 45 | 55 | 55 | 190 |
| 6 | Unavailable | 30 | 30 | 30 | 30 | 120 |
| | Available | 50 | 40 | 40 | 40 | 170 |
| | Field cap. | 80 | 70 | 70 | 70 | 290 |
| | Free space | 30 | 40 | 40 | 40 | 150 |

Process (4) is leaching, which is described in accordance with the mass transfer model presented in section 7.6.2.

Process (5) is the denitrification, $N_d$, which consists of a denitrification pulse, $N_1$ and diffuse denitrification, $N_2$:

$$N_d = N_1 + N_2 \tag{7.235}$$

$$N_1 = \alpha \times N_F \tag{7.236}$$

$N_1$ is included only as pulse and $\alpha$ is dependent on the soil type.
$N_F$ is the amount of fertilizer applied.
$N_2$ occurs when the following three criteria all are satisfied:

$$I:\ T_s \geqslant \ 5°C \tag{7.237}$$

where $T_s$ is the soil temperature

II: $\dfrac{(\Theta - \Theta_{WP})}{(\Theta_{FC} - \Theta_{WP})} \geq 0.9$ (7.238)

III: $N_S > 0.6 \text{kgNha}^{-1}\text{cm}^{-1}$ (7.239)

$N_2$ is described by a first order reaction: $N_2 = -\, dN_s/dt = -\, K_{\dot{N}}N_S$ (7.240)

**TABLE 7.32**

**Residues in the Field after different Crops and Harvest Methods.**

| Method : | Farm harvest | | | | Experimental harvest | | | |
|---|---|---|---|---|---|---|---|---|
| Fertilizer: | No nitrogen | | N-fertilized | | No nitrogen | | N-fertilized | |
| Crop | t DM | kg N | t DM | kg N | t DM | kg N | t DM | kg N |
| Grass[1] | 1.5 | 20 | 1.5 | 25 | 1.5 | 20 | 1.5 | 25 |
| Grass[2] | 3 | 40 | 3 | 50 | 3 | 25 | 3 | 50 |
| W-wheat | 3 | 40 | 3 | 50 | 2 | 30 | 2 | 35 |
| S-barley | 3 | 40 | 3 | 50 | 2 | 30 | 2 | 35 |
| S-rape | 3 | 40 | 3 | 50 | 2 | 30 | 2 | 35 |
| F-beet | 3 | 40 | 3 | 50 | 1 | 20 | 1 | 20 |

1) First year in autumn established in a cereal cover crop 2) Second and following years in autumn.

In the model inorganic nitrogen is considered either to be ammonium or nitrate. Ammonium might be adsorbed or bound by ion exchange. This is process (6), which is described as presented in section 7.9.2.

The oxidation of ammonium to nitrate, nitrification, is process (7), which can be modelled by use of:

$$\dfrac{d\left(NH_4^+ - N\right)}{dt} = -k_{NH_4^+} \cdot \left(NH_4^+ - N\right)$$ (7.241)

where $\left(NH_4^+ - N\right)$ is the ammonium concentration in a given soil layer and $k_{NH_4^+}$ is a rate constant.

**TABLE 7.33**

**Residue in the Field after Grass (After Jensen, 1980)**

| Fertilizer applied, kg N/ha | 0 | 500 | 750 |
|---|---|---|---|
| Stubble, t DM/ha | 1.2 | 1.2 | 1.2 |
| Root, t DM/ha | 2.0 | 1.4 | 1.5 |
| N in stubble, % | 0.74 | 2.19 | 2.27 |
| N in root, % | 0.87 | 1.49 | 1.54 |
| N in stubble, kg/ha | 8.3 | 25.4 | 27.7 |
| N in root, kg/ha | 17.7 | 21.4 | 23.3 |
| N in stubble and root, kg/ha | 26.0 | 46.8 | 51.0 |

Process (8) corresponds to volatilization of ammonia. The concentration of ammonia in the upper layer is found by an equilibrium expression:

$$\log \frac{(NH_3 - N)}{\left(NH_4^+ - N \ - \ NH_3 - N\right)} = pH - pK \tag{7.242}$$

where $(NH_3 - N)$ is the concentration of ammonia and $\left(NH_4^+ - N\right)$ is the total concentration of ammonia and ammonium, pH is pH of soil and pK = - log K (pK = 9.3).
The volatilization rate, $V_{NH_3}$, is found from the following expression:

$$V_{NH_3} = D_{NH_3} \cdot (NH_3 - N) \tag{7.243}$$

where $D_{NH_3}$ is the diffusion coefficient for ammonia in soil (see also section 7.9.2).
The uptake of nitrogen, which means inorganic nitrogen (ammonium and nitrate) ions found from the model presented in section 7.9.5. It is included in the model as process (9). This process corresponds to a sink of nitrogen in the layers corresponding to the root zone, which is dependent on the plant species.

## PROBLEMS, CHAPTER 7

1. Two alternatives exist for improving the visual quality of Lake X
1) Increase the dilution (flushing) rate and 2) decrease the concentration of nutrients in the inflow by waste water treatment. The present detention time is 8 months and the average inflow of phosphorus, which is considered the most limiting nutrient is 120 $\mu$g $1^{-1}$. The lake can be considered a completely mixed reactor. Which alternative would you choose and why?

2. A toxic nonbiodegradable (conservative) substance is released into a river at point X continuously and at a constant rate. The flow rate of the river is 0.5 m $s^{-1}$ and the dispersion coefficient is 25 m² $s^{-1}$. The concentration of the toxic substance at point X is denoted z. What will be the concentration 2 km downstreams from X? What will be the concentration if dispersion is considered negligible?

3. The average flow velocity of a stream is 0.7 m $s^{-1}$ and the average depth is 1.5 m. Estimate the rate of oxygen tranfer from the atmosphere to the water at 12°C, 15°C and 20°C.

4. A stream has the following characteristics during a low flow period: flow rate 70 m³ $s^{-1}$ and 0.4 m $s^{-1}$, temperature 24°C, depth 2 m, dissolved oxygen 85% and $BOD_5$ 2 mg/1 at point X, How many kg of $BOD_5$ can be discharged into the stream at point X, if a minimum of 5 mg/1 is to be maintained in the stream? Rate constants can be considered averagely. Nitrification is negligible.

5. A 30 km (width 10 km depth 30 m) fjord receives freshwater with a salinity of 0.1% at a rate of 50 m³ $s^{-1}$: The salinity of the sea is 3%. What is the salinity 10 km, 15 km and 20 km from the open sea, if it is estimated that the dispersion coefficient is 30 m² $s^{-1}$?

# 8. APPLICATION OF ECOLOGICAL MODELS IN ENVIRON- MENTAL MANAGEMENT

This chapter deals with the special problems related to the application of environmental models in management. The first section overviews the characteristic features of these types of models, while the second section discusses the relationship between the environmental problem and the models including the generality of management models.

Three illustrative examples have been selected to demonstrate the wide spectrum of problems, which can be solved by use of models. The three examples are: a eutrophication problem, i.e. validation of a prognosis set up by use of the previously presented models, see section 7.4.4., an evaluation of human carrying capacity of a recreational area by use of threshold considerations, an energy model and finally selection of management strategies by use of a wetland forest model. The problems are all three characteristic for the types of environmental problems,which can be solved to-day by use of models, and the structure of the three models presented demonstrates simultaneously the wide range of modelling possibilities, which are offered to modellers to-day.

## 8.1. ENVIRONMENTAL MANAGEMENT MODELS

There is no principal difference between scientific and environmental management models. Environmental management models have, on the other side, some characteristic features, which will be presented in this chapter.

The management problem to be solved can often be formulated as follows: if certain forcing functions (management actions) are varied, what will be the influence on the ecosystems state? The model is used to answer this question or in other words to predict, what will change in the system, when forcing functions are varied with space and time.

The term **control functions** is used to indicate forcing functions, that can be controlled by man such as consumption of fossil fuel, regulation of water level in a river by a dam, discharge of pollutants or fishery policy.

A certain class of environmental management models is named control models. They differ from other such models by the content of the following two elements:

1) a quantitative description of control processes
2) a formalization of objectives and evaluation of achievements.

The difference between control models and other environmental management models can best be illustrated by an example. The eutrophication model presented in section 7.4.4 can be used as a management model. If we find the model response to various input of nutrients, we get the corresponding scenarios as model output. Among these scenarios the manager can select the one, that he prefers from an ecological-economic viewpoint. The model is used as an environmental management tool, but it is not a control model, which needs to formulate a goal f.inst. that we want to achieve, within a certain period of time, a certain transparency of lake water. Furthermore, we must introduce into the model a variable input of nutrients and find by use of the model the relation between the transparency and the nutrients input. The manager find directly from this relation which nutrients inputs, he must select to achieve his goal. It requires with other words some additional equations and concepts are required to be introduced into the model to construct a control model. In many cases this is quite feasible, but it adds to the complexity of the model. In cases where the control function can be varied continuously the advantages of control models are often sufficient to justify the additional complexity of the model, but if only a few possibilities are available, it hardly pays to construct a control model.

In the case of eutrophication there are only a few methods able to reduce the nutrient inputs, which can be varied easily in accordance with the known efficiency of these methods or combinations of methods. The management problem in this case is: which method to select among a few possibilities? A question, which can be answered simply by comparison of the corresponding scenarios.

In the case, where objectives are multiple, not all the formulated goals might be achieved simultaneously. Some of the goals might even be contradictory. Several methods are available in operation research to solve such multi goal problems: linear transformation, use of control indices, use of metrics in goal function space, Pareto methods etc. (see Haines, Hall and Freedman, 1975). Nevertheless, the final selection of a control function may ultimately be determined by subjective criteria, such as aesthetics which cannot be formulated. The final decision is in other words political.

A further step in complexity is the construction of ecological-economic models. As we gain more experience in the construction of ecological and economic models, more and more ecological-economic models will be developed. It is often feasible to find a relation between a control function and the economy, but it is in most cases quite difficult to assess a relationship between the economy and the ecosystem state. What is f.inst. the economic advantages of an increased transparency? Ecological-economic models are useful in some cases, but should at the least be used with much precaution and the relation between economy and environmental conditions should be critically evaluated, before the results are applied.

This presentation of control and ecological-economic models could give the impression, that environmental models always are more complex than scientific models. This is not the case. Environmental management models have often more clear formulation of the objectives of the model than scientific models, which might render it more easy to select the complexity of the model in the first hand. Knowledge as to the needed predictive value of an environmental management model might also enable the modeller to reduce its complexity , while the scientific use of the model implies that the modeller rigorously questions the possibilities of complexity reductions.

Data collection is the most costly part of model construction. For many water quality models it has been found that the data collection has amounted to 80-90% of the total modelling costs. As complex models require far more data than simple ones, the selection of the complexity of environmental management models should be closely related to the environmental problem to be solved.

It is, therefore, not surprising that the most complex environmental management models have been developed for large ecosystems, where the economic involvment is great.

The predictive capability of environmental models can always be improved in a specific case by expansion of the data collection program and by increased complexity of the model, provided, of course, that the modellers are sufficiently skilled to know in which direction further expansion of the entire program has to develop. However, the relation between the economy of the project and the accuracy of the model is somewhat as shown in fig. 8.1. The reduction in discrepancy between model and reality is lower for the next dollar invested in the project. But it is also clear from the shape of the curve in fig. 8.1 that the error will never be completely eliminated : all model predictions have a standard deviation. This is not surprising to scientists, but it is often not realized by the decision makers, to whom the modeller has to present his results.

A validation, if carried out properly, can be used to determine the standard deviation. Model results used in environmental decisions should always be accompanied by an indication of the standard deviation on the prognosis and it is important to clarify the meaning of this standard deviation to the decision maker. The modeller should even give recommendations as to how to use results and standard deviations in their proper context. Decision makers have often used standard deviations wrongly, when they were presented to them: as numbers, which indicate how much the costs could be reduced without any effect on environmental quality.

Engineers use safety factors to assure that a building or a bridge can last for a certain period of time with very low probability of breakdown to occur, even under extreme conditions. Decision makers would in such a case never interfere with the use of safety factors to account for the standard deviations of the engineer's model computations.

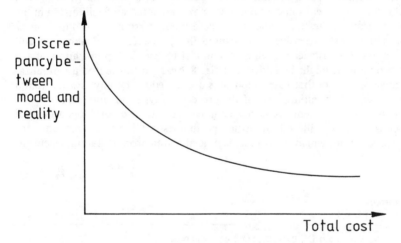

Fig.8.1: The more a skilled modeller invests in a model and in data collection the closer he will come to realistic predictions, but he will always gain less for the next dollar invested and he will never be able to give completely accurate predictions.

Nobody would propose to use a smaller or no safety factor at all to save some concrete and reduce the cost. The reason is obvious: nobody want to take the responsibility for even the smallest probability of a building or bridge collapsing.

When decision makers are going to take decisions on environmental issues, the situation is strangely enough different. Here he wants to use the standard deviation to save money not to assure under all circumstances a high environmental quality. It is therefore the modellers duty to explain to the decision makers the consequences of the various decision possibilities. A standard deviation of a prognosis for an environmental management model can, however, not always be translated into a probability, because we do not know the probability distribution. It might hardly be one of the common distribution functions, but it is possible to use the standard deviation qualitatively or semiquantatively and translate by use of words the meaning of the results. The civil engineers are actually more or less in the same situation and they have succeeded in convincing their decision makers. Why should the environmental modeller not be able to do the same?

It is often advantageous to attack an environmental problem in the first hand by use of simple models. They require very few data and can give the modeller and the decision maker some priliminary results. If the modelling project is stopped at this stage for one or another reason, a simple model is still better than no model, because it will at least give a survey of the problem.

The simple model is, furthermore, a good starting point for the construction of more complex models. In many cases the construction of a model is carried out as an iterative process, as mentioned in section 2.2. In fig. 8.2 is shown how a stepwise development of a complex model might take place. The first model step is here as in fig. 2.2 a conceptual model. It is used to get a survey of the processes and state variables in the system. The next step is a simple model, which is even calibrated and validated. It is used to set up a data collection program for a more comprehensive program, which is used for a model, which is close to the final selected version. Often, however, as is also shown in fig. 8.2 the third model will in use reveal some weaknesses, which are attempted eliminated in the fourth version. It seems at the first glance to be a very cumbersome procedure, but as data collection is the most expensive part of modelling, as mentioned above, it will require less resources to construct a preliminary model to use for optimization of the data collection program. It might be added that fig. 8.2 can be considered a formalization of the iterative procedure that many modellers are forced to use anyhow and that a planning of these steps at an initial phase of the project always is advantageous.

A simple mass balance scheme is also recommendable to use for biogeochemical models. The mass balance will indicate what posibilities we have for reduction or increase of a concentration, which, of course, is crucial for environmental management.

**TABLE 8.1**

**Examples of Sources.**

| Source | Examples |
|---|---|
| Point sources | Waste water (N, P, BOD), $SO_2$ from fossil fuel, discharge of toxic substances from industries. |
| Nonpoint man made sources | Agricultural use of fertilizers, deposition of lead from vehicles, contaminants in rain water. |
| Non point natural sources | Run-off from natural forests, deposition on land of salt originated from the sea. |

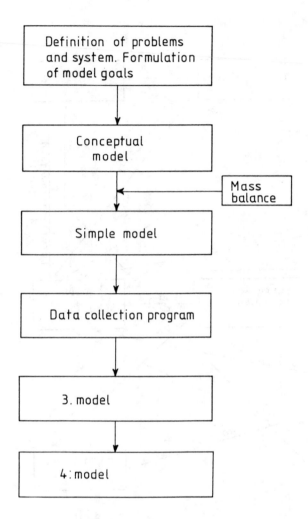

Fig.8.2:Scheme for development of a complex management model.

Point sources are usually more easy to control than man made non-point sources, which again usually can be controlled better than natural sources. Examples of these three types of sources are given in Table 8.1. We can also distinguish between local, regional and global sources. As the mass balance indicates the quantities it is possible to conclude as to which sources to concentrate on first. If f.inst. a non-point regional source of pollutants is dominating it might be useless to eliminate small local point sources first, unless, of course, this might have some political influence on the regional decisions.

Fig. 8.3 shows a mass balance model. From the model can be seen that diffuse contami-

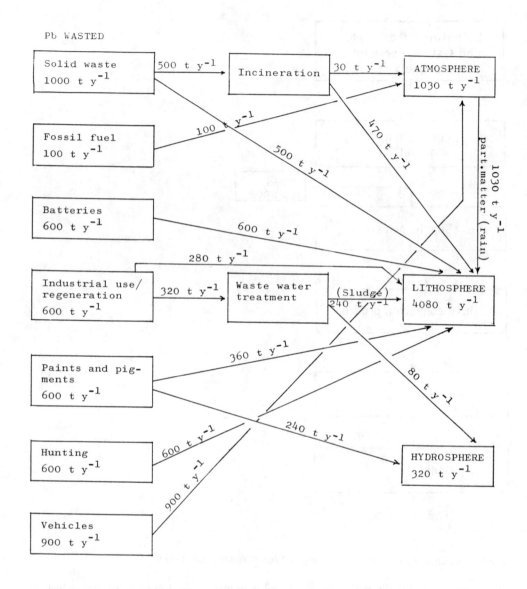

Fig.8.3: The dispersion of lead losses to the environment. (Denmark 1969).

nation of lead is caused by lead in gasoline. Because of the diffuse character the lead will contaminate uncontrollably, whereas most of the lead in waste water can be removed by a proper treatment method or even recovered. As lead is toxic it is harmful in food items and water. It is therefore understandable that lead has been banned in gasoline in many countries or at least the allowed concentration reduced considerably.

It has been touched upon already that the modeller and the dicision maker should understand each other. It is here recommendable to let the decision maker follow the model construction from the very first phase to get acquainted with its strength and the shortcomings of the model. It is also important that the modeller and the decision maker together formulate the objectives of the model and interpret the model results. Holling (1978) has demonstrated how such a teamwork develops and run phase by phase. The recommendations given in this reference will not be referred here, but before the start of a modelling management team, it would be a good idea to get acquainted with the procedure experienced by Holling. The conclusions are, however, clear: The modeller and the decision maker should work together in all phases, while a model procedure, where the modeller first build the model and the decision maker then use it with a small report on the model in hand, cannot be recommended.

Communication between the decision maker and the modeller can be facilitated in many ways, and often it is in the hand of the modeller to do so. If the model is built as a menu system as presented by Mejer (1983) it might be possible in few hours to teach the decision maker, how to use the model and that will, of course, increase his understanding of the model and its results. If an interactive approach as presented by Fedra (1982) is applied it is furthermore possible for the decision maker to visualize a wide range of possible decisions. The effect of this approach is increased by the use of various graphic methods to give the best possible decision as to what happens in the system by use of various management strategies. It is under all circumstances recommendable to invest time in a good graphic presentation of the results to the decision maker. Even he has been currently informed about the model project in all its phases, he will hardly understand the background and assumptions of all model components. Therefore it is important that the results including the main assumptions, shortcomings and standard deviations are carefully presented by an illustrative method.

## 8.2. ENVIRONMENTAL PROBLEMS AND MODELS

We are not yet that far advanced in environmental modelling, that model results can be used solely. The model should never be used as decision maker, but should rather be considered as a very useful tool in the decision process. This implies that the modelling results should be clearly and illustratively presented and be considered an important component in the discussion of which decision to select. Side effects, the interpretation of the prognosis, implications of the prognosis accuracy etc are elements which might be considered in such a discussion.

If a good environmental model is available, it is a very powerful tool in the decision-making process. A wide range of environmental problems has been modelled during the last 10-15 years. They have all been of important assistance to the decision makers and with the rapid growth in the use of environmental models, the situation will only improve in the nearest future. Obviously, we do not have achieved the same level of experience for all environmental problems. Table 7.2 gives a survey of some important environmental problems, where models are available as a decision tool. An attempt is

327

made in the table to indicate the present level of modelling for the various problems. A scale from 0 to 5 is used to indicate a) the quality of to-day's model, b) the depth of the experience or you might say the number of case studies. Such an evaluation is naturally subjective but it will give the reader a feeling of how far advanced we are to-day with some of the most crucial environmental problems. Note that we need far more experience in environmental modelling of large scale problems, such as the regional distribution of airborne pollutants and the water quality of the sea. Our knowledge and experience about the distribution and effect of toxic substances and the stability of ecosystems including the significance of species are, generally, limited. Some case studies of toxic substance distribution and effect are well examined, but our experience is still much too limited when the great number of toxic substances is taken into account.

The use of models in environmental management is growing. They are widely used in several European Countries, in North America and in Japan, but more and more countries take up the application of models by environmental agencies. Through the journal "Ecological Modelling" and the ISEM, International Society for Ecological Modelling, it is possible to follow the progresses in the field. This "infrastructure" of the field facilitates the communication and accelerate the exchange of experience and thereby the growth of the entire field of ecological modelling. Soon there will be a need for a model bank, where the user can obtain information about existing models, their use and characteristics. It is not possible to transfer a model from one case study to another as stated throughout the book and it is difficult to transfer a model from one computer to another, unless it is exactly the same type of computer, but it is often a great help to get the experience gained by somebody else in modelling a similar situation at some other place in the world.

This touch on the problem of generality of models. Few models have been used on several case studies to give a wide experience on this important matter. The eutrophication model presented in section 7.4.4 has been used on a wide range of case studies included lakes in the temperate and the tropical zone, shallow and deep lakes and even fjords. The experience gained by these case studies is illustrative, but does not necessarily represent general properties of models.

Tables 7.14 and 7.15 give some impressions of the generality of the examined eutrophication model. As seen some modifications must be carried out from case to case and different parameters must be applied. On the other side are the parameters not more different, at least the most sensitive parameters, than it is possible to explain it by site effects. The modifications, that were needed, do not throw doubt on the general applicability of the model, but they can also be explained, as expected and natural differences from ecosystem to ecosystem. If this experience is valid generally, it implies that a model can never be used generally, but that it is necessary to make modifications, when it is used on other ecosystems. On the other hand it is a clear advantage to have experience from other similar ecosystem and a major part of the model can probably be used also in the next case study. This calls for the construction of flexible models, which can easily be modified or which even might contain alternative descriptions, which can be selected from. Furthermore, it should be feasible to include processes, which can be switched in and out as needed in the individual case studies. In this way it is possible to make the model applicability more general, although it is necessary to take the ecosystem characteristics into account, when selecting the right model version. This will lead to general models for a specific environmental problem f.inst. eutrophication, oxygen deficit, DDT pollution in soil etc. Various modifications will be used on each ecosystem. It will be much more difficult to construct a general model for a specific type of ecosystem, which due to its flexible structure could be modified to deal with several environmental problems. Each environmental problem touches different ecological processes. This approach will therefore due to the enormous complexity of nature lead to too complex models and too large modifications from problem to problem. It seems possible to conclude that a certain generality is and will be further developed on the environmental problem level, but not on the ecosystem level.

## 8.3. MANAGEMENT EXAMPLES

8.3.1. Introduction. Three examples are presented here to illustrate the application of ecological models in environmental management. The three examples have been selected to give the reader, in a few pages, an impression of different approaches and thereby demonstrate the wide range of possibilities.

The first example shows a validation of a prognosis, which has been used as management tool. The model has already been presented in section 7.4.4 and cope with the eutrophication problem. The generality of the model has briefly been discussed in section 8.2. The validation of the prognosis shows that the model gives acceptable results, but reveals also that a possible description of algal succession would improve the model results. These problems are further discussed in section 9.2 and possible solutions are presented in section 9.3.

This example illustrates one of the few models, in which the prognosis was first published and then validated afterwards, which makes the result interesting in management context.

The second example focuses on the human carrying capacity of a recreational area: Lake Tahoe Basin. While the first model is concerned with mass flows this model focuses on energy flows. The model is not ideal to solve the problem of the assessment of the human carrying capacity, but it touches on some essential concepts of regional planning. However, the model does give workable results and represents the state of the art in integrated regional planning.

The third example deals with management of forest succession. It concerns a population model, which describes the growth in number and size of tree species. It focus with other word on management of natural ressources, which is another important area of environmental management modelling.

The three examples selected demonstrate thereby typical and different applications of ecological models for environmental managemnent: Control of pollution discharge, regional planning and management of natural resources. Three different types of models are applied: a biogeochemical model, an energy flow model and a model of population dynamics.

All three models are rather complex. It is therefore not possible to give all details of all three, but the major features are given together with a discussion of the management results.

### 8.3.1. Validation of a Prognosis, based upon a Management Model

Prognosis for different removal efficiencies for phosphorus, nitrogen or phosphorus and nitrogen simultaneously have been made. It has been stated that removal of nitrogen has little or no effect, while removal of phosphorus would give substantial reduction in the phytoplankton concentration.

The results of the two case studies are summarized in fig. 8.4 and Table 8.2.

Case A: The treated waste water has a concentration of 0.4 mg P l$^{-1}$, corresponding to an about 92 percent removal efficiency, which should be achieved by chemical precipitation.

Case B: The treated waste water has a concentration of 0.1 mg P l$^{-1}$, corresponding to an about 98 percent removal efficiency, which will require chemical precipitation in combination with, for example, ion exchange. This case might correspond to conveyance of the waste water, too.

As seen in Table 8.2, the water quality will improve significantly in accordance with the prognosis. Case B, 98% removal of phosphorus, must be prefered.

The third year case B will give a reduction in production from 1100 g C/m$^2$ y (Table 8.2) to 500 g C/m$^2$ y and the transparency is increased from a minimum value of 20 cm (table 8.2) to 60 cm. The ninth year would even reduce the production to 320 g C/m$^2$ y, which corresponds to a mesotrophic lake. An acceptable improvement for a shallow lake situated in an agricultural area. The prognosis predicts a pronounced effect of 98% phosphorus removal, which therefore could be recommended to the environmental authorities. Further improvements after nine years should not be expected (compare the three last years in fig. 8.4).

Conveyance of the waste water was also considered but has the following disadvantages:

1) It is slightly more expensive than the Case B solution, taking interests, depreciation and running costs into consideration;
2) The phosphorus is not removed but only transported to the downstreams Susaa River, where its effects not have been considered;
3) The sludge produced at the biological treatment plant will be less valuable as a soil conditioner, since the phosphorus concentration will be lower than when phosphorus removal is included; and
4) The fresh water is not retained in the lake, from where, after storage for some time, it could have been reclaimed, if needed. Fresh-water is not at present a problem in this area, but it is foreseen that it might be in 20 to 40 years.

In spite of these arguments the community has chosen to convey waste water to the Susaa River due to a preference for traditional methods. Construction of the pipeline occurred in 1980, and it began operation in April 1981, which has enabled a validation of the presented prognosis.

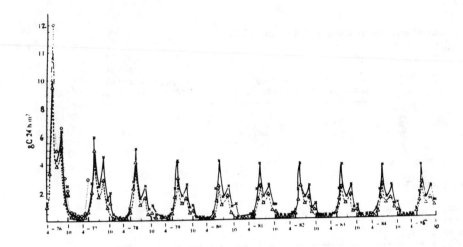

Fig.8.4: Productivity g C/m² 24 h in the case that (a) phosphorus is not removed from the waste water o, (b) waste water with 0.4 mg P/1 is discharged (x),(c) waste water 0.1 mg P/l is discharged. All data based upon the model presented in tables 7.3 - 7.5 (version II).

**TABLE 8.2**

**Predictions by means of Model in two Cases for Concentration of treated Waste Water: A: 0.4 mg P/l; B: 0.1 mg P/1**

|  | Third year | | Ninth year | |
|---|---|---|---|---|
|  | Case A | Case B | Case A | Case B |
| g C/m² year | 650 | 500*) | 500 | 320*) |
| Minimum transparency (cm) | 50 | 60 | 60 | 75 |

*) an error of 3% on this value could be expected if the validation results hold, see R for production in Table 7.13.

# TABLE 8.3

## Comparison of Prognosis Data and measured Data

| | Prognosis approximately (Case A, 92% P reduction) | Measurement approximately (88% reduction) |
|---|---|---|
| Minimum tranparency first year | 20 cm | 20 cm |
| Minimum transparency second year | 30 - | 25 - |
| Minimum transparency third year | 45 - | 50 - |

| | Prognosis approximately (Case A, 92% P reduction) | Measurement approximately (88% P reduction) |
|---|---|---|
| g C/24 h m$^2$ maximum first year | 9.5 ± 0.8 | 5.5 ± 0.5 |
| g C/24 h m$^2$ maximum second year | | |
| (spring) | 6.0 ± 0.5 | 11 ± 1.1 |
| (summer) | 4.5 ± 0.4 | 3.5 ± 0.4 |
| (autumn) | 2.0 ± 0.2 | 1.5 ± 0.2 |
| g C/24 h m$^2$ maximum third year | | |
| (spring) | 5.0 ± 0.4 | 6.2 ± 0.6 |

| | Prognosis approximately (Case A 92% P reduction) | Measurement approximately (90% P reduction) |
|---|---|---|
| (Chlorophyll (spring) maximum mg/m$^3$ first year | 750 ± 112 | 800 ± 80 |
| Chlorophyll (spring) maximum mg/m$^3$ second year | 520 ± 78 | 550 ± 55 |
| Chlorophyll (spring) maximum mg/m$^3$ third year | 320 ± 48 | 380 ± 38 |

**TABLE 8.4**

**Validation of the Prognosis.**

|       | Version A | Version B |
|-------|-----------|-----------|
| Y     | 0.79 *)   | 0.72 *)   |
| SDPC  | 0.18      | 0.08      |

*) Phytoplankton, soluble and total nutrient concentrations were considered.

Lake Glumsoe was ideal for these studies due to the limited depth and size as indicated section 7.4.4, but also because a reduced nutrient input to the lake could be foreseen. The limited retention time (about six months) makes it realistic to obtain a validation of a prognosis within a relatively short time interval (a few years). On April 1, 1981 the input of waste water directly to the lake was stopped. As the capacity of the sewage system is still too small, a minor input of mixed rain water and waste water is, however, from time to time discharged through an upstream tributary of the lake. The phosphorus loading is therefore not reduced by 98% but rather only by 88%, (determined by a phosphorus balance). It means that the prognosis case A should be used for comparison. During the third year after the reduction in loading had taken place a pronounced effect was observed. Table 8.3 compares some of the most important data of the prognosis. This table also includes data obtained recently during the first two months of the third year. In the Table errors are indicated as ± for g C/24h m² and chlorophyll maximum mg/m³. For the prognosis values the results are used from Table 7.13 - values for production 8% resp. phytoplankton concentration 15% - to determine errors. For the measured values an error of 10% are estimated. A comparison between the prognosis and measured values is illustrated in figs. 8.5 - 8.7. As seen from Table 8.3, the prognosis has given an almost correct production in the third year for maximum spring production and phytoplankton concentration, but the maximum concentration of phytoplankton occurs about 1. of April, while the prognosis predicts in the beginning of May (fig. 8.7). Previously, the lake was dominated by Scenedesmus, but now diatoms, which have a lower optimum temperature and therefore bloom earlier in the spring than Scenedesmus, are dominant in March and April. This seems to explain the discrepancy between prognosis and measurements on this point. Consequently the model might improve its predictions, if it was possible to account for shifts in species composition. Results published by Jørgensen (1981) and Jørgensen and Mejer (1981) indicate that this would be possible by the introduction of a maximum growth rate of phytoplankton, which is variable and currently determined as the value, that gives the highest exergy (for further explanations see these papers and section 9.3. This type of model is named flexible structure models. However, since diatoms take up silica, it would probably also be necessary to introduce a silica cycle into the model. The other production and chlorophyll values are well predicted except the spring production the second year (Table 8.3). The predictions on minimum transparency are acceptable as they are given with a difference of 5 cm or less (Table 8.3).

The general trends in the nutrient concentrations, fig. 8.5 and 8.6, give good accordance between predicted and measured values, although the fluctuations in phosphorus concentration were not well predicted.

The prognosis was validated by use of Y (see section 7.4.4.) and the average standard deviation of the predicted and measured maximum phytoplankton concentration, named SDPC. The results are shown in Table 8.4 for model version A, which was used up to 1979 and model version B, which includes the improvement introduced 1979-83 (detail on version A and B, see section 7.4.4).

The Y values are 72% and 79% compared with 31% for the validation under unchanged loading. The increased standard deviation, Y, between model values and measured values is due to the above mentioned shift in algal composition. The maximum phytoplankton concentration is predicted with an error of 8% (version B Table 8.4), which is acceptable.

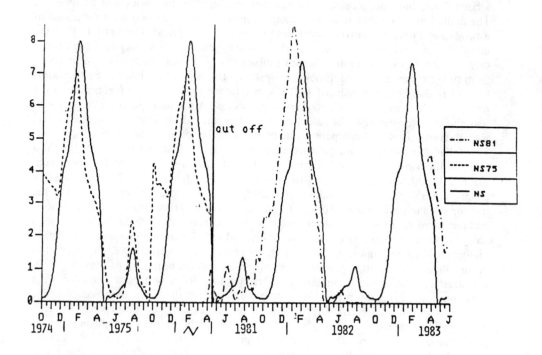

Fig.8.5: Prognosis validation.

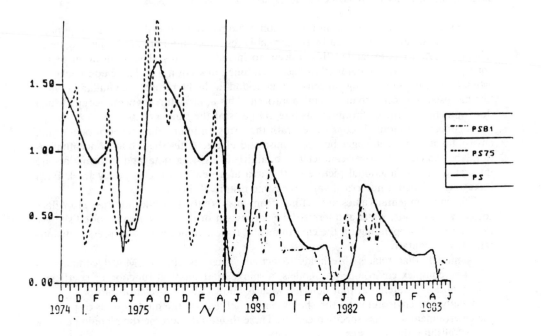

Fig.8.6: Prognosis validation.

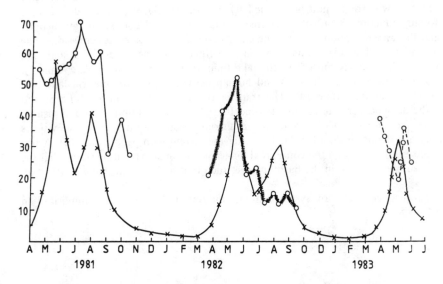

Fig.8.7: Phytoplankton concentration versus time. Prognosis validation.
o corresponds to measured values.
x corresponds to model output.

### 8.3.2. Model for evaluating Human Carrying Capacity of a Recreational Area

Carrying capacity is originated in population ecology, where it is defined as the population level, that can be sustained by the available resources in an area over the long term. Gilliland (1983) has expanded this concept to include humans. The application of the concept in population dynamics has been mentioned in section 6.2. Three approaches to estimate the human carrying capacity can be found in the literature, see Gilliland (1983). She has used the "eco-urban system" approach. This approach recognizes people, urban development and the environment as integral parts of the same system.

A conceptual model is constructed with the focus on interaction between people and environment and on linkages between cause and effect. After the model conceptualization, the systems historical character is quantified. Existing data and information are integrated into a functional picture of the area at a new and useful policy level. Each linkage is assigned a numerical value for some historical time period.

The environmental values are defined from answers to such questions as: How does society want the air, water and terrestrial environment of the study area to look like over the long term? Depending on the environmental issues, values may be expressed quantitatively or qualitatively.

A model, that relates thresholds to actual values, is chosen. Model components might be complex environmental models, a simple mathematical function or coefficient or a qualitative statement, that describe what is known about a relationship.

Finally, the environmental thresholds can be defined via the models so as to preserve the environmental values specified earlier. These thresholds become the planning criteria for establishing the area's carrying capacity.

All steps are presented below for a model developed by Gilliland (1983) for a case study of the Lake Tahoe Basin, California-Nevada.

The conceptual model is shown in fig. 8.8. Circles indicate forcing functions. They are divided into two groups: natural (to the left) and people related (to the right). People related forcing functions (control functions) include purchased goods, energy, people, dollars and the transportation system. The energy consumption produces emissions (line 4 and 11) and heat (line 13). Some emissions are filtered by the terrestrial system and stored here or show up as run off to the lake (line 5). People consume water (line 7), which produces waste water (line 12) and human activities consume natural land and convert it to urban land (line 14). The transfer of land affects run-off (line 6) and increases the demand for external inputs (line 10). The resident population and visitors use the lake resources (line 15) and the terrestrial resources (line 16). The visitors eventually leave the basin (line 17). Dollars enter the system with people (line 10). A few more processes are shown on the conceptual diagram, but in short can be stated that the four major subsystems within the basin interact such that a change in one will affect all others.

The historical character of systems inputs and internal flows are summarized in Tables 8.5 - 8.8.

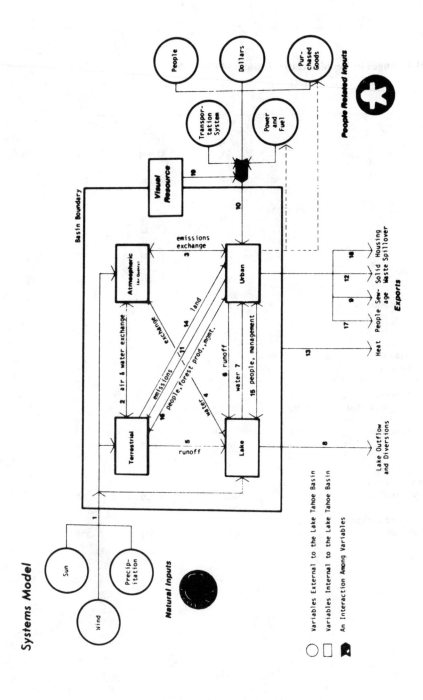

Fig.8.8: The Lake Tahoe Basin Systems Model.

337

# TABLE 8.5

## Summary of Changes in People related Inputs: 1970-1978.

| Input | 1970 | 1978 | Percent change |
|---|---|---|---|
| People (number) | | | |
| Residents | 33.600 | 73.200 | 118 |
| Tourists | 46.600 | 87.900 | 89 |
| Total | 80.000 | 161.100 | 101 |
| Money (million 1975 dollars) | | | |
| Federal | 21 | 39 | 88 |
| State | 7 | 14 | 100 |
| Private | 55 | 148 | 169 |
| Visitor | 287 | 484 | 69 |
| Total | 370 | 685 | 85 |
| Goods and Services (million 1975 $) | 120 | 280 | 133 |
| Energy | | | |
| Electric Power Transmission Capacity (kilovolts) | 420 | 720 | 71 |
| Peak Electricity Demand (Mwe) | 82.5 | 125.2 | 52 |
| Electric Power Consumption (thousand megawatt/hours) | 363.7 | 574.2 | 58 |
| Natural Gas Consumption (million cubic meters per yr) | 38.6 | 105.7 | 174 |
| External Transportation Traffic on Entry Roads (Average daily traffic) | 25.620 | 46.150 | 80 |
| Commercial Airline Passengers (number per year) | 123.426 | 589.103 | 377 |

# TABLE 8.6

## Increases in Tahoe's Urban Components:   1970-1978.

| Component | 1970 | 1978 | Percent increase |
|---|---|---|---|
| Recreation - Visitation | | | |
| (million visitor-days) | | | |
| Gaming | 4.6 | 10.4 | 126 |
| Outdoor | 5.3 | 5.8 | 9 |
| Other | 1.9 | 6.1 | 221 |
| **Total** | **11.8** | **22.8** | **93** |
| Recreation - Facilities | | | |
| Gaming - | | | |
| (Number of games and devices) | 4,164 | 8,220 | 97 |
| Outdoor - | | | |
| (Number of campground sites) | 1,774 | 2,080 | 17 |
| Housing (Number of dwelling units) | 20,263 | 36.043 | 78 |
| Employment | 18.420 | 38,060 | 107 |
| Transportation | | | |
| Daily Traffic: Stateline | 29,000 | 55,000 | 90 |
| Public Utilities | | | |
| Water Use (billion liters per yr) | 17.3 | 21.1 | 22 |
| Sewage Flows (Million liters per average day) | 23.6 | 32.5 | 37 |
| Solid Waste Generated (1,000 cubic meters per year) | 118.0 | 179.0 | 51 |
| Urban Land Use (hectares) | 6,435 | 11,453 | 78 |

**TABLE 8.7**

**Changes in Variables Affecting Lake Tahoe.**

| Variable | 1970 | 1978 | Percent change |
|---|---|---|---|
| Water Run-off (billion liters/yr) | 495.4 | 501.9 | 1 |
| Sediment Loading (1.000 kg/yr) | 53.370 | 70.330 | 31 |
| Nutrients in Run-off (kg/yr) | | | |
| Nitrogen | 132.500 | 156.600 | 18 |
| Phosphorus | 69.800 | 86.000 | 23 |
| Biomass (micrograms per liter fresh weight)a | 90 | 225 | 150 |
| Primary Productivity (grams C/m²/yr) | 50.2 | 80.5 | 50 |
| Clarity (Secchi disk depth in meters) | 29.982 | 5.95 | -6 to -13b |

a  These data represent 1969 and 1975

b  Annual average values for 1970 and 1978 show a 13% decline in clarity; the 4 year average for 1968-1971 compared with the 4 year average for 1975-1978 shows a 6% decline.

**TABLE 8.8**

**Decrease in Tahoe'a Wildlife Habitats:  1970-1978.**

| Habitat | Hectares 1970 | 1978 | Percent Decrease |
|---|---|---|---|
| Forest | 64.482 | 60.133 | 7 |
| Shrub | 5.728 | 5.499 | 4 |
| Stream Environ- ment Zone | | | |
| Riparian | 527 | 469 | 11 |
| Meadow | 1.815 | 1.514 | 17 |
| Marsh | 280 | 192 | 31 |

# TABLE 8.9

## Environmental Value Definitions and the Form of environmental Thresholds for the Lake Tahoe Basin.

| Air | Water | Terrestrial |
|---|---|---|
| **VALUES** | | |
| 1. CO concentrations<br><br>　8 hr:　　6 ppm<br>　1 hr:　25 ppm<br><br>2. Ozone concentrations<br><br>　1 hr:　0.10 ppm<br><br>3. Non-degradation for criteria pollutants and visibility | 1. Protect the exceptional and unique recreational and ecological characters of Lake Tahoe via:<br>　a. Existing federal and state numerical and narrative standards.<br>　b. The non-degradation requirements.<br>2. Protect native fish resources in Lake Tahoe and its tributary streams; provide fisheries for recreation; protect the public right to fish in public waters. | 1. Protect threatened, endangered, rare, and sensitive plant and animal species for their irreplaceability and scientific value.<br><br>2. Protect wetlands, including riparian, meadows, marshes, beaches, and lake shorelines, for their nutrient and sediment filtration values, high value wildlife habitat, and other biological, aesthetic, and recreational benefits. |
| **THRESHOLDS** | | |
| Establish limits on: | Establish limits on: | Establish limits on: |
| 1. CO emissions in grams/mile/hour.<br><br>2. HC and $NO_x$ emissions in grams/day | 1. Loading rates of sediment, nitrogen, phosphorus, and iron in tons per year<br><br>2. Disturbance in fish spawning areas; stream flow requirements in l/s; and stream sediment load in mg/l. | 1. Development in occupied habitat in hectarees, use in adjacent habitat.<br><br>2. Development in and adjacent to wetlands in hectares |

# TABLE 8.10

## Verbal Description of the Carrying Capacity Model.

| SYSTEM INPUTS AND STATE VARIABLES | INTERNAL MODELS | | DIFFERENTIAL EQUATIONS |
|---|---|---|---|
| People-Related Inputs | Combine Information on: | To determine: | $Q = K_1 Q_2 J_1 - K_2 Q_1 J_2 J_3$ |
| $J_1$ Mitigation Dollars | A. Population Activity ($J_3$), development dollars ($J_2$), and natural land ($Q_1$) | Additions to the amount of urban land ($Q_2$) | $Q_2 = K_2 Q_1 J_1 J_3 - K_1 Q_2 J_1$ |
| $J_2$ Urban Development Dollars | | | |
| $J_3$ People | B. Mitigation Expenditures ($J_1$), urban land development ($Q_2$) | Additions to the amount of natural land ($Q_1$) | |
| Natural Inputs | C. Land development ($Q_1$ and $Q_2$) and precipitation quantity and quality ($J_4$). | Runoff quantity and quality | $Q = Q_1 Q_2 J_4 [C][D] - K_3 Q_3$ |
| $J_4$ Precipitation | | | $Q_4 = Q_1 Q_2 J_3 [E]$ |
| $J_5$ Meteorological Conditions | D. Runoff quantity and quality, precipitation quantity and quality ($J_4$), and the hydrological/biological/chemical character of Lake Tahoe ($Q_3$). | Water Quality of Lake Tahoe ($Q_3$) | $Q_5 = Q_2 J_2 J_5 [F][G] -$<br>$\quad K_4 Q_5 J_5 [H]$ |
| State Variables | | | |
| $Q_1$ Amount and location of natural land | | | |
| $Q_2$ Amount, type, and location of urban land | E. Land development ($Q_1$ and $Q_2$) and population activity ($J_3$). | Wildlife populations and wetland characteristics ($Q_4$) | |
| $Q_3$ Lake Tahoe | F. Urban land development ($Q_2$) and population activity ($J_2$). | Emission rates (Transportation models). | |
| $Q_4$ Terrestrial environmental quality | G. Emission rates and meteorological conditions ($J_5$) | Air Quality ($Q_5$) (Air dispersion models). | |
| $Q_5$ Air Quality | H. Meteorological conditions ($J_5$) and air quality ($Q_5$). | Export of air pollutants (air mass stability measures). | |

**TABLE 8.11**

**People related embodied Energy consumed in Support of the Urban Acitivty in the Lake Basin in 1978,by Type of Recreational Activity. (109 kilocalories).**

| | Recreational Activity | | | | Total Basin |
|---|---|---|---|---|---|
| | Gaming | Outdoor | Other[a] | Total | |
| "Embodied" in | | | | | |
| Goods | | | | | |
| Goods[b] | | | | | |
| | | | | | |
| Construction | | | | | |
| Matls | 773 | 104 | 642 | 1.519 | 1.730 |
| Retail | 1.980 | 727 | 883 | 3.590 | 3.686 |
| | | | | | |
| Subtotal | 2.753 | 831 | 1.525 | 5.109 | 5.416 |
| | | | | | |
| Fuels[c] | | | | | |
| Buildings | 1.664 | 261 | 840 | 2.765 | 3.074 |
| Transportation | 524 | 184 | 221 | 929 | 959 |
| | | | | | |
| Subtotal | 2.188 | 445 | 1.061 | 3.694 | 4.033 |
| | | | | | |
| Food/Fiber[d] | 160 | 33 | 4 | 235 | 252 |
| | | | | | |
| Total | 5.101 | 1.309 | 2.629 | 9.039 | 9.701 |

a  Other recreation includes second homes, hotel/motel and general rest and relaxation.
b  Represents the fuel consumed to manufacture these goods.
c  Evaluated as the enthalpy of the fuel forms with electricity converted to its thermal equivalent at 2645.8 kcal/kwh.
d  Evaluated as 3000 kcal/person/day for food and 1.6 million kcal/resident/year of paper.

A preliminary list of environmental values was developed at a workshop in September 1980, involving people from various levels of government and interest groups. Some of the values are listed in table 8.9. Some of the values are quantified others are not. Consequently the models that relate the values to thresholds vary from quantitative to qualitative depending on the relationship. For air quality there are well documented models, that link CO-emissions to CO-concentrations in the air and that link hydrocarbon and nitrogen oxide emissions to ozone concentrations. In the case of water quality and the terrestrial environment models are used, which loosely quantify the relationship between sediment, nitrogen, phosphorus and iron loading to the quality of the lake water and between impact and species population and loss of habitat.

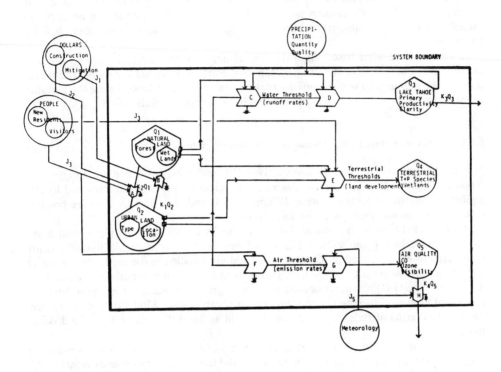

Fig.8.9: Threshold - Carrying Capacity Model of the Lake Tahoe Basin. (Interactions are described in Table 8.10)

A model for establishing consistency among the thresholds for the three environmental subsystems and for testing alternative carrying capacities is given in fig. 8.9. Environmental values are state variables: Q3, Q4 and Q5. Models that link these state variables are indicated by the arrow shaped symbol and the letters A through E. Mathematical formulations exist for these models to a certain extent. Thresholds are represented by the lines that link land use to the environmental values. The carrying capacity is expressed as the amount type and location of land development in Tahoe's forested areas and its wetlands (Q1). Table 8.10 defines all variables, describes the functions of the models and provide the differential equations required to simulate the model. The carrying capacity becomes the maximum population and associated urban activity, that a region can accommodate without exceeding either environmental thresholds limits or mitigation cost constraints.

The limitation of this approach are implicit in threshold limits and the carrying capacity, as they are only valid for a given kind of development, a given mix of population activity and existing technologies and policies. Odum (1976) has suggested to use the energy unit as a common unit to measure carrying capacity. In Table 8.11 is shown Gilliland's translation of inputs into energy units. She suggests to use (as an example) 7.0 x 10¹² kilocalories per year as input limit. Based upon this approach it is easily possible to test innovation alternatives. It is also interesting to compare the influence of the distribution of recreation time. F.inst. will one visitor day of gaming require 490,480 kilocalories per year of embodied energy support, while one visitor day of outdoors recreation require only 225,690 kilocalories per year. The use of energy units to express carrying capacity renders it possible to account for changes in technology, which can alter the energy efficiency.

It can be concluded based upon this experience that the energy is easy to use and interprete as a measure for the carrying capacity. Furthermore that the model is a useful tool for the planners to test various alternatives. In the case of Lake Tahoe the carrying capacity seems to be close to the present loading or maybe slightly below.

### 8.3.3. Simulation of Management Alternatives in Wetland Forests

The model, presented here, is named SWAMP and was developed by Phipps and Applegate (1983). It is intended to describe interrelationships between trees and hydrological conditions in wetland forests (Phipps, 1979) and it is able to simulate possible scenarios for management of forest succession.

Growth in the model is controlled in the model by crowding, overtopping and depth to water table in organic soils. Additions of new and removal of old individuals are considered to be controlled in part by chance and are handled in the model by the use of a random number generator. The model consists of a series of subroutines, see fig. 4.10. The subroutine GROW determines the amount of growth of each tree per year, and it is calculated as annual increments of basal area of breast height. Model output is tree size, described in terms of basal area per size class and as basal diameter (dbh) of individual trees.

This growth form results in constant increments of basal area if annual increments of height growth and environmental conditions of the tree, including those associated with crowing and shading, remain constant (Phipps, 1967). Three-dimensional growth of the main stem is assumed to follow a paraboloidal shape.

The basic growth rate, B, was calculated from tree-ring widths using the following equation:

$$B = \frac{1}{n}\left(r_t^2 - r_{t-n}^2\right) \tag{8.1}$$

where $r_t$ is radius at outside of ring formed in year t and $r_{t-n}$ is radius at inside of ring formed n years previously.

Average annual basal area (BA) is obtained by multiplying basic growth rate by $\pi$:

$$BA = \pi \cdot B \tag{8.2}$$

The basic growth rate of each tree is modified each year by effects of crowding, overtopping and depth to water table. The modified growth rate G is furthermore multiplied by a water table factor, H. Thus calculation of basal area at year, t, is found from:

344

$$BA_t = \pi\, G_t \cdot H_t$$

G is found by subroutine CROWD and H by subroutine WATER. Subroutine C R O W D accounts for the effect of crowding and overtopping on growth rate.

**TABLE 8.12**

**Basal Diameter Limits and stocking Constants by Size Class.**

| Size class | Basal diameter<br>dbh (cm) | Stocking constant<br>S (cm²) |
|---|---|---|
| Small tree | 3-10 | 780 |
| Sub-canopy | 10-20 | 4.700 |
| Canopy | >20 | 12.500 |

It calculates actual stocking, s, of each layer and compare these values with the stocking constants, S (table 8,12). A crowding factor, C, is determined for each tree layer such that :

$$\text{if}\quad s \geq S, \text{then}\quad C = \left(\frac{S}{s}\right)^2, \tag{8.4}$$

or

$$\text{if}\quad s < S, \text{then}\quad C = 2 - \left(\frac{s}{S}\right)^2. \tag{8.5}$$

The crowding factor, C, can theoretically have any value between zero and 2, being larger when the stand is more open and smaller when the stand is crowded. In keeping with the assumption that more shaded understory layers grow at reduced rates, the crowding factor is adjusted by multiplying the crowding factor of any layer with the factors of higher layers. The adjusted crowding factors, M, are intended to account for both crowding and shading:

$$\text{Small tree layer:} M_1 = C_1 \times C_2 \times C_3 \tag{8.6}$$

$$\text{Sub-canopy layer:} M_2 = C_2 \times C_3 \tag{8.7}$$

$$\text{Canopy layer:} M_3 = C_3 \tag{8.8}$$

An assumption is made that the greater the stocking of the stand, the more dense the foliage, and hence the more dense the shade. When the actual stocking of the stand exceeds the sum of the stocking constants of all 3 layers, an adjustment is made for shade tolerance. The growth rate, G, is calculated such that if $(s_1 + s_2 + s_3) < (S_1 + S_2$

+ S$_3$) then:

$$G = MB \tag{8.9}$$

Subroutine W A T E R deals with the relationship between growth rate and depth to water table:

$$H = 1 - 0.05 \, (D - W)^2 \quad \text{where} \tag{8.10}$$

H is growth multiplier for water table effects, D = actual depth to water table and W is the optimum depth for a given species, see Table 8.13.

Subroutine B I R T H adds new individuals to the plot by simulating natural forest regeneration. For details about this submodel see Phipps (1979) and Bedinger (1971). 3 cm dbh is used in the submodel as minimum size of trees entering the model. The submodel PLANT provides a means of simulating the addition of planted trees to the plot. It includes a lag time between the year of planting and the year when they are large enough to enter calculations.

Subroutines K I L L and C U T provide means of removing trees from the plot. Age alone is not assumed to kill a tree. However, when the growth rate is less than a specified minimum, the probability that the tree will not survive should greatly increase. Cambial activity at a basal height in a tree is defined as the annual increment of basal area relative to the average circumferential length of the cambium (Duff and Nolan, 1857). The cambial activity, CA of the nth ring, can be found from:

$$CA_w = \pi \left( r_n^2 - r_{n-1}^2 \right) / \pi (r_n + r_{n+1}) = r_n - r_{n-1} \tag{8.11}$$

$r_n$ - $r_{n-1}$ is the width of the $n^{th}$ ring, and ring width can be considered direct measure of CA.

If $CA_w$ is less than 0.1 mm per year, the program generates a random number between 1 and 1000 and if the number is between 1 and 369 then KILL removes the tree. Subroutine CUT simply provides a means by which the program operator may simulate lumbering.

The presented model has been used to manage wetland forest. The questions raised are such as: how are these and these tree species influenced by this and this strategy and water level. It was found by simulations that the species composition is pronouncedly influenced by the strategy selected. Some of the results are shown in figs. 8.10 and 8.11 to illustrate this statement.

**TABLE 8.13**

**Examples of Growth Adjustment for Depth to Water Table.**

|  |  | Growth factor (H) | |
| --- | --- | --- | --- |
|  | Optimum depth (W) | Depth (T) = 1 m | Depth (T) = 2.5 m |
| Species A | 0.5 m | 0.9875 | 0.8000 |
| Species B | 2.5 m | 0.8875 | 1.0000 |

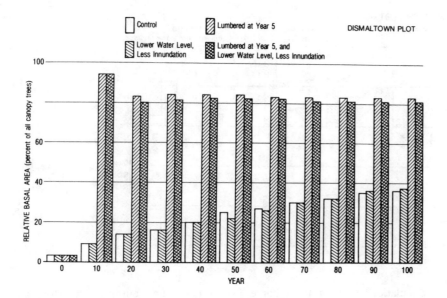

Fig.8.10: Graph of relative basal area of Acer rubrum at 10-year intervals as simulated for different management treatments. Simulations are of the Dismaltown plot.

347

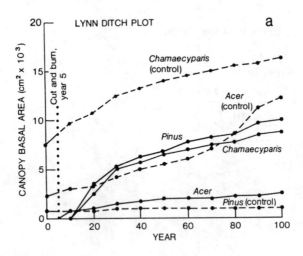

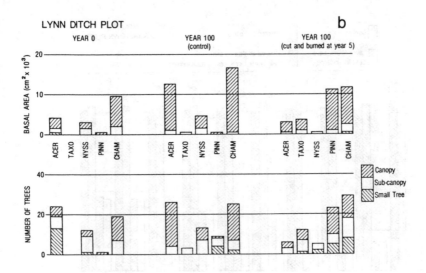

Fig.8.11: Simulation resulting from harvesting overstory and burning understory at Lynn Ditch plot. Control simulations, included comparison, are without harvesting or burning. (a) Graph of basal area data at 10-year intervals for the more important species in the canopy layer. (b) Graphs of basal area and density (number of individuals) of each species by vegetative level at years 0 and 100. Basal area values are to nearest 0.5 cm$^2$·1000. Data for understory Magnolia and Ilex not included.

# 9. ECOSYSTEMS CHARACTERISTICS AND MODELS

Models of ecosystems attempt to capture the characteristic features of ecosystems. However, ecosystems differ from most other systems by beeing adaptive and by having an enormous amount of feedback mechanisms,which are able to regulate the result of the dominating forcing functions. This chapter deals with the possibilities of modelling these ecosystems characteristics. The models presented in the previous chapters of the book account only to a limited extent for these characteristic features of ecosystems, but recent modelling approaches attempt to include them.

The two first sections of the chapter present the characteristic features of ecosystems, while the two last sections discuss two workable approaches to include them in the modelling effort. The two methods described are the application of goal functions and catastrophy theory. They are useful but do also have shortcomings. It is expected that new and better approaches will appear in the nearest future and that improvement of the present methods will increase their applicability.

## 9.1. CHARACTERISTIC FEATURES OF ECOSYSTEMS

Models are constructed on the basis of examinations of the considered ecosystem. The model will therefore reflect the processes and the structure of the ecosystem in the period of investigations. Through the use of the model, we attempt to learn about the reactions of the system including reactions to changes in external factors (forcing functions).

However,an ecosystem has on all levels several feedback mechanisms, which attempt to meet the changes in external factors with the smallest possible internal changes in function as a result. There are many biochemical and physiological feedback mechanisms in the cells, which maintain the functions of the cells and thereby the reactions of species.

The ability of the ecosystem to maintain its function (almost) independent of changes or fluctuations in external factors is expressed by use of different stability criteria. In the ecological literature can be found description of such stability criteria as persistence, resistance, resilience and ecological buffer capacity (see for this last concept also section 2.9). All these criteria attempt to capture the ability of the ecosystem to be soft, not rigid and to be able to adapt to new circumstances, when needed to maintain its function.

These properties of stability originate from many regulation- and feedback mechanisms. They are organized in a hierarchy. They have an increasing regulating effect (and are more and more long-term based, up through the hierarchy, see Table 9.1).

The two first mentioned mechanisms in Table 9.1 are considered in most ecological models. The two next mechanisms - adaptation - have been taken into account in some ecological models. F.inst. the adaptation to light of the photosynthesis can be considered by a current change of the Michaelis-Menten's constant in the following expression:

$$\text{PHOT} = \mu \frac{I}{K_r + I} \tag{9.1}$$

where

PHOT is the rate of photosynthesis, $\mu$ is the growth rate coefficient, I irradiance and $K_r$ a Michaelis-Menten's constant.

$K_r$ is changed in accordance with the radiance of the preceeding days. The change in DDT resistance of fruit fly populations, see fig. 9.1, can like other similar adaptions be considered by a description of the resistance in the first generation by use of a distribution function, which will indicate, which organisms will survive and what resistance, they will pass on to the next generation a.s.o.

The limited use of adaptation processes in models to a certain extent makes all ecological models "grey". Many biologist may use this as an excuse for not using models, but as the most important adaptation processes (but of course not all) can be considered in to-days models, this excuse is not any longer valid.

The last three mechanisms are possible reactions to changes in external factors, but they might also be observed as a result of a more general, long-term, development and evolution of the entire ecosystem.

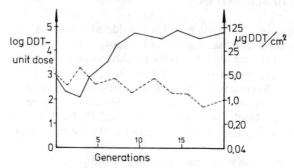

Fig.9.1: Change in DDT resistance of fruit fly populations exposed to directional selection for high (-) and low resistance (---). The left hand scale = log of dose tolerated. The right hand scale = the actual dose, which could kill half a population sample in standard exposure time.

**TABLE 9.1**

**Hierarchy of Regulation- and Feedback Mechanisms.**

1. Regulation of rates e.g. uptake of nutrient by algae.

2. Feedback regulation of rates e.g. by high nutrient concentration in algae the uptake rate will slow down.

3. Adaptation of process rates e.g. by changing the dependence of nutrient uptake rate on temperature.

4. Adaptation of species to new conditions e.g. adaptation of insects to DDT.

5. Shift in species composition. Species better fitted to new conditions will be more dominant.

6. A more pronounced shift in species compositions causing a shift in the structure of the ecosystem.

7. Change in the genetic pool available for selection.

The question, that is treated in this chapter, is: What are the possibilities of building these properties of ecosystems into models. We are mainly concerned with adaptation in its broadest sense and changes in species composition and foodweb structure.

## 9.2. ECOSYSTEM DYNAMICS

This paragraph attempts to give a system oriented presentation of the ecosystem properties presented in section 9.1. Straskraba (1980) operates with system dynamics of four orders:

1)Dynamics of the first order.
Ecosystem models are commonly expressed as systems of ordinary time variant equations with fixed parameters. In matrix notation:

$$(9.2)$$

$$S = f(S_i \, S_o \,, \bar{p}, \bar{z}, t)$$

where

$$
\begin{aligned}
S_i &= \epsilon \, S \\
x_{i,j} &= \epsilon \, S_i \\
x_{i,j} &= \Sigma \, r_{i,j,k} \\
r_{i,j,k} &= f(S, p, \bar{z}, t) \\
x_i &= \text{trophic levels} \\
x_{ij} &= \text{species} \\
r_{i,j,k} &= \text{subprocesses} \\
p &= \text{parameters} \\
z &= \text{driving variables}
\end{aligned}
$$

The dynamics represented be equation (9.2) is characterized by a system, which is non-adaptive and which has fixed parameters and fixed structure. The trajectory of the system is solely determined by the system state, given its state in time $t = 0, S_o$, vector of fixed parameters p and time t. This is covered by far the most simulating ecosystem models.

2)Dynamics of the second order.
In this case the systems parameter vector is dependent on a control vector, u, which

might represent either external variables, state variables or other control variables. In matrix notation:

$$S = f\left(S_1 S_0, \bar{p}(\bar{u}), \bar{u}, \bar{z}, t\right)$$ (9.3)

The structure of the system described by equation (9.3) remains constant, but the function changes in accordance with adaptation of parameters. It opens the possibilities for inclusion of self-adaptation on three different levels of organization:

a) The level of process,
b) the levels of individuals or population of species
c) the level of compartments.

In all instances the general numerical realization will be via adaptive parameters, although the nature of the adaptation and its description will, of course, be different. An example is the light adaptation of photosynthesis due to variation in the chlorophyll content or shifting of temperature optima and maximum due to changes in the environmental temperature pattern. (Groden, 1977 and Fedra, 1979). It is noticeable that such types of adaptation are accompanied with an increase in the internal organization of the system, (Jørgensen and Mejer, 1979), (see section 2.9.) measured by use of exergy or negative entropy, which are related by the following equation:

(9.4)

$$Ex = T(S^{eq} - S) \text{ where}$$

$S^{eq}$ is the entropy at thermodynamic equilibrium with the environment. It is, in other words, possible to describe this self-adaptation by use of exergy as goal function, as it has been attempted by Jørgensen and Mejer, 1981a.

3. Dynamics of the third order.
In this case the structure of the system becomes dependent on its state:

$$S = f\left(S(\bar{u}) S_0, \bar{p}, \bar{u}, \bar{z}, t\right)$$ (9.5)

This description implies that some relations between elements are ruled out, while others are entering according to the control information signals. Correspondingly, some elements vanish and others, not introduced before, become essential. Such a system will be denoted a self-organizing system or a variable structure system.

In ecosystems such changes in structure are observed as the presence of different species at the same trophic level in different situations. The numerical realization is possible by two methods:

a) by use of adaptation of parameters
b) by use of models with variable structure.

The latter type of models is represented by use of the catastrophy theory in ecosystem modelling, see section 9.4. In this case a mathematical analysis of the model reveals that for some values of the control variables two states are possible. The switch from one state to another is dependent on previous states, which implies that the model shows a hysteresis effect. This is only a pure mathematical consequence of the equation system.

The first mentioned method using adaptation of parameters has been used by Radtke and Straskraba (1980) and is also illustrated in Jørgensen and Mejer (1981a) and Jørgensen (1982), see section 9.3

4) Dynamics of fourth order.

In this case the control of the system will become dependent on the system state. We could call this dynamics of the systems evolution. In matrix notation it is:

$$S = f\left(S(\bar{u})S_0\bar{p}(\bar{u})\bar{u}(S)\bar{z}t\right)$$

(9.6)

The goal function or the strategy and control performance change in accordance with the state of the system. This is caused in the ecosystem by changes in the genetic pool.

## 9.3. ECOLOGICAL MODELS WITH GOAL FUNCTIONS

Ecosystem dynamics of fourth order, the change in the genetic pool of the system, are not needed to be included in short-term models. It is, however, crucial for the use of models to set up prognoses to be able to simulate changes in species composition, see section 8.3.2.

Straskraba (1979) uses a maximization of the biomass as a governing principle. The model computes currently the biomass and adjust one or more selected parameters to give the maximum biomass at every instance. The model has a routine, which computes the biomass for all possible combinations of parameters within a given realistic range. The combination of parameters which gives the maximum biomass is selected for the next time step a.s.o.

Jørgensen and Mejer (1977, 1979, 1981) have suggested the use of thermodynamic function exergy as a goal function. They show that the ecosystem reacts to changes in the external factors by changing the structure or composition in such a way that the ecosystem becomes better buffered to meet such changes. They introduce a new concept called ecological buffer capacity, $\beta$, defined in section 2.9.

There are, in accordance with this definition, an infinite number of buffer capacities for all combinations of all possible forcing functions and all possible state variables, but the exergy, Ex, seems to be related to the buffer capacities by:

$$Ex = \sum_{i=1}^{i=n} \beta_i k_i$$

(9.7)

The principle introduced can therefore be formulated as follows: Changes in external factors will create new conditions for the ecosystems that the system will meet by changing the structure or composition in such a way that the exergy under these new circumstances is maximized. Exergy measures the ability of the ecosystem to be buffered against changes in the system caused by changes in external factors. Thermodynamically exergy also measures the organization or order of the system. The idea behind such models is illustrated in fig. 9.2 (taken from Jørgensen and Mejer, 1981a).

The final results of these additional attempts to improve the model will often be a model which gives slightly better validation results than the model developed at first.

The possibilities of the ecosystem to change the structure is dependent on the genetic pool, as shown in the figure. The wide spectrum of species on earth to-day is fitted to almost all natural conditions, which implies that there will be species in all ecological niches contributing to the total exergy of the ecosphere. The thermodynamics of evolution states that an ecosystem evolves towards maximum biomass. It can easily be shown that this principle is included in the hypothesis on development (or evolution) towards maximum of exergy.

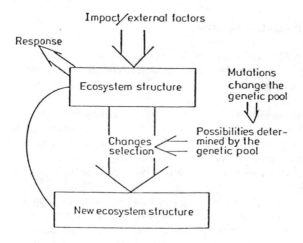

Fig.9.2: Principle of the presented ecological theory.

This implies that the response of ecosystems to new external conditions is linked to the evolution of ecosystems. It is a necessity for an ecosystem to meet perturbations by changing the structure in such a way that the ecological buffer capacity, that is, the ability to meet the perturbations, is increased, and that corresponds to a higher exergy level. This constant change in the external factors will, of course, also change the selection pressure on the species.

The selection is, however, not only serving the survival of the best fitted species but also the prevailing composition and structure of the ecosystem. The relatively quick changes in composition are caused by alterations in the external factors, which again are modifying the selection pressure. However, the pool of genetic material is simultaneously growing slowly and being modified. This opens avenues for a new combination of selection pressure and possibilities of meeting this pressure. Many mutations will not be better fitted to the steadily changing external factors, but there will always be a probability that some mutations, better fitted to a set of external factors valid at a given time, will occur.

As everything is linked to everything in an ecosystem, the evolution of species must work hand in hand with the selection of ecological composition and structure. As seen from this discussion, the principle of development or evolution towards higher exergy is able to explain how ecosystems react to perturbation as well as to evolution in the Darwinian sense.

The use of a goal or control function for determination of the variations of essential parameters has up to now been quite limited, but an example should be given to illustrate the possibilities.

A current modification of the maximum growth of phytoplankton in a eutrophication model (Jørgensen, 1976; Jørgensen, Mejer, Friis, 1978), see section 7.4.4, was attempted. The exergy for a wide range of values for the maximum growth rate of phytoplankton was computed and the value which gave the highest exergy was selected. The model was applied on a hypereutrophic lake and a 99% reduction of the phosphorus input was simulated. It was found that along with decreased phosphorus concentration and eutrophication, the selected maximum growth rate increased, see Table 9.2.

**TABLE 9.2**

**Maximum Growth Rate for Algae.**

| Case | Max growth rate at highest exergy |
|------|-----------------------------------|
| Oligotrophic Lake $P_{total} < 0.05$ mg/l$^{-1}$ | 3.3 day$^{-1}$ |
| Eutrophic Lake $P_{total} - 0.5$ mg/l$^{-1}$ | 2.2 day$^{-1}$ |
| Hypereutrophic Lake $P_{total} - 1.5$ mg/$^{-1}$ | 1.6 day$^{-1}$ |

This is in accordance with observations that phytoplankton species in oligotrophic lakes are generally smaller, i.e. the specific surface is higher, giving a higher growth rate, than in eutrophic lakes.

Other possible goal functions are summarized in Table 9.3. It is expected that more and more models in the nearest future will account for the flexibility of ecosystem structure by application of goal functions and as already mentioned this is possibly absolutely necessary, when drastic changes in forcing functions are expected.

**TABLE 9.3**

**Goal Functions.**

| Proposed for | Principle (Objevtive function) | Reference |
|---|---|---|
| All systems at all temporal spatial, and organizational levels of resolution | Maximum energy flow | Lotka (1922; Odum & Pinkerton (1955) |
| | Minimum entropy | Glansdorff & Prigogine (1970) |
| | Maximum exergy | Mejer & Jørgensen (1979) |
| Ecological systems at organizational level of resolution greater than populations | Optimum ascendency | Ulanowicz (1980) |
| | Maximum persistent organic matter | Whittaker & Woodwell (1971) O'Neill et al.(1975) |
| Ecological systems; applicable scale | Maximum biomass; maintenance metabolism ration | Margalef (1968) |
| Economic systems | Maximum profit (specific growth rate) | Various authors |

## 9.4. APPLICATION OF CATASTROPHY THEORY TO ECOLOGICAL MODELLING

Another attempt to include higher order dynamics in ecological models is the application of catastrophy theory, which considers the existence of multiple stable equilibrium states.

Catastrophy theory was introduced by the french mathematician R, Thom (1973). Several papers dealing with the application of this theory in ecological modelling have been published: Jones and Walthers (1976), Jones (1977) Duckstein, Casti and Kempf (1977), Dubois (1979), Duckstein, Casti and Kempf (1979) and Kempf (1980).

As an illustration of this approach will be presented a river model, developed by Dubois (1979).

## ILLUSTRATION 9.1

The change in oxygen concentration in a stream can be expressed by use of the following equation:

$$\frac{dC_t}{dt} = \text{rate of exch:air/water} + \text{production by photosynthe-sis-consumption by respiration.} \tag{9.8}$$

$C_O$ is oxygen concentration at time t.

The consumption of oxygen $C_O$, can be given by a Michaelis-Menten expression:

$$C_0 = k_1 \cdot \frac{C_t}{k_1 + C_t} \tag{9.9}$$

where

$C_O$ is oxygen concentration at time t.   $k_1$ and $k_2$ are constants.

The production of oxygen by photosynthesis, PP, is given by the logistic equation:

$$\tag{9.10}$$
$$PP = k_3 \cdot C_t (1 - \alpha \cdot C_t)$$

where $k_3$ and $\alpha$ are constants.

The reaeration, RA, is described as:

$$\tag{9.11}$$
$$RA = K_a (C_S - C_t)$$

where $K_a$ is the reaeration constant and $C_S$ oxygen concentration at saturation, see section 7.2.1

Equation (9.8) can now be transformed to:

$$\frac{dC_t}{dt} = K_a \cdot C_s - K_a \cdot C_t + k_3 C_t (1 - \alpha C_t) - k_2 \cdot \frac{C_t}{k_1 + C_t} \tag{9.12}$$

$C_S$ is a function of temperature.

If we use the following symbols:

X      $= C_t / k_1$
$X_s$      $= C_s / k_1$ (function of temperature T)
a      $= K_a \cdot C_S$ (function of temperature T)
b      $= k_1 - K_a$
c      $= \alpha k_3 \cdot k_1 / b$  and
d      $= k_2 / k_1$

Equation (9.12) might be written as:

$$\frac{dx}{dt} = a(T) + bx(1 - cx) - d\frac{x}{1 + x}$$

The stationary solution of this equation is given by $\frac{dx}{dt} = 0$.                Fig.9.3.

shows dx/dt + a versus x for particular values of b, c and d.(b = 1, c = 0.1 and d = 4).

a(T) varies with the temperature and depending on the value of a, one, two or three stationary states exist for $x$. When only one stationary state exists, the state is asymptotically stable. When two stationary states exist, one state is asymptotically stable, while the other state is unstable. When three stationary states exist two are asymptotically stable and one is unstable, see fig. 9.4, which shows the state of x for different values of a(T). Arrows in the figure show how x will evolve. S1 and S2 are attractors, as x values above the attractor points give a negative dx/dt and below the attractor points a positive dx/dt. If we vary a from 0.5 to 1.3, x will jump from S1 to S2, which can be seen by comparison of the figures. If we on the other hand decrease a from 1.3 to 0.5, x will jump from S2 to S1. The critical jump value for the first jump is 1.2 for the second one 0.75. Thus the two jumps of x do not occur for the same value of a and consequently not for the same temperature. It means that between the two a values 0.75 and 1.3 the stable state of x is dependent on the history of the system. In fig. 9.5 is shown the pattern of stable x-values versus a. -the hysteresis effect is clearly shown.

If we consider an explicit temperature dependent relation for a(T), f.inst.:

(9.13)

a(T) = B - C sin(wt)

where

B, C and w are constants, fig. 9.5 is transformed into fig. 9.6 where the hysteresis effect is demonstrated for two different values of w.

Similar results are obtained for other set of equations, which possess the same mathematical properties. It is, however, not only a mathematical property, but observations in ecosystems support the results. In the spring, when the temperature is increasing, it means that a(T) is decreasing, the jump to a lower oxygen concentration in polluted rivers is observed at a higher temperature (lower a-value) than the jump back to the high oxygen concentration in the autumn.

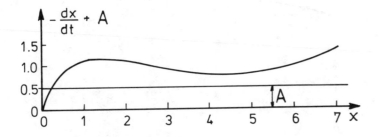

Fig.9.3: dx/dt + a plotted versus x.

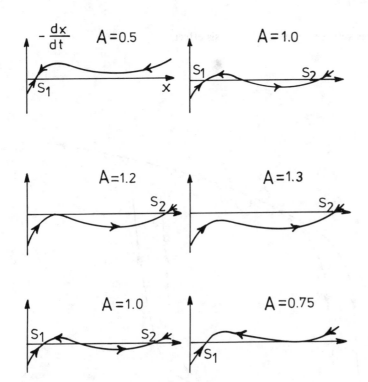

Fig.9.4: dx/dt versus x for 6 different a-values. Arrows show how x will evolve. $S_1$ and $S_2$ are attractors.

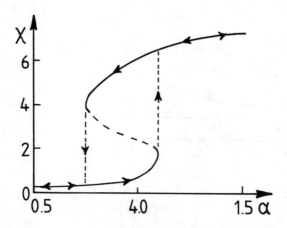

Fig.9.5: Stable x-values versus a. Note the hysteresis effect.

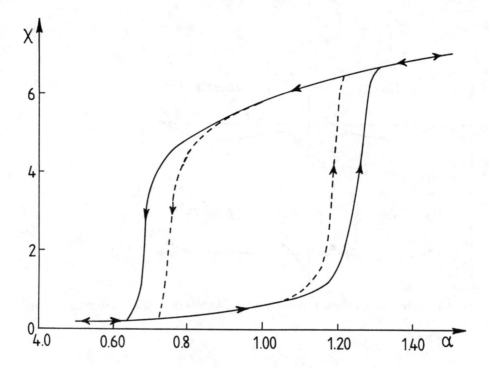

Fig.9.6: Pattern of stable x-values versus a.

# 10. REFERENCES

*Abou-Donia, M.B., and Preissig, S.H., 1976. Delayed neurotoxicity from continuous low-dose oral administration of leptophos to hens. Toxicology and Applied Pharmacology 38: 595-608.

*Ahlgren, I., 1973. Limnologiska studier av Sjoen Norrviken. III. Avlastningens effekter. Scripta Limnologica Upsaliencia No. 333.

*Allen, T.F.H., and Starr, T.B., 1982. Hierarchy: Perspectives for Ecological Complexity. University of Chicago Press, Chicago. pp. 310.

*Andersen, K.P. and Ursin, E., 1977. A multispecies extension to the Beverton and Holt theory of fishing, with account of phosphorus circulation and primary production. Meddr. Danm. Fisk. - og Havunders. N.S. 7: 319-435.

*Anderson, J.M., 1973. The eutrophication of lakes. Towards Global Equilibrium. ed. D. Meadows, *Cambridge. MA:MIT Press, pp. 117-140.

*Anon, 1931. Report of the Ad.Hoc. working group on the use of effort data in assessments. ICES C.M. 1981/G:5 (mimeo.).

*Aoyama, I, Yos. Inoue and Yor. Inoue, 1978. Simulation analysis of the concentration process of trace heavy metals by aquatic organisms from the viewpoint of nutrition ecology. Water Research 12: 837-842.

*Armstrong, F.A.J. and Schindler, D.W., 1971. Preliminary chemical characterization of waters in the experimental lakes area, Northwestern Ontario. J. Fish. Res. Board Can., 28: 171-187.

*Armstrong, N.E., 1977. Development and Documentation of Mathematical Model for the Paraiba River Basin Study, Vol 2 - DOSAGM: Simulation of Water Quality in Streams and Estuaries. Technical Report CRWR-145. Center for Research in Water Resources, The University of Texas at Austin, Austin, Texas.

*Arp, P.A., 1983. Modelling the Effects of Acid Precipitation on Soil Leachates: A Simple Approach. Ecological Modelling 19: 105-117.

*Baly, E.C.C., 1935. The kinetics of photosynthesis. Proc.Roy.Soc.

London 117B: 218-239.

*Bartell, S.M., Gardner, R.H. and O'Neill, R.V., 1984. The Fates of Aromatics Model. Ecol. Modelling. 22: 109-123.

*Beddington, J.R. and May, R.M., 1980. Maximum sustainable yields in systems subject to harvesting at more than one trophic level. Math. Biosci., 51: 261-281.

*Bedinger, M.S., 1971. Forest Species as Indicators of Flooding in the lower White River Valley, Arkansas. U.S. Geol. Surv. Prof. Pap. 750-C. USDI, Washington, D.C., pp. 248-253.

*Bergstrand, E., and Cawse, P.A., 1979. The deposition of trace elements and major nutrients in dust and rainwater in Northern Nigeria. Sci. Total Environ 13: 263-274.

*Betzer, S.B., and Pilson, M.E.Q., 1974. The seasonal cycle of copper concentration in Bysycon canaliculatum. Biological Bulletin 142: 165-175.

*Beyer, J. and Sparre, P., 1983. Modelling Exploited Marine Fish Stocks. Application of Ecological Modelling in Environmental Management, Part A. Elsevier Scientific Publishing Company. Amsterdam-Oxford-New York. Edited by S.E. Jørgensen.

*Bhumralkar, M.B., Johnson, R.L., Mancuso, R.L., and Wolf, D.E., 1979. Regional patterns and transfrontier exchanges of airborne sulphur pollution in Europe. Final Report, SRI project 4797, SRI International.

*Bierman, V.J., Verhoff, F.H., Poulson, T.C., and Tenney, M.W., 1974. Multinutrient dynamic models of algal growth and species competition in eutrophic lakes. Modeling the Eutrophication Process. eds. E. Middlebrooks, D.H. Falkenberg and T.E. Maloney, Ann Arbor, MI: Ann Arbor Science, pp. 89-109.

*Bonner, J.T., 1965. Size and Cycle. An Essay on the Structure of Biology. Princeton University Press, Princeton, NJ, 219 pp.

*Bosserman, R.W., 1980. Complexity measures for assessment of environmental impact in ecosystem networks. In: Proc. Pittsburgh Conf. Modelling and Simulation. Pittsburgh, PA, April 20-23, 1980.

*Bosserman, R.W., 1982. Structural comparison for four lake

ecosystem models. In: L. Troncale (Ed.), A General Survey of
Systems Methodology. Proceedings of the Twenty-sixth Annual
Meeting of the Society for General Systems Research. Washingron,
DC, January 5-9, 1982, pp. 559-568.

* Botkin, D.B., Janak, J.F. and Wallis, J.R., 1972. Some ecological
consequences of a computer model of forest growth. J. Ecol.,
60: 849-872.

* Brandes, M., Chowdry, N.A. and Cheng, W.W., 1974. Experimental
Study on Removal of Pollutants from Domestic Sewage by
Underdrained Soil Filters. National Home Sewage Disposal
Symposium. Agric. Eng., Chicago, Ill.

* Bro-Rasmussen, F. and Christiansen, K., 1984. Hazard assessment
- a summary of analysis and integrated evaluation of exposure
and potential effects from toxic environmental chemicals.
Ecol. Modelling. 22: 67-85.

* Broqvist, S., 1971. Matematisk modell for ekosystemet i en sjo.
Forskningsgruppen for Planeringsteori, Matematiske Institution,
Tekniska Hogskolan, Stockholm.

* Brown, S.L., 1978. A comparison of cypress ecosystems in the
landscape of Florida. Ph.D. Diss., University of Florida,
Gainsville, FL, 569 pp.

* Bryan, G.W., 1976. Some aspects of heavy metal tolerance in
aquatic organisms. In: "Effects of Pollutants on Aquatic
Organisms". (A.P.M. Lockwood, ed.) 7-34. Cambridge University
Press, Cambridge.

* Burns, L.A. and Taylor, R.B., 1979. Nutrient-uptake model in
marsh ecosystems. Proc.Am.Soc.Civ.Eng. J. Tech.Counc., 105:
177-196.

* Canale, R.P., DePalma, L.M., and Vogel, A.H., 1976. A plankton-based
food web model for Lake Michigan. In: Modelling Biochemical
Processes in Aquatic Ecosystems. ed. R.P. Canale, Ann Arbor
Science, Michigan, pp. 33-74.

* Canale, R.P., DePalma, L.M., and Vogel, A.H., 1976. A plankton-based
food web model for Lake Michigan. Modeling Biochemical
Processes in Aquatic Ecosystems. ed. R.P. Canale, Ann Arbor,
MI: Ann Arbor Science, pp. 33-74.

* Chen, C.W., 1970. Concepts and utilities of ecological models.
Proceedings of the American Society of Civilengineers,
Journal of the Sanitary Engineering Division 96(SA5): 1085-1097.

*Chen, C.W., and Orlob, G.T., 1975. Ecologic simulation of aqatic environments. Systems Analysis and Simulation in Ecology vol. 3. ed. B.C. Patten, New York, NY: Academic Press, pp. 476-588.

*Chen, C.W., Dean, J.D., Gherini, S.A., and Goldstein, R.A., 1982. Acid rain model: Hydrologic module. The Jour. of Environ. Eng. ASCE, EE3, 108: 455-472.

*Chester, P.F., 1982. Acid rain, catchment characteristics and fishery status. Int. Conf. on Coal Fired Power Plans and the Aquatic Environment, Copenhagen 16-18 Aug. 1982. Proceedings Water Qual. Inst., Denmark, 447-457.

*Chiou, C.T., Freed, V.H., Schmedding, D.W., and Kohnert, R.L., 1977. Partition coefficient and bioaccumulation of selected organic chemicals. Environmental Science and Technology 11: 475-478.

*Christophersen, N., and Weight, R.F., 1980. Sulfate flux and a model for sulfate concentration in stream-water at Birkenes, a small forested catchment in southernmost Norway. Water Resour. Res., in press (also SNSF-project, IR70/80).

*Cloern, J.E., 1978. Simulation model of Cryptomonas ovata population dynamics in Southern Kootenay Lake, British Colombia. Ecological Modelling 4: 133-150.

*Colwell, R.K., 1973. Competition and coexistence in a simple tropical community. Amer. Matur. 107: 737-760.

*Costanza, R. and Sklar, F.H.(Baton Rouge, LA. USA), 1985. Articulation, accuracy and effectiveness of mathematical models: a review of freshwater wetland applications. Ecol. Modelling. 27: 45-69.

*Cowardin, L.M., Carter, V., Golet, F.C. and LaRoe, F.T., 1979. Classification of wetlands and deepwater habitats of the United States. U.S. Fish and Wildlife Service Pub. FWS/OSB-79/31 Dept. of the Interior, Washington, D.C., 103 pp.

*Cridland, C.C., 1960. Laboratory experiment on the growth of Tilapia spp. Hydrobiologia, 15: pp. 135-160.

*Cridland, C.C., 1962. Laboratory experiments on the growth of Tilapia spp. Hydrobiologia 20: pp 155-166.

*Daan, N., 1975. Consumption and production in the North cod, Gadus morhua,: An assessment of the ecological status of the stock. Neth. J. Sea Res. 9: 24-55.

* Dahl-Madsen, K.I. and Strange-Nielsen, K., 1974. Eutrophication
  models for ponds. Vand, 5: 24-31.

* Dahl-Madsen, K.I., and Strange-Nielsen, K., 1974. Eutrophication
  models for ponds. Vand 5: 24-31.

* Deevey, E.S., Jr., 1947. Life tables for natural populations of
  animals. Quart. Rev. Biol. 22: 283-314.

* Dennis, R.L., 1983. Dispersion Models for Management of Urban
  Air Quality. Application of Ecological Modelling in Environmental
  Management, Part B. Edited by S.E. Jørgensen and Mitsch,
  W.J. Elsevier Scientific Publishing Company, Amsterdam-Oxford-New
  York-Tokyo.

* Dillon, P.J., and Kirchner, W.B., 1975. The effects of geology
  and land use on the export of phosphorus from watersheds.
  Water Res., 9: 135-148.

* Dillon, P.J., and Rigler, F.H., 1974. A test of a simple nutrient
  budget model predicting the phosphorus concentration in lake
  water. J. Fish. Res. Board Can., 31: 1771-1778.

* Dobbins, W.E., 1964. BOD and Oxygen Relationship in Streams.
  Journal of Sanitary Engineering Division. Proceedings ASCE
  90, SA 3, 53.

* Dodson, I.S., 1975. Predation rates of zooplankton in arctic
  ponds. Limnology and Oceanography 20(3): 426-433.

* Dubois, D.M., 1979. Catastrophy Theory Applied to Water Quality
  Regulation of Rivers. State-of-the-Art in Ecological Modelling.
  Proceedings of the Conference on Evcological Modelling,
  Copenhagen, Denmark, 28 August - 2 September 1978. Editor S.E.
  Jørgensen, International Society for Ecological Modelling.

* Dubois, D.M., 1978. Modelisation ecologique d'une riviere en vue
  de l'optimisation de l'epuration. Proceedings of the Colloque
  International (Liege, 16-19 May 1978). University of Liege,
  Environmental Engineering School, 45: 1-3.

* Duckstein, L., Casti, J., and Kempf, J., 1977. Modelling phytoplankton growth in small eutrophic ponds with catastrophe
  theory. Proceedings, 13th American Water Resources Conference,
  Tuscon, Arizona.

* Duff, G.H., and Nolan, N.J., 1957. Growth and Morphogenesis in

the Canadian Forest Species. II. Specific Increments and
their Relation to the Quantity and Activity of Growth in
Pinus resinosa Alt. Can. Jour. Bot., 35: 527-572.

* Dugdale, R.C., 1975. Biological modelling. Modelling of Marine
  Systems. ed. J.C. Nihoul, Amsterdam: Elsevier, pp. 187-206.

* EAFRO East African Fishery Research Organization (Jinja).

* Eckenfelder, W.W., Jr., O'Connor, D.J., 1961. Biological Waste
  Treatment. Pergamon Press, New York.

* Eckenfelder, W.W., Jr., 1970. Water Quality Engineering for
  Practicing Engineering. Barnes and Noble, Inc., New York.

* Ehrenfeld, D.W., 1973. Biological Conservation. Holt, Rinehart
  and Winston, New York.

* Eliassen, A. 1978. The OECD study of long range transport of air
  pollutants: Long range transport modelling. Atmos. Environ.
  12: 479-487.

* Eliassen, A., and Saltbones, J., 1982. Modelling of long range
  transport of sulphur over Europe: A two-year model run and
  some model experiments. EMEO/MSCW Report 1/82, Norwegian
  Meteorological Institute. To appear in Atmospheric Environment.

* Emlen, J.M., 1973. Ecology: An evolutionary approach. Addison-Wesley,
  Reading, Mass. 493 pp.

* Evans, R.B., et al., 1966. Principles of Desalination. N.Y. In:
  Spiegler, K.S. (ed).

* Fagerstroem, T., and Aasell, B., 1973. Methyl mercury accumulation
  in an aquatic food chain. A model and implications for
  research planning. Ambio 2: 164-171.

* Fedra, K, 1983. A Modular Approach to Comprehensive System
  Simulation: A Case Study of Lakes and Watersheds. Analysis
  of Ecological Systems: State-of-the-Art in Ecological Modelling.
  Edited by William K. Lauenroth (Natural Resource Ecology
  Laboratory), Gaylord V. Skogerboe (Agricultural and Chemical
  Engineering Department) and Marshall Flug (National Park
  Service, Water Resources Laboratory), Colorado State University,
  Fort Collins, CO 80523, U.S.A. Elsevier Scientific Publishing
  Company. Amsterdam-Oxford-New York.

*Fenchel, T., 1974. Intrinsic rate of natural increase: the
relationship with body size. Oecologia, 14: 317-326.

*Findeisen, W., Iastebrov, A., Lande, R., Lindsay, J., Pearson,
M. and Quade, E.S., 1978. A sample glossary of systems
analysis. Working Paper WP-78-12 (Laxenburg, Austria: International Institute for Applied Systems Analysis).

*Fisher, B.E.A., 1984. Long-range transport of air pollutants and
some thoughts on the state of modelling. Atmospheric Environment
18(3): 553-562.

*Foree, E.G., 1976. Reaeration and velocity prediction for small
streams. Proceedings of American Society of Civil Engineers,
Journal of Environmental Engineering Division 102(EE5): 937-952.

*Forrester, J.W., 1961. Industrial Dynamics. MIT Press, Cambridge.

*Gallegos, A.F., and Whicker, F.W., 1972. Radio cesium retention
by rainbow trout as affected by temperature and weight.
Report of National FNF Series 100, 115642: 1-25.

*Gardner, R.H., Huff, D.D., O'Neill, R.V., Mankin, J.B., Carney,
J. and Jones, J., 1980. Application of error analysis to a
marsh hydrology model. Water Resour. Res., 16: 659-664.

*Gargas, E., 1976. A three-box eutrophication model of a mesotrophic
Danish Lake. Water Quality Institute, Hoersholm, Denmark.

*Garrod, D.J., 1960. A review of Lake Victoria fishery service
records 1954-1959. East African Agricultural and Forestry
Journal, Vol. XXVI, pp. 42-48.

*Gause, G.F., 1934. The Struggle of Existence. New York, NY:
Hafner, p. 133.

*Gillett, J.W., et al., 1974. A conceptual model for the movement
of pesticides through the environment. National Environmental
Research Center, US Environmental Protection Agency, Corvallis,
OR Report EPA 660/3-74-024, p. 79.

*Gilliland, M.W., 1983. Models for Evaluating Human Carrying
Capacity: A Case Study of the Lake Tahoe Basin, California-
Nevada. Application of Ecological Modelling in Environmental
Management, Part B. Edited by S.E. Jørgensen and W.J. Mitsch.
Elsevier. Amsterdam-Oxford.New York-Tokyo.

*Greenwood, P.H., 1965. Two new species of Haplochromis (Pisces:
Cichlidae) from Lake Victoria. Ann. Mag. Natur. Hist,, Vol.

8 (89/90), pp. 303-318.

* Gromiec, M.J.,1983. Biochemical Demand - Dissolved Oxygen.
Application of Ecological Modelling in Environmental Management,
Part A. Elsevier Scientific Publishing Company, Amsterdam-
Oxford-New York.Editor S.E. Jørgensen.

* Gromiec, M.J., and Gloyna, E.F., 1973. Radioactivity transport
in water. Final Report No. 22 to US Atomic Energy Commission,
Contract AT(11-1)-490.

* Haimes, Y.Y.,Hall, W.A., Freedman, H.T., 1975. Multiobjective
Optimization in Water Resources Systems. The Surrogate Worth
Trade-off Method. Elsevier Scientific Publishing Company.
Amsterdam-Oxford-New York.

* Halfon, E., 1983. Is there a best model structure? II. Comparing
the model structures of different fate models. Ecol. Modelling,
20: 153-163.

* Halfon, E., Unbehauen, H. and Schmid, C., 1979. Model order
estimation and system identification theory to the modelling
of 32P kinetics within the trophogenic zone of a small lake.
Ecol. Modelling, 6: 1-22.

* Halfon, E.(Burlington, Ont., Canada) and Reggiani, M.G.(Rome,
Italy), 1978. Adequacy of wcosystem models. Ecol. Modelling.
4: 41-51.

* Halfon, E., 1984. Error analysis and simulation of Mirex behaviour
in Lake Ontario. Ecol. Modelling. 22: 213-253.

* Hamblyn, E.L., 1966. The food and feeding habits of Nile Perch
Lates niloticus (Linne) (Pisces: Centropomidae).

* Hansen, H.H., Frankel, R.J., 1965. Economic Evaluation of Water
Quality. A Mathematical Model of Dissolved Oxygen Concentration
in Freshwater Streams. Second Annual Report, Sanitary Engineering
Laboratory Report No. 65-11, Sanitary Engineering Research
Laboratory, University of California, Berkeley.

* Harlemann, D.R.F., 1978. Tech. Rep. MZT TR 227. Parsons Lab. MIT.

* Harris, J.R.W., Bale, A.J., Bayne, B.L., Mantoura, R.C.F.,
Morris, A.W., Nelson, L.A., Radford, P.J., Uncles, R.J.,
Weston, S.A. and Widdows, J. A Preliminary model of the
dispersal and biological effect of toxins in the Tamar
estuary, England. Ecol. Modelling. 22: 253-285.

* Henriksen, A., and Seip. H.M., 1980. Strong and weak acids in surface waters pf southern Norway and southwest Scotland. Water Res., 14: 809-813, also SNSF-project, FR17/80.

* Holling, C.S., 1959. Some characteristics of simple types of predation and parasitism. Canad. Entomol. 91: 385-398.

* Holling, C.S., 1966. The functional response of invertebrate predators to prey density. Mem. Entomol. Soc. Canada 48: 1-87.

* Hopkinson, C.S., Jr. and Day, J.W., Jr., 1980. Modelling hydrology and eutrophication in a Louisiana swamp forest ecosystem. Environ. Manage., 4: 325-335.

* Huff, D.D. and Young, H.L., 1980. The effect of a marsh on runoff. I.A. water-budget model. J. Environ. Qual., 9: 633-640.

* Huff, D.D., Koonce, J.F., Ivarson, W.R., Weiler, P.R., Dettmann, E.H. and Harris, R.F., 1973. Simulation of urban run off, nutrient loading, and biotic response of a shallow eutrophic lake. In: E.J. Middlebrooks, D.H. Falkenberg and T.E. Maloney (Editors), Modelling the Eutrophication Process. Ann Arbor Science, Ann Arbor, MI, pp. 33-55.

* Hutchinson, G.E., 1970. The biosphere. Scient. Amer., 223 (3): 44-53.

* Imboden, D.M., 1974. Phosphorus model for lake eutrophication. Limnology Oceanography 19: 297-304.

* Imboden, D.M., 1979. Modelling of vertical temperature distribution and its implication on biological processes in lakes. State of the Art in Ecological Modelling, ed. S.E. Jørgensen (Copenhagen: International for Ecological Modelling), pp. 545-561.

* Imboden, D.M., and Gachter, R., 1978. A dynamic lake model for trophic state prediction. Ecological Modelling 4: 77-98.

* Jackson, P.B.N., 1970. The African Great Lakes Fisheries: Past, present and future. Afr. J. Hydrobiol. Fish.

* Jacobsen, O.S., and Jørgensen, S.E., 1975. A submodel for nitrogen release from sediments. Ecological Modelling 1: 147-151.

* Jansson, B.O., 1972. Ecosystem approach to Baltic problem. Swedish Natural Science Research Council Bulletins from Ecological Research Committee No. 16.

* Jeffers, N.R.J., 1978. An Introduction to Systems Analysis with

Ecological Applications. E. Arnold.

*Jørgensen, S.E., 1976. A model of fish growth. J. Ecol. Model., 2: 303-313

*Jørgensen, S.E., (editor-in-chief; editorial board: M.B. Friis, J. Hendriksen, L.A.Jørgensen, S.E. Jørgensen and H.F. Mejer), 1979. Handbook of Environmental Data and Ecological Parameters. International Society of Ecological Modelling, Copenhagen.

*Jørgensen, S.E., 1981. A Holistic Approach to Ecological Modelling by Application of Thermodynamics. In Systems and Energy edited by W. Mitsch et al., 1982, Ann Arbor.

*Jørgensen, S.E., Kamp Nielsen, L., Jørgensen, L.A. and Mejer, H., 1982. An Environmental Management Model of the Upper Nile Lake System. ISEM Journal, 4: 5-72.

*Jørgensen, S.E., Mejer, H., 1977. Ecological buffer capacity. J. Ecol. Model., 3: 39-61.

*Jørgensen, S.E., Mejer, H. and Friis, M., 1978. Examination of a lake model. J. Ecol. Model., 4: 253-279.

*Jørgensen, S.E., Jacobsen, O.S. and Hoei, I., 1973. A prognosis for a lake. Vatten, 29: 382-404.

*Jørgensen, S.E., 1976. A eutrophication model for a lake. Ecol. Model., 2: 147-165.

*Jørgensen, S.E., Jørgensen, L.A., Kamp Nielsen, L., and Mejer, H.F., 1981. Parameter Estimation in Eutrophication Modelling. Ecol. Model., 13: 111-129.

*Jørgensen, S.E., Mejer, H.F., 1979. A Holistic Approach to Ecological Modelling. Ecol. Model., 7: in press.

*Jørgensen, S.E., Mejer, H.F. and Friis, M., 1978. Examination of a lake model. Ecol. Model., 4: 253-279.

*Jørgensen, S.E., 1982. Modelling the eutrophication of shallow lakes. In: D.O. Logofet and N.K. Luckyanov (Editors), Ecosystem Dynamics in Freshwater Wetlands and Shallow Water Bodies, Vol. 2. UNEP/SCOPE, U.S.S.R. Academy of Sciences, Moscow, pp. 125-155. *Jørgensen, S.E., 1983. Eutrophication Models of Lakes. Application of Ecological Modelling in Environmental Management, Part A. Elsevier Scientific Publishing Company. Amsterdam-Oxford-New York. Edited by S.E. Jørgensen.

* Jørgensen, S.E., 1983. Modelling the Distribution and Effect of Toxic Substances in Aquatic Ecosystems. Application of Ecological Modelling in Environmental Management, Part A. Elsevier Scientific Publishing Company. Amsterdam-Oxford-New York. Edited by S.E. Jørgensen.

* Jørgensen, S.E., 1984. Parameter estimation in toxic substance models. Ecol. Modelling. 22: 1-13.

* Jørgensen, S.E. and Mejer, H.F., 1983. Trends in Ecological Modelling. Analysis of Ecological Systems: State-of-the-Art in Ecological Modelling. Edited by William K. Lauenroth (Natural Ecology Laboratory), Gaylord V. Skogerboe (Agricultural and Chemical Engineering Department) and Marshall Flug (National Park Service, Water Resources Laboratory), Colorado State University, Fort Collins, CO 80523, U.S.A. Elsevier Scientific Publishing Company, Amsterdam-Oxford-New York.

* Jørgensen, S.E., Kamp-Nielsen, L. and Jacobsen, O.S., 1975. A submodel for anaerobic mudwater exchange of phosphate. Ecological Modelling 1: 133-146.

* Jørgensen, S.E., Jacobsen. O.S., and Hoi, I., 1973. A prognosis for a lake. Vatten, 29: 382-404.

* Jørgensen, S.E., 1981. A Holistic Approach to Ecological Modelling by Application of Thermodynamics. In: Systems and Energy. ed. W. Mitsch et al., 1982, Ann Arbor.

* Jørgensen, S.E., and Mejer, H.F., 1981a. Application of Exergy in Ecological Models. In: Progress in Ecological Modelling, edited by D. Dubois, Liege, p. 311-47.

* Jørgensen, S.E., and Mejer, H.F., 1981b. Exergy as Key Function in Ecological Models. In: Energy and Ecol. Modelling, edited by W. Mitsch et al., p. 587-590.

* Jørgensen, S.E., 1979. Modelling the distribution and effect of heavy metals in an aquatic ecosystem. Ecological Modelling 6: 199-223.

* Jørgensen, S.E., 1979. Modelling the distribution and effect of heavy metals in aquatic ecosystems. J. Ecol. Model., 6: 199-223.

* Jørgensen, S.E., Kamp-Nielsen, L, and Jørgensen, L.A., 1985. Examination of the Generality of Eutrophication Models. Ecol. Modelling, in press.

* Jørgensen, S.E. (The Royal Danish School of Pharmacy Dept. of

Pharmaceutical Chemistry AD, 2 Universitetsparken, DK-2100
Copenhagen, Denmark), 1985. Structural Dynamic Model. Ecol.
Modelling, in press.

* Jørgensen, S.E., Kamp-Nielsen, L., Christensen, T., Windolf-Nielsen,
J., and Westergaard, B., 1985. Validation of a Prognosis
based upon a Eutrophication Model. Ecol Modelling, in press.

* Johnson, W.B., Wolf, D.E., and Mancuso, R.L., 1978. Long term
regional patterns and transfrontier exchanges of airborne
sulphur pollution in Europe. Atmos. Environ. 12: 511-527.

* Jones, D.D. and Walters, C.J., 1976. Catastrophe theory and
fisheries regulation. J. Fish. Res. Bd. Can. 33: 2829-2833.

* Jones, D.D., 1977. Catastrophe theory applied to ecological
systems. Simulation 29: 1-15.

* Jones, R., 1978. Further observations on the energy to the major
fish species in the North Sea. ICES C.M. 1978 Gen: 6. (mimeo.).

* Jost, J.L., Drake, I.F., Frederickson, A.G., and Tsandriya,
H.M., 1973. Interactions of Tetrahymena pyriformis, Escherichia
coli, Az tobacter vinelandiii and glucose in minimal medium.
Journal of Bacteriology 113: 834-840.

* Kamp-Nielsen, L., 1983. Sediment-Water Exchange Models. Application
of Ecological Modelling in Environmental Management, Part A.
Elsevier Scientific Publishing Company. Amsterdam-Oxford-New
York. Edited by S.E. Jørgensen

* Kamp-Nielsen, L., 1975. A kinetic approach to the aerobic sediment-water exchange of phosphorus in Lake Esrom. Ecological
Modelling 1: 153-160.

* Kamp-Nielsen, L., 1974. Mudwater exchange of phosphate and other
exchange rate. Arch. Hydrobiol., 2: 218-237.

* Kauppi, P., Posch, M., Matzner, E., Kauppi, L., and Kamari, J.,
1984. A model for predicting the acidification of forest
soils: application to acid deposition in Europe. (Forthcoming
IIASA Research Report).

*Kempf, J., 1980. Multiple Steady States and Catastrophes in Ecological Models. ISEM-Journal 2: 55-80.

*Kenaga, E.E., and Goring, C.A.I., 1978. Relationship between water solubility, soil sorption, octanol-water partitioning, and concentration of chemicals in biota. In "Aquatic Toxicology", J.G. Eaton, P.R. Parrish and A.C. Hendricks, eds. Special Technical Publications, No.707: 78-113. American Society for Testing and Materials, Philadelphia.

*Killus, J.P. et al., 1980. Continued Research in Mesoscale Air Pollution Simulation Modelling: Volume IV --Refinements in Numerical Analysis Transport, Chemistry, and Pollutant Removal. Final Report to the U.S. Environmental Protection Agency, prepared by Systems Applications, Inc., EF77-142R

*Kirchner, T.B. and Whicker, F.W, 1984. Validation of PATHWAY, a simulation model of the transport of radionuclides through agroecosystems. Ecol. Modelling. 22: 21-45.

*Kohlmaier, G.H.., Sirre', E.O., Brohl, H., Killian, W., Fishbach, U., Plochl, M., Muller, T. and Jiang, Y., 1984. Ecol. Modelling. 22: 45-67.

*Kohlmaier, G.H., Sire, E.O, Brohl, H., Kilian, W., Fischbach, U., Plochl, M., Muller, T., and Yunsheng, J., 1984. Dramatic development in the dying of German spruce-fir forests: In search of possible causeeffect relationships. Ecological Modelling 22: 45-65.

*Kramer, B.M., 1979. Air Quality Modeling: Judicial, Legislative and Administrative Reactions. Colombia Journal of Environmental Law, 5: 236-263.

*Krenkel, P.A., and Orlob, G.T., 1962. Turbulant diffusion and the reaeration coefficient. Proceedings of American Society of Civil Engineers, Journal of Sanitary Engineering division 88(SA2): 53-83.

*Lam, D.C.L. and Simons, T.J., 1976. Computer model for toxicant spills in Lake Ontario. Environmental Biogeochemistry vol. 2 Metals transfer and ecological mass balances. ed. J.O. Nriago (Ann Arbor, MI: Ann Arbor Science), pp. 537-549.

*Lam, D.C.L., and Simons, T.J., 1976. Numerical computations of advective and diffusive transport of chloride in Lake Erie. J. Fish Res. Canada, 33: 537-549.

*Lamanna, C, and Malette, M.F., 1965. Basic Bacteriology. Baltimore, MD: Williams and Wilkins.

*Lamb, R.G., 1975. The Calculation of Long Term Atmospheric Pollutant Concentration Statistics Using the Concept of a Macro-Turbulence. Proc. of the Seminar of Air Pollution Modelling, Venice, Italy, November 27-28.

*Lamb, R.G., and Durran, D.R., 1978. Eddy Diffusitivities Derived from a Numerical Model of the Convective Planetary Boundary Layer. Il Nuovo Cimento, 1: 1-17

*Lamb, R.G., 1984. Air pollution models as descriptors of cause-effect relationships. Atmospheric Environment 18(3): 591-606.

*Lappalainen, K.M., 1975. Phosphorus loading capacity of tubes and a mathematical model for water quality prognoses. Proceedings of 10th Nordic Symposium on Water Research, Vaerloese, May 20-22, 1974 (Helsinki:"Entrofierung" NORFORSK.).

*Larsen, D.P., Mercier, H.T. and Malveg, K.W., 1974. Modeling algal growth dynamics in Shagawa Lake, Minnesota. Modeling Eutrophication Process, eds. E.J. Middlebrooke, D.H. Falkenberg and T.E. Maloney, Ann Arbor, MI: Ann Arbor Science, pp. 15-33.

*Larsen, D.P., Mercier, H.T., and Malveg, K.W., 1974. Modeling algal growth dynamics in Shagawa Lake, Minnesota. Modeling the Eutrophication Process, eds. E.J. Middlebrooks, D.H. Falkenberg and T.E. Maloney, Ann Arbor, MI: Ann Arbor Science, pp. 15-33.

*Lassen, H., and Nielsen, P.B., 1972. Simple mathematical model for the primary production as a function of the phosphate concentration and the incoming solar energy applied to the North Sea. Danmarks Fisker- og Havundersoegelser. International Council for the Exploration of the Sea. Plankton Committee 1972.

*Lassiter, R.R., 1978. Principles and constraints for predicting exposure to environmental pollutants. U.S. Environmental Protection Agency, Corvallis, OR Report EPA 118-127519

*Lassiter, R.R., 1975. Modeling dynamics of biological and chemical components of aquatic ecosystems. EPA-660/3-75-012, U.S. Environmental Protection Agency, Washington, DC.

*Lassiter, R.R., and Kearns, D.K., 1974. Phytoplankton population changes and nutrient fluctuations in a simple aquatic ecosystem model. Modeling the Eutrophication Process, eds. E.J. Middlebrookes, D.H. Falkenberg, and T.E. Maloney, Ann Arbor, MI: Ann Arbor Science, pp. 131-138.

*Lau, L.Y., 1972. Prediction equation for reaeration in open-channel flow. Proceedings of American Society of Civil Engineers. Journal of Sanitary Engineering Division 96(SA6): 1063-1068.

*Laws, R.M., 1962. Some effects of whaling on the southern stocks of baleen whales. In The Exploitation of Natural Animal Populations. Editor Le Cren, E.D. and Holdgate, M.W., 242-59. Blackwells, Oxford.

*Lehman, J.T., Botkin, D.B., and Likens, G.E., 1975. The assumptions and rationales of a computer model of phytoplankton population dynamics. Limnology and Oceanography 3: 343-364.

*Leung, D.K., 1978. Modeling the bioaccumulation of pesticides in fish. Center for Ecological Modeling Polytechnic Institue, Troy, NY Report 5.

*Liss, P.S. and Slater, P.G., 1974. Flux of gases across the air-sea interface. Nature, 247: 181-184.

*Loehr, R.C., 1974. Characteristics and comparative magnitude of nonpoint sources. J. Wat. Poll. Cont. Fed., 46: 1849-1872.

*Løenholdt, J., 1973. The BOD5, P and N content in raw waste water. Stads- og Havneingenioeren, 7: 1-6.

*Løenholdt, J., 1976. Nutrient Engineering WMO Training Course on Coastal Pollution (DANIDA): 244-261.

*Lorenzen, M.W., Smith, O.J. and Kimmel, L.V., 1976. A long-term phosphorus model for lakes: Application to Lake Washington. Modeling Biochemical Processes in Aquatic Ecosystems, ed. R.P. Canale, Ann Arbor, MI: Ann Arbor Science, pp. 75-92.

*Lotka, A.J., and Harleman, D.R.F., 1975. A real-time model of nitrogen-cycle dynamics in an estuarine system. MIT Department of Civil Engineering. R.M. Parsons Laboratory Report 204.

*Lotka, A.J., 1956. Elements of mathematical biology. Dover, New York. 465 pp.

*Lotka, A.J., 1922. Contribution to the energetics of evolution. Proc. Nat. Acad. Sci. 8:147-150.

*Louma, S.N., and Bryan, G.W., 1978. Factors controlling the availability of sediment-bound lead to estuariane bivalve

Scrobicularia plana. Journal of the Marine Biological Association
of the United Kingdom 58: 793-802.

*Lu, J.C.S. and Chen, K.Y., 1977. Migration of trace metals in
interface of seawater and polluted surficial sediments.
Environmental Science and Technology 11: 174-182.

*Lu, P.-Y., and Metcalf, R.L., 1975. Environmental fate and
biogradability of benzene derivatives as studied in a model
aquatic ecosystem. Environmental and Health Perspectives 10:
269-284.

*Mackay, D. and Cohen, Y., 1976. Prediction of Volatilization
Rate of Pollutants in Aqueous Systems. Symposium on Non-bio-
logical Transport and Tranformation of Pollutants on Land
and Water, May 11-13. National Bureau of Standards, Gaithersburg,
Maryland.

*Margalef, R., 1968. Perspectives in Ecological Theory. University
Chicago Press. Chicago, 112 pp.

*Matis, J.H., and Patten, B.C., 1981. Environ analysis of linear
compartmental systems: the static, time invariant case.
Proc. 43nd Session, Int. Stat. Inst., Manila, Philipines,
Dec. 4-14, 1979, in press. *May, R.H., 1973. Stability and Complexity in Model
Ecosystems.

*Mejer, H. and Jørgensen, L.A., 1981. Model Identification Methods
applied to two Danish Lakes. Proc. of Task Force Meeting,
November 1979. IIASA.

*Mejer, H.F. and Jørgensen, S.E., 1979. Energy and ecological
buffer capacity. State of the Art in Ecological Modelling,
ed. S.E. Jørgensen, Copenhagen: International Society for
Ecological Modelling, pp. 829-846.

*Mejer, H.F., 1983. A Menu Driven Lake Model. ISEM-Journal 5: 45-50.

*Mertz, D.B., 1970. Notes on methods used in life-history studies.
Pp. 4-17 in J.H. Connell, D.B. Mertz, and W.W. Murduch
(eds.), Readings in ecology and ecological genetics. Harper
& Row, New York, 397 pp.

*Metcalf, R.L., Sangha, G.K., and Kopoor, I.P., 1975. Model
ecosystem for the evaluation of pesticide biodegradability
and ecological magnification. Environmental Science and
Technology 5: 709-713.

*Miller, D.R., 1979. Models for total transport. Principles of Ecotoxicology Scope vol. 12, ed. G.C. Butler, 1979, New York, NY: Wiley, pp. 71-90.

*Miller, J.G., 1978. Living Systems. McGraw-Hill, New York. 1102 pp.

*Miller, P.C., Stoner, W.A. and Tieszen, L.L., 1976. A model of stand photosynthesis for the wet meadow tundra at Barrow Alaska. Ecology, 57: 411-430.

*Mitsch, W.J., 1976. Ecosystem modeling of waterhyacinth management in Lake Alice, Florida. Ecol. Modelling. 2: 69-89.

*Mitsch, W.J., 1983. Ecological Models for Management of Freshwater Wetlands. Application of Ecological Modelling in Environmental Management, Part B. Edited by S.E. Jørgensen and W.J. Mitsch. Elsevier. Amsterdam-Oxford-New York-Tokyo.

*Mogensen, B., 1978. Chromium pollution in a Danish fjord. Licentiate Thesis. Royal Danish School of Pharmacy, Copenhagen.

*Mogensen, B., and Jørgensen, S.E., 1979. Modelling the distribution of chromium in a Danish firth. Proceedings of 1st International Conference on State of the Art in Ecological Modelling, Copenhagen, 1978. ed. S.E. Jørgensen, Copenhagen: International Society for Ecological Modelling, pp. 367-377.

*Morowitz, 1968. Energy Flow in Biology. Ac. Press.

*Muniz, I.P., and Seip, H.M., 1982. Possible effects of reduced Norwegian sulphur emissions on the fish populations in lakes in Southern Norway, SI-report 81 03 13-2, 28 s.

*Neely, W.B., Branson, D.R., and Blau, G.E., 1974. Partition coefficient to measure bioconcentration potential of organic chemicals in fish. Environmental Science and Technology 8: 1113-1115.

*Nihoul, J.C.J., 1984. A non-linear mathematical model for the transport and spreading of oil slicks. Ecol. Modelling. 22: 325-341.

*Nyholm, N., 1978. A simulation model for phytoplankton growth and nutrient cycling in eutrophic, shallow lakes. Ecol. Modelling, 4: 279-310.

*Nyholm, N., Nielsen, T.K. and Pedersen, K., 1984. Modeling heavy metals transport in an arctic fjord system polluted from mine tailings. Ecol. Modelling. 22: 285-325.

*Nyholm, N., 1976. Kinetics studies of phosphate-limited algae growth. Thesis, Technical University of Copenhagen.

*Nyholm, N., 1978. A simulation model for phytoplankton growth and nutrient cycling in eutrophic, shallow lakes. Ecological Modelling 4: 279-310.

*OECD, 1977. The OECD programme on long range transport of air pollutants. Measurements and findings. OECD, Paris.

*O'Brien, J.J., and Wroblewski, J.S., 1972. An ecological model of the lower marine trophic levels on the continental shelf of West Florida. Geophysical Fluid Dynamics Institute, Florida State University, Tallahassee, FL Technical Report, 170 pp.

*O'Brien, J.J., 1970. A Note on the Vertical Structure of the Eddy Exchange Coefficient in the Planetary Boundary Layer. Journal of Atmospheric Sciences, 27: 1213-1215.

*O'Connor, D.J., and Dobbins, W.E., 1956. The mecahnism of reaeration in natural streams. Proceedings of American Society of Civil Engineers. Journal of Sanitary Engineering Division 96 (SA2): 547-571.

*O'Connor, D.J., 1962. The effect of Stream Flow on Waste Assimilation Capacity. Proceedings of 17th Purdue Industrial Waste Conference, Lafayette.

*O'Connor, D.J., 1967. The Temporal and Spatial Distribution of Dissolved Oxygen in Streams. Water Resources Research 3, 1, 65.

*O'Connor, D.J., DiTorro, D.M., 1970. Photosynthesis of Oxygen Balance in Streams. Journal of Sanitary Engineering Division, Proceedings ASCE 96, SA 2, 547.

*O'Melia, C.R., 1974. Phosphorus cycling in lakes. North Carolina Water Resources Research Institute, Raleigh Report 97, 45 pp.

*O'Neill, R.V., W.F. Hanes, B.S, Ausmus and D.E.Reichle. 1975. A theoretical basis for ecosystem analysis with particular reference to element cycling. pp. 28-40. In, F.G.Howell, J.B.Gentry and M.H.Smith (eds.) Mineral Cycling in Southeastern Ecosystems. NTIS pub. CONF-740513.

*Octavia, K.A.H., Jirka, G.H. and Harleman, D.R.F., 1977. Vertical

Heat Transport Mechanisms in Lakes and Reservoirs. MIT Dept.
of Civil Eng., R.M. Parsons Laboratory for Water Resources
and Hydrodynamics, Tech. Report no. 227.

* Odum, E.P., 1959. Fundamentals of Ecology (2nd Edition). Saunders,
Philadelphia, PA.

* Odum, H.T. and Pinkerton, R.C. 1955. Time's speed regulator: The
optimum efficiency for maximum power output in physical and
biological systems. Amer.sci. 43:331-343.

* Odum, H.T., 1972. An energy circuit language. Systems Analysis
and Simulation in Ecology vol. 2, ed. B.C. Patten, New York,
NY: Academic Press, pp. 139-211.

* Odum, H.T., 1971. Environment, Power, and Society. Wiley Interscience, New York. 331 pp.

* Odum, H.T., 1983. Systems Ecology. Wiley Interscience, New York,
644 pp.

* Olson, M.P., and Voldner, E.C., 1981. Documentation of the
Atmospheric Environment Service long-range transport of air
pollutants model. Work Group 2 Report 2-5 AES, Toronto.

* Ondok, J.P. and Pokorny, J., 1982. Models of the O2 and CO2
regimes in shallow ponds. In: D.O. Logofet and N.K. Luckyanov
(Editors). Ecosystem Dynamics in Freshwater Wetlands and
Shallow Water Bodies, Vol. 2. UNEP/SCOPE, U.S.S.R. Academy
of Sciences, Moscow, pp. 174-189.

* Orlob, G.T., Hrovat, D. and Harrison, F., 1980. Mathematical
model for simulation of the fate of copper in a marine
environment. American Chemical Society, Advances in Chemistry
Series 189: 195-212.

* Park, R.A., Groden, T.W., and Desormeau, C.J., 1979. Modification
to model CLEANER. requiring further research. Perspectives
on Lake Ecosystem Modeling. eds. D. Scavia and A. Robertson,
Ann Arbor, MI: Ann Arbor Science, pp. 87-108.

* Parker, R.A., 1972. Estimation of ecosystem parameters. Verhandlung
Internationale Vereinigung Limnologie 18: 257-263.

* Parker, R.A., 1974. Empirical functions relating metabolic
processes in aquatic systems to environmental variables. J.
Fish. Bd.Can. 31: 1550-1552.

* Paschal, J.E., Soneshine, D.E. and Richardson, J.H., 1979. A

simulation model of a peromyscus leucopus population in an area of the great dismal swamp. In: P.W. Kirk, Jr. (Editors), The great dismal swamp. University Press of Virginia, VA, pp. 277-296.

*Pattee, H.H., 1973. Hierarchy Theory: The Challenge of Complex Systems. Braziller, New York.

*Patten, B.C., 1983. On the Quantitative Dominance of indirect Effects in Ecosystems. Analysis of Ecological Systems: State-of-the-Art in Ecological Modelling. Edited by William K. Lauenroth (Natural Resource Ecology Laboratory), Gaylord V. Skogerboe (Agricultural and Chemical Engineering Department) and Marshall Flug (National Park Service, Water Resources Laboratory), Colorado State University, Fort Collins, CO 80523, U.S.A.. Elsevier Scientific Publishing Company, Amsterdam-Oxford-New York.

*Patten, B.C., Egloff, D.A., and Richardson, T.H., 1975. Total ecosystem model for a cove in Lake Texoma. Systems Analysis and Simulation in Ecology vol. 3. ed. B.C. Patten, New York. NY: Academic Press, pp. 206-423.

*Patten, B.C., 1971-1976. Systems Analysis and Simulation in Ecology, Vols. 1-4. Academic Press, New York.

*Patten, B.C., 1982. Environs: relativistic elementary particles for ecology. Am. Nat. 119: 179-219.

*Patten, B.C., and Auble, G.T., 1981. System theory of the ecological niche. Am. Nat. 118: 345-369.

*Patten, B.C., Bosserman, R.W., Finn, J.T., and Cale, W.G., 1976. Propagation of cause in ecosystems. Pages 457-579 in System analysis and simulation in ecology, vol. 4, B.C. Patten, ed. Academic Press, New York, 593 pp.

*Phipps, R.L. and Applegate, L.H., 1983. Simulation of Management Alternatives in Wetland Forests. Application of Ecological Modelling in Environmental Management, Part B. Edited by S.E. Jørgensen and W.J. Mitsch. Elsevier. Amsterdam-Oxford-New York-Tokyo.

*Phipps, R.L., 1979. Simulation of wetlands forest vegetation dynamics. Ecological Modelling, 7: 257-288.

*Phipps, R.L., 1967. Annual Growth of Suppressed Chestnut Oak and Red Maple, a Basis for Hydrologic Inference. U.S. Geol. Surv. Prof. Pap. 485C. USDI, Washington, D.C. 27 pp.

* Phipps, R.L., 1979. Simulation of Wetlands Forest Vegetation
  Dynamics. Ecol. Modelling 7: 257-288.

* Pielou, E.C., 1969. An introduction to mathematical ecology.
  Wiley-Interscience, New York. 286 pp.

* Platt, T., Subba Rao, D.V., 1975. Primary production of marine
  microphytes.In: Cooper, J.P.(ed.): Photosynthesis and Productivity in Different Environments, pp. 249-280. Cambridge
  University Press.

* Puccia, C.J., 1983. Qualitative models for east coast benthos.
  p. 719-724. In: W.K. Lauenroth, G.V. Skogerboe, and M. Flug
  (eds.) Analysis of Ecological Systems: State-of-the-Art in
  Ecological Modelling. Elsevier Scientific, Amsterdam.

* Radtke, E., and Straskraba, M., 1980. Self-Optimization in a
  Phytoplankton Model. Ecol. Modelling 9: 247-268.

* Raytheon Company, Oceanographic Environmental Services, 1973.
  REBAM - A Mathematical Model of Water Quality for the Beaver
  River Basin. U.S. Environmental Protection Agency, Washington,
  D.C.

* Reverton, R.J.H. and Holt, S.J., 1957. On the dynamics of exploited
  fish populations. Fishery Invest., Ser.2. 19: 1-533.

* Reynolds, S.D. et al., 1976. Continued Research in Mesicale Air
  Pollution Simulation Modelling--Volume II: Refinements in
  the Treatments of Chemistry, Meteorology, and Numerical
  Integration Procedures. Prepared by Systems Applications,
  Inc. for U.S. EPA (EPA-600/4-76-016b).

* Rich, L.G., 1973. Environmental Systems Engineering, p. 5.
  McGraw Hill, U,S.A.

* Richey, J.E., 1977. An empirical and mathematical approach
  toward the development of a phosphorus model of Castle Lake.
  In: C.S. Hall and J.W. Day, Jr. (Editors). Ecosystem Modeling
  in Theory and Practice. Wiley, Sons, New York, N.Y, pp. 267-288.

* Richey, J.E., 1977. An empirical and mathematical approach
  toward the development of a phosphorus model of Castle Lake.
  Ecosystem Modeling in Theory and Practice. eds. C.A.S. Hall
  and I.W. Day, Jr., New York, NY: Wiley-Interscience, pp. 267-287.

* Ricker, W.E., 1954. Stock and recruitment. J. Fish. Res. Board
  Canada 11: 559-623.

* Roberts, T.R., 1974. Geographical distribution of African
  fresh-water fishes. Zool. J. Limn. Soc., Vol. 57: pp. 249-319.

* Ross, J., 1967. Systema uravnenii dlya opisaniya kolichestvennogo
  rosta rastenii (System of equations describing the quantitative
  plant growth),In: Fytoaktinometricheskie Issledovaniya
  Rastitelnogo Pokrova, pp. 64-88. Izd. Valgus, Tallin.

* Sakamoto, M., 1966. Primary production by phytoplankton community
  in some Japanese lakes and its dependence on lake depth.
  Arch. Hydrobiol., 62: 1-28.

* Schwarzenbach, R.P. and Imboden, D.M., 1984. Modelling concepts
  for hydrophobic pollutants in lakes. Ecol. Modelling. 22:
  171-213.

* Seip, K.L., 1983. Mathematical Models of the Rocky Shore Ecosystem.
  Application of Ecological Modelling in Environmental Management,
  Part B. Edited by S.E. Jørgensen and W.J. Mitsch. Elsevier.
  Amsterdam-Oxford-New York-Tokyo.

* Seip, K.L., 1978. Mathematical model for uptake of heavy metals
  in benthic algae. Ecological Modelling 6: 183-198.

* Sklar, F.H., 1983. Water budget, benthological characterization
  and simulation of aquatic material flows in a Louisisana
  freshwater swamp. Ph.D.Diss., Louisisana State University,
  Baton Rouge, L.A., 280 pp.

* Smale, S., 1966. Am. J. Math. 87, 491-496.

* Smidth, F.L./MT, 1973. Report on the eutrophication of Lake Lyngby.

* Smith, F.E., 1963. Population dynamics in Daphina magna and a
  new model for population growth. Ecology, 44: 651-663.

* Smith, J.H. et al., 1977. Environmental Pathways of Selected
  Chemicals in Freshwater Systems. Part I, EPA 600/7-77-113.

* Snodgrass, W.J., and O'Melia, C.R., 1975. Predictive model for
  phosphorus lakes. Sensitivity analysis and applications.
  Environmental Science and Technology 9(10): 937-944.

* Sparre, P., 1979. Some rematks on the application of yield/recruit
  curves in estimation of maximum sustainable yield. ICES C.M.
  1979/G:41.

* Ssentongo, G.W., 1972. Yield Isoplets of Tilapia esculenta
  Graham 1928 in Lake Victoria and Tilapia nilotica (linnaeus)

1957 in Lake Albert. Afr. J. Trop. Hydrobiol. fish, Vol. 2(2): pp. 121-128.

*Steele, J.H., 1974. The Structure of the Marine Ecosystems. Blackwell Scientific Publications, Oxford, 128 pp.

*Steele, J.H., 1974. The structure of Marine Ecosystems. Oxford: Blackwell, pp. 74-135.

*Steele, J.H., 1962. Environmental control of photosynthesis in the sea. Limnology and Oceanography 7: 137-150.

*Stone, J.H. and McHugh, G.F., 1979. Hydrologic effects of canals in coastal Louisiana via computer simulations. In: Proc. 1979 Summer Computer Simulation Conference. AFIPS Press, Montvale, NJ, pp. 339-346.

*Straskarba, M., 1980. Cybernetic-Categories of Ecosystem Dynamics. ISEM-Journal 2: 81-96.

*Straskraba, M., 1976. Development of an analytical phytoplankton model with parameters empirically related to dominant controlling variables. Umweltbiophysik, eds. R. Glaser, K. Unger and M. Koch. Berlin: Akademie-Verlag, pp. 33-65.

*Straskraba, E., Gnauck, A., 1983. Aquatische Okosysteme. Modiellierung and Simulation. VEB Gustav Fischer Verlag, Jena.

*Straskraba, M., 1979. Natural control mechanisms in models of aquatic ecosystems. Ecol. Model. 6: pp. 305-322.

*Streeter, H.W., Wright, C.T., and Kehr, R.W., 1936. Measures of natural oxidation in polluted streams, Part III, An experimental study of atmosphere reaeration under stream flow conditions. Sewage Work Journal 8(2): 282-316.

*TRACOR, Inc., 1971. Estuarine Modeling. An assessment. EPA, WQO. 16070 DZV 02/71. Texas.

*Texas Water Development Board, 1970. DOSAG-I Simulation of Water Quality in Streams and Canals. Program Documentation and User's Manual. Prepared by Systems Engineering Division.

*Thackston, E.L., and Krenkel, P.A., 1969. Reaeration prediction in natural streams. Proceedings of American Society of Civil Engineers, Journal of Sanitary Engineering Division 95(SA1): 65-94.

* Thom, R., 1973. Stabilite structurelle et morphogenese: essai d'une theorie generale des modeles, W.A. Benjamin Inc., Mass., U.S.A.

* Thomann, R.V., 1984. Physio-chemical and ecological modeling the fate of toxic substances in natural water systems. Ecol. Modelling. 22: 145-171.

* Thomann, R.V., et al., 1974. A food chain model of cadmium in westwen Lake Erie. Water Research 8: 841-851.

* Thomas, A.H., Jr., 1961. The dissolved Oxygen Balance in Streams. Proceedings, Seminar on Waste Water Treatment and Disposal, Boston Society of Civil Engineering.

* Tinkle, D.W., 1967. The life and demography of the side-blotched lizard. Uta stansburiana. Misc. Publ. Mus. Zool., Univ. Mich. No. 132-182 pp.

* Tisvoglou, E.C., 1967. Tracer measurement in stream reaeration.Water Pollution Control Administration, US Department of Interior, Washington, DC Report.

* Uchrin, C.G., 1984. Modeling transport processes and differential accumulation of persistent toxic organic substances in groundwater systems. Ecol. Modelling. 22: 135-145.

* Ulanowicz, R.E. 1980. Anhypothesis of the development of natural communities. J. Theor. Biol. 85:223-245.

* Ulanowicz, R.E., 1981. A Unified Theory of Self-Organization. Energy and Ecological Modelling. Edited by W.J. Mitsch and J.M. Klopatek. Elsevier Scientific Publishing Company. Amsterdam-Oxford-New York-Tokyo.

* Ursin, E., 1967. A mathematical model of some aspects of fish growth, respiration and mortality. J. Fish. Res. Bd. Can. 13: 2355-2453.

* Ursin, E., 1979. On multispecies fish stock and yield assessment in ICES. A Workshop on multispecies approaches to fisheries management advice. St. John's, November 1979.

* Ursin, E., 1979c. Principles of growth in fishes. Symp.Zool.Soc. London. No 44: 63-87.

* Usher, M.B., 1972. Developments in the Leslie matrix model. In Mathematical Models in Ecology. Editor Jeffers J.N.R.,

29-60. Blackwells, Oxford.

* Veith, G.D., Defoe, D.L., and Bergstedt, B.V., 1979. Measuring
  and estimating the bioconcentration factor of chemicals in
  fish. J. Fish. Res. Board Can., 36: 1040-1048.

* Vollenweider, R.A., 1969. Moglichkeiten und Grenzen elementarer
  Modelle der Stoffbilanz von Seen. Archiv Hydrobiologia 66: 1-36.

* Vollenweider, R.A., 1975. Input-output models with special
  reference to the phosphorus loading concept in limnology.
  Schweizeriche Zeitschrift fur Hydrologie 37: 53-83.

* Vollenweider, R.A., 1968. The Scientific Basis of Lake and
  Stream Eutrophication, with Particular Reference to Phosphorus
  and Nitrogen as Eutrophication Factors. Tech. Rep. OECD,
  Paris, DAS/DSI/68, 27: 1-182.

* Vollenweider, R.A., 1965. Calculation models of photosynthesis-depth
  curves and some implications regarding day rate estimates in
  primary production. Memorie dell Istituto Italiano di Idro-
  biologia 18 Suppl.; 425-457.

* Volterra, V., 1926. Fluctuations in the abundance of a species
  considered mathematically. Nature 188: 558-560.

* WMO, 1975. Intercomparison of Conceptual Models used in Operational
  Hydrological Forecasting. Geneva, 160 pp.

* Walsh, J.J., and Dugdale, R.C., 1971. Simulation model of the
  nitrogen flow in the Peruvial upwelling system. Investigacion
  Pesquera 35: 309-330.

* Walters, C.J., Park, R.A. and Koonce, J.F., 1980. Dynamic models
  of lake ecosystems. In: E.D. LeCren and R.H. Lowe-McConnell
  (Editors), The functioning of Freshwater Ecosystems. Cambridge
  University Press, pp. 455-479

* Wangersky, P.J., and Cunningham, W.J., 1956. On time lags in
  equations of growth. Proc. Nat. Acad. Sci., 42: 699-702.

* Wangersky, P.J., and Cunningham, W.J., 1957. Time lag in population
  models. Cold Spring Harbor Symp. Quant. Biol., 42: 329-338.

* Wangersky, P.J., and Cunningham, W.J., 1956. On time lags in
  equations of growth. Proc. Nat. Acad. Sci. 42: 699-702.

* Water Resources Engineers, Inc., 1973. Computer Program Documentation

for the Stream Quality Model QUAL-II. Prepared for U.S.
Environmental Protection Agency, Systems Analysis Branch,
Washington, D.C.

* Weinberg, G.M., 1975. An Introduction to General Systeme Thinking.
Wiley, New York.

* Welcomme, R.L., 1970. Studies of the effect of abnormally high
water levels on the ecology of fish in certain shallow
regions of Lake Victoria. J. Zool. London, Vol. 160: pp 405-436.

* Wheeler, G.L., Rolfe, G.L. and Reinbold, K.A., 1978. A simulation
for lead movement in a watershed. Ecol. Modelling. 5: 67-76.

* Wheeler, G.L., Rolfe, G.L., and Reinhold, K.A., 1978. A simulation
model for lead movement in a watershed. Ecological Modelling
5: 67-76.

* White, G.C., Adams, L.W. and Bookhout, T.A., 1978. Simulation of
tritium kinetics in a freshwater marsh. Health Physics, 34:
45-54.

* Whittaker, R.H., and Mitchell, R., 1973. Ecology of Yellowstone
thermal effluent systems. Hydrobiologia 41. p. 251-271.

* Whittaker, R.H. and G.M. Woodwell, 1971. Evolution of natural
communities. pp. 137-159. In J.A. Weins (ed.), Ecosystem
Structure and Function. Oregon State Univ. Press, Corvallis.

* Whitten, G.Z., Killus, J.P., and Hogo, H., 1980. Modeling of
Simulated Photochemical Smog with Kinetic Mechanisms. Final
Report prepared by Systems Apllications, Inc. for the Envi-
ronmental Sciences Research Laboratory of the U.S. EPA,
Research Triangle Park, N.C.

* Wiegert, R.G., 1971. Simulation modelling of the algal-fly
components of a thermal ecosystem: effects of spatial heterogeneity, time delays, and model condensation. In: B.C.
Patten (Editor), Systems Analysis and Simulation in Ecology,
Vol. III. Academic Press, New York, NY, pp. 157-181.

* Wolfe, N.L., Zepp, R.G., Baughman, G.L., Fincher, R.C. and
Gordon, J.A., 1975. Chemical and Photochemical Tranformation
of Selected Pesticides in Aquatic Systems. U.S. Environmental
Protection Agency Research Report No. EPA-600/3-76-067.

*Wolfe, N.L. et al., 1977. Methoxychlor and DDT degradation in water rates and products. Env. Science & Tech., 11: 1077-1081.

*Yanni, M., 1970/1971. Biochemical studies on some Nile Fish, 1. Fat and water contents of Anquilla vulgaris, Synodontis schall, and Clarias Lazfra. Zool. Soc. Egypt. Bull., Vol. 23: pp. 90-101.

*Zeigler, B.P., 1976. Theory of Modelling and Simulation. Wiley, New York, NY, 435 pp.

*Zepp, R.G. et al., 1977. Photochemical tranformation of DDT and methoxychlor degradation products DDE and DMDE by sunlight. Arc. Env. Contam. Toxicol., 6: 305-314.

*Zitko, V., and Carson, W.G., 1976. The mechanism of the effect of water hardness on the lethality of heavy metals to fish. Chemisphere, 299-303.

*Zweifel, R.G., and Lowe, C.H., 1966. The ecology of a population of Xantusia vigilis, the desert night lizard. Amer. Mus. Novitates 2247: 1-57.

# INDEX

Abbreviations ........................................ 121
Acidic rain ................................... 270, 273
Adaption, processes ........................... 120
Aggregation ......................................... 39
Agriculture systems .......................... 139
Algae .......................................... 57, 59
Algae, max.growth rate ..................... 358
Algal culture ..................................... 161
Algal growth ........................................ 10

Benthic animals .................................. 264
Beverton-Holt ..................................... 190
Biochemical functions ......................... 64
Biochemical processes .......................... 79
Biogeochemical ................. 193, 196, 204
Biological concentration factor .............. 71
Biological processes ....................... 79, 90
Biological programme (IBP) .................. 15
BOD$_5$ ............................................... 40
BOD/DO ......................... 193, 196, 201
Boolean powers .................................... 37
Boundary conditions ............................. 77
Buffer capacity ............................... 30, 34

C-cycle, global .......................... 131, 136
Cadmium ......................................... 267
Calibration ................................... 13, 48
Calibration, ecological model ................. 73
Carbon bond mechanisms ..................... 282
Carrying capacity ................. 160, 166, 239
Catabolism ....................................... 113
Chemical processes .................. 79, 90, 102
Chemostats .......................................... 10
Chromium ......................................... 259
Classification of models ........................ 22
Computer flowchart ........................... 129
Concentration, zinc, copper, lead ........... 154
Conceptualization ...................................
    computer ...................................... 134
    diagram ............... 12, 44, 92, 132, 136, 168
    matrix ......................................... 128
Conservation .........................................
    energy ........................................... 76
    mass ............................................. 76
    momentum ..................................... 76
Control functions ............................... 321
CSMP programme ................................ 85

Data collection ................................. 322
Decomposition ........................... 118, 119
Denitrification ................................... 119
Diffusion coefficients ........................... 95
Dispersion coefficients .......................... 95
Dispersion, horizontal ......................... 284
    vertical ....................................... 286
Dynamic structure ................................ 19
Dynamics .......................................... 193

Ecological buffer capacity ........... 30, 82, 353
Ecosystem ......................................... 195
Ecosystem, total ................................ 193
Eddy diffusivity profile ....................... 280
Energy flow ................................ 114, 329
Energy conservation .............................. 65
Energy flow diagram .............................. 31
Environment, simulation ........................ 84
Environmental factors ........................... 63
Esrom, Lake ...................................... 237
Estuarine ......................................... 211
Eutrophication, prediction of ......... 232, 329
Eutrophication, parameters from 12 studies ... 244

Exergy .............................................. 81
Faaborg Firth .................................... 259
Feedback mechanisms ............................ 63
Finite difference methods ..................... 214
Florida cypress dome .......................... 247
Flowchart ........................................... 60
Food chain ........................................ 259
Forcing functions ........................ 12, 123
Freundlich's adsorption isotherm ............. 98
Fundamental property ............................ 65

Goal functions ................................... 359
Glumsoe, Lake .............................. 243, 333
Growth coefficient ............................. 162
Growth rate, max. ................................ 54

Hardanger Fjord ................................. 128
Henry's law ......................................... 96
Hierachical levels .............................. 124
Hydrodynamics ................................... 204

Identification of models ........................ 23
Initial conditions ................................ 77
Intrinsic rate ...................................... 67
Ion in soil/water ............................... 296

Langmuir's adsorption isotherm .............. 98
Lead, losses to the environment ............. 326
Lotka-Volterra ............................ 163, 168

Magnification factor ............................. 70
Mathematical equations ......................... 12
Metabolic action .................................. 66
Michaelis-Menten .............. 166, 268, 348
Model structure ................................... 35

Models ................................................
    Air chemistry ........................... 270, 281
    Air pollution ................................ 271
    ARP's ......................................... 278

    Biodemographic ................................ 24
    Biogeochemical .............................. 329
    Black box ...................... 21, 126, 131
    BOD/DO ........................................ 74
    Bogs .......................................... 246
    Box ...................................... 125, 131

    Carrying capacity ........................... 341,
    Compartment .......................... 19, 143
    Complex ................................ 123, 324
    Complex, management ..................... 325
    Conceptual ................. 18, 123, 324, 336
    Control ....................................... 321
    Copper, simple ............................. 258
    Crop production ............................ 292
    Cypress dome ............................... 247

    DDT , simple ................................ 258
    Deterministic ............................ 19, 59
    Dynamic ....................................... 58

    Ecological-economic ........................ 322
    Ecotoxicological ............................ 257
    Effects ....................................... 278
    EMEP .......................................... 273
    Energy circuit .............................. 137
    Energy flow .................................. 329
    Energy/nutrient ............................. 246
    EULERIAN multiple box ............... 271, 277
    EUROMAP/ENAMAP .......................... 272
    Eutrophication . 33, 51, 124, 193, 200, 224, 232, 321, 328

    Fishery ....................................... 177

Fish growth ...................................................... 65
Flexible .......................................................... 328
Flexible structure .......................................... 333
Forested swamps ............................................ 246

Glumsoe, Lake ................................................ 243
Grey ............................................................... 127
Growth ...................................... 157, 162, 170, 234, 193
Grussian plume .................................... 271, 282, 284

Harvest ........................................................... 183
Holistic ........................................................... 19
Hydrological .................................................... 124

Initial ............................................................. 123
Input/output ............................................. 127, 131

Lagrangian trajectory ..................................... 271
Linear ............................................................. 144
Long-range ...................................................... 273

Marshes ........................................................... 246
Mass transfer ................................................... 291
Mathematical ................................................... 18
Matrix ...................................................... 19, 131, 170
MIT reservoir .................................................. 217
Multidimensional ............................................ 213

Nine-compartment ........................................... 270
Nitrogen ................................................... 143, 146
Numerical ................................................. 213, 271

Oyster reef ............................................... 129, 138

Phosphorus, simple .......................................... 42
Phytoplankton .............................. 124, 129, 234
Picture ..................................................... 125, 131
Plant growth ................................................... 308
Plumme dispersion ................................... 270, 282
Population ................................................ 156, 174
Population, dynamics ...................................... 329
Process ........................................................... 246

Reductiomistic ................................................. 19
Response ......................................................... 151

Signed diagraphs ...................................... 131, 155
Simple ...................................... 152, 195, 324
Soil ................................................................. 273
Static ............................................................. 143
Statistical ....................................................... 276
Stochastic ....................................................... 19
Stratified lake ................................................ 215

Temperature .................................................... 116
Temperature, profile ...................................... 219
Three-compartment ........................................ 268
Toxicological .................................................. 265
Toxic substances ..................................... 64, 252
Tree growth .................................................... 246
Tundra ............................................................ 246
Two-film ......................................................... 95
Two-layer ........................................................ 276

Victoria, Lake ................................................. 178

White box ....................................................... 127
WORD ....................................................... 125, 131

MONOD, expression ......................................... 119
Mortality .................................................. 157, 186

Natality .......................................................... 156
Nitrogen cycle ................................................ 14
Nitrogen, soil processes .................................. 314
North Atlantic ................................................ 188
North Sea ................................................... 176, 184
Nutrient ..........................................................
    concentration in lakes .................................. 277

    loadings of lakes ......................................... 225
    prediction of concentration .......................... 227

Okefenokee ..................................................... 148
Optimum, harvest ............................................ 247
Oxygen ........................................................... 200

Parameter values, comparison ......................... 62
Paratisism ....................................................... 167
Penman-Monteith ............................................ 299
pH effect ........................................................ 277
pH in soil ....................................................... 277
Photosynthesis, equations .............................. 108
    rate ........................................................... 105
Phytoplankton ........................................ 264, 335
Prognosis, data and measured data comparison .......... 332

Regression equation ........................................ 72
Reproductive, value ........................................ 160
Retention curve .............................................. 298
River-stream ............................................. 202, 204
Saprobic system .............................................. 210
Sensitivity analysis ........................................ 48
Settling rates ................................................. 263
Silica cycle ..................................................... 223
Silver Springs ................................................. 31
Simulation language features ........................... 86
Soil pollution ................................................. 193
Solubility constants ....................................... 104
Solubility of metals ....................................... 104
State variables ......................................... 12, 16, 58
Static ............................................................. 20
Storage and pathways, definitions for models ......... 250
Submodels ....................................................... 53
Summary of model characteristics ................... 28
Susaa, River ................................................... 330
Symbols .......................................................... 181
Systematic perbutation ................................... 61

Tahoe, River .............................................. 329, 336
Tentative modelling procedure ........................ 17
Thermodynamics .............................................. 78
Toxic material .......................................... 193, 259
    concentration .............................................. 266
Transport processes ........................................ 93

Upper Nile Lake System .................................. 125
Urban development .......................................... 336
Urbanization ................................................... 9

Validation ................................................ 15, 333
    numerical .................................................... 242
Verhulst-Pearl ................................................ 160
Verification .............................................. 15, 41
Vertical transport ........................................... 270
Victoria, Lake ........................................... 30, 178

Washington, Lake ........................................... 231
Wetland .................................................... 193, 246

Zooplankton ................................................... 264